The Development of Flowers

THE DEVELOPMENT OF FLOWERS

Richard I. Greyson

with a contribution by
Carl N. McDaniel

New York Oxford
OXFORD UNIVERSITY PRESS
1994

Oxford University Press

Oxford   New York   Toronto
Delhi   Bombay   Calcutta   Madras   Karachi
Kuala Lumpur   Singapore   Hong Kong   Tokyo
Nairobi   Dar es Salaam   Cape Town
Melbourne   Auckland   Madrid

and associated companies in
Berlin   Ibadan

Copyright © 1994 by Oxford University Press, Inc.

Published by Oxford University Press, Inc.,
200 Madison Avenue, New York, New York 10016

Oxford is a registered trademark of Oxford University Press

All rights reserved. No part of this publication may be reproduced,
stored in a retrieval system, or transmitted, in any form or by any means,
electronic, mechanical, photocopying, recording, or otherwise,
without the prior permission of Oxford University Press.

Library of Congress Cataloging-in-Publication Data
Greyson, Richard I.
The development of flowers / Richard I. Greyson.
including a contribution by Carl N. McDaniel.
p.   cm.
Includes bibliographical references and index.
ISBN 0-19-506688-X
1. Flowers.   2. Inflorescences.
3. Angiosperms—Development.
I. Title
QK653.G75   1994   582.13′043—dc20   93-22316

2 4 6 8 9 7 5 3 1

Printed in the United States of America
on acid-free paper

Preface

Plant development has often been presented as a descriptive study of macrostructure—cells, tissues and organs. Increasingly, it becomes obvious that development must be viewed as a multidimensional process involving many different levels of interaction which are open to analysis and experimentation as well as description. The developmental biology of flowers, a traditionally undernourished segment of this discipline, is on the verge of breaking out of this stereotype due, in large part, to the discoveries of and experimentation with a wealth of genetically based variability. Opportunities now exist, in some materials, for serious experimental explorations from which to assess the relative contributions of different levels of organization to the overall process of flower formation.

For those who pursue these newer studies, a text identifying, summarizing and evaluating the status of knowledge of flower development has been an obvious need. Despite the fact that information about flower development is scattered throughout the literature of disparate disciplines, data exist at a variety of levels, in a range of species which bear on flower development. My objective has been to bring some of these observations, views and data together into a single volume. I have concentrated my attention on the information that bears on our questions about the cellular and organismal aspects of development and have left serious examination of the genetic and molecular aspects to future commentators.

Because plant growth and development can be variable and plastic, great care must be given to the design of investigations both descriptive and experimental, and to their presentations, to insure accuracy, precision and reproducibility. One set of guidlines to meet these objectives includes: (a) the articulation of precise attainable objectives, considering the material available and the resolving power of the analysis; (b) the selection of defined genetic or clonal stocks of experimental and control materials; (c) an assessment of the optimum environmental and cultural conditions for the questions under investigation and the materials available; (d) the maintenance of records of the environmental and cultural conditions actually achieved; (e) the execution of the description or experiment with concern for reproducibility and statistical significance; (f) accuracy and detail concerning all these items in published reports.

These concerns dictated my decisions to cite, in some cases, specific numerical data, to use the taxonomic binomial (and occasionally to identify the cultivar or genotype), to use graphical and tabular descriptions of experimental results, to describe the experimental conditions used in the studies and to provide numerous citations. Obviously, it was neither possible nor desirable to follow this format in all cases, and my comments are not intended to substitute for the primary research data of the original papers cited.

The illustrations, tables and their labels are considered to be an integral part of the manuscript, and where possible they stand alone without much repetition of the data in the text. In most cases the reproductions of graphs and histograms do not include data points: once again, readers should consult original papers for details.

Throughout most of the manuscript, the discussion focusses on development in a limited number of species. The attributes of those highlighted include: (a) they possess known genetic variability in flower structure; (b) some data on flower physiology are available; (c) seed sources of genetically based variants are available; and (d) the life cycle and structure of the mature plant are amenable to experimental conditions. Our selection includes many domesticated species which can be cultured and studied under convenient laboratory conditions. Thus, perennials, trees and others with complexities of culture and life cycle have not made my short list of model species. Species that presently approach these criteria include: *Arabidopsis, Antirrhinum,* L*ycopersicon, Nicotiana, Petunia, Pisum* and *Zea*. Other species which have interesting developmental traits but do not exist as well-developed genetic collections include: *Begonia, Brassica, Cucumis, Helianthus, Lolium, Mercurialis, Nigella, Oryza, Pharbitis, Sinapis, Spinacia, Triticum* and *Xanthium*.

I hope that it is apparent that many questions about flower development remain unanswered. Many opportunities for original discoveries are available at many different levels. It is not too far-fetched to speculate that studies of flower development may contribute to basic insights about plant development in general.

To those who use this book to assist in their explorations of flowers and plant development and especially to those who begin their studies here, may I wish you "happy hunting."

London, Ontario R. I. G.
March 1993

Acknowledgments

It is a pleasure to acknowledge those researchers who lent original negatives or prints from which we prepared the photographic plates. These include G. Bernier, V. Bommineni, P. Cheng, G. Cusset, J. Hill, L. Hufford, P. Leins, E. Lord, R. Lyndon, C. McDaniel, E. Meyerowitz, M. Moncur, E. Nishino, J. Palmer, P. Polowick, A. Ritterbusch, R. Sattler, V. Sawhney, C. Sekhar, M. Sundberg, S. Tepfer, S. Tucker, W. van Heel, J. Verbeke. To all these and others who gave permission to copy their tables or data or line drawings, I am most grateful and wish to thank each one for his/her generosity and encouragement extended in support of this project.

Permission to use published figures and data was requested from all publishers and many individual authors. Those items used are individually acknowledged to the individual authors and/or publishers, as requested. I am grateful to all the publishers who gave permission to use items from their publications.

Those reviewers who read portions or all of different drafts of the manuscript and spotted errors, offered comments and suggested revisions—Venkat Bommineni, Lloyd Evans, Gary Hicks, Arnold Karpoff, Elliot Meyerowitz, Rolf Sattler, Vipen Sawhney, David Walden—I want to thank most sincerely for their help. Each made helpful suggestions, many of which have been adopted. The final responsibility for all opinions, interpretations, omissions and errors is, however, mine. I would appreciate hearing from researchers who wish to draw attention to particular errors of fact or interpretation.

I wish especially to thank Carl McDaniel for preparing Chapter 3 and for offering positive ideas and encouragement at different times during the preparation of the manuscript. I am also grateful to Peter Greyson for the preparation of line drawings, Alan Noon for the preparation photographic copies and prints, Maureen Greyson for proofreading and Stefani Tichbourne for secretarial assistance and preparation of the final draft.

It is a pleasure to acknowledge, as well, sectors of the University of Western Ontario which assisted in many ways: the Interlibrary Loan, Circulation and Reference departments of the Allyn and Betty Taylor Library; Vice-President's Research Special Competition; and the Department of Plant Sciences. I wish also to acknowledge the Natural Sciences and Engineering Research Council of Canada (NSERC) for financial support throughout my research career.

Finally, I wish to dedicate my efforts on this project to the memory of my late wife, Dorothy, who provided editorial and artistic advice and enthusiastic encouragement during its preparation. Unfortunately, she was unable to see the final product.

Contents

Abbreviations, xi

1. A Developmental Point of View, 3
2. The Angiosperm Flower and Life Cycle, 13
3. Photoperiodic Induction, Evocation and Floral Initiation by Carl N. McDaniel, 25
4. The Perianth, 44
5. The Androecium, 87
6. The Gynoecium, 130
7. Inflorescences, 171
8. Flowers and Inflorescences of Grasses, 195
9. Flower Organogenesis: Summary and Prospect, 231

Glossary, 259

Appendix 1: Culture Protocols for Rearing Research Plants, 261

Appendix 2: Sources of Information about Genetic Stocks of Model Species for Flower Studies, 264

References, 267

Index, 309

Abbreviations

The following abbreviations for frequently used words and phrases are used throughout the text. Abbreviations and symbols within figure and table legends are individually defined in each case. Abbreviations of units of measure (length, weight, time, concentration) follow the *CBE Style Manual,* Fifth Edition (1983).

ABA	abscisic acid
CH	chasmogamous
CL	cleistogamous
c-mst	cytoplasmic male sterility
DAG	days after germination
DNP	day neutral plant
DW	dry weight
FW	fresh weight
g-mst	genic male sterility
GA_3	gibberellic acid
GA(s)	gibberellin(s)
GSI	gametophytic self-incompatibility
IAA	indole acetic acid
LD	long day
LM	light microscope
L/S	longitudinal section
mtDNA	mitochondrial DNA
PGS	plant growth substance(s)
SD	short day
SEM	scanning electron microscope or scanning electron microscopy
SSI	sporophytic self-incompatibility
TEM	transmission electron microscope
T/S	transverse or cross section

The Development of Flowers

1

A Developmental Point of View

> The basic problem of the developmental biologist is to account, in a mechanistic sense, for the manner in which the single-celled zygote is elaborated into the morphologically more complex multicellular adult.
>
> Raff and Kaufman (1983), with permission.

> At the beginning of its development the young plant, as it grows from a fertilized egg or from some larger embryonic mass, is relatively simple and homogeneous. A characteristic feature of the developmental process, however, is the origin of differences in the amount, character and location of growth which lead to differences between the various parts of an individual.
>
> Reproduced from *Plant Morphogenesis* by Sinnott (1960). Copyright © 1960. McGraw-Hill.

Explicit developmental views such as those expressed above are rare in discussions of flower structure. This is not because little is known about flower ontogeny: much pertinent information is available and many details of flower development are well understood. The data on development, however, are scattered and the intellectual motivation behind many studies of flower morphology and anatomy frequently differs markedly from that of the developmental biologist.

A Developmental View of the Flower

The elements of the problem of flower development are implicit in the three views of the development of a *Nigella* flower in Fig. 1-1A–C (see color insert). The floral meristem, frequently begins as a mound of structurally undifferentiated cells from which the organs initiate in patterns at the margins of the meristem (Fig. 1-1A). Subsequently, organs of different types initiate towards the center. The individual organs grow and differentiate (Fig. 1-1A and B) in a coordinated manner so that at flower maturity (Fig. 1-1C) the individual parts of the flower contribute to the final event—the production of viable seed.

Our problem as developmental biologists is to uncover and unravel the interrelated sequences of cellular activity at various levels of organization and ultimately to understand how these different processes interact to produce the complex of flower organs. An imperfect, though useful, reminder of the complexity of the potential levels of interaction involved, is illustrated in Fig. 1-2. Until recently, only the outermost levels of flower organization have been available for analysis. It is obvious that real insight about flower development will only come from a concerted, multifaceted research thrust encompassing all relevant levels.

The term **development** is used frequently in modern biology, with a variety of meanings, though it is rarely defined. It is used here: (a) as a synonym for embryogenesis and ontogeny; (b) to include all the molecular and physiological events and stages that connect the gene to the phenotype; (c) as an adjective to describe the variable expressions of a single "developmental sequence." Thus, as used here, "development" could include all the sub-disciplines that are included in modern treatments of developmental biology (Lash and Whittaker 1974; Browder 1980). It could also include all those events and processes that are involved in pattern generation and in the morphogenesis of the individual flower organs.

Similarly, the term "developmental sequence" can have a number of meanings. It is used here to indicate either a series of stages of expression or a temporal series of developmental events (Alberch 1985). In molecular developmental studies, "sequence" is generally reserved for the base sequence of individual genes. Further discussion of levels of organization and attempts to identify causal relationships is left for Chapter 9.

Calls for such an explicitly multidisciplined assault on the questions of flower ontogeny (Wardlaw 1956–57; Sattler 1988a) have only recently (Drews and Goldberg 1989) attracted the serious attention of developmental biologists. The fortuitous coincidence of the current needs of agriculture with the arrival of modern molecular technologies makes the serious exploration of flower development both intellectually attractive and potentially resolvable.

What then should be the specific objectives for a developmental analysis of the flower? At the risk of oversimplification and omission, the salient stages in the development of the flower and its organs are identified below, and some questions that should be applied at each stage are asked.

It is convenient to identify the main cellular events (Fig. 1-3) in the development of each organ type. Simplistically these are **alteration of meristem fate, initiation, growth** (cell division, cell elongation), **differentiation and maturation, maturity** and **senescence.** As indicated, these processes overlap considerably within the organ, between organs within the flower and in some cases within individual cells.

The approach taken here is from "the outside to the inside" or in Brenner's (1975) terminology "top–down analysis." The alternative approach, "bottom–up" analysis, which extends the logic and technology of molecular genetics from the gene towards the phenotype, has begun and is currently yielding fascinating details. The eventual synthesis of these approaches, which may reveal novel levels of complexity and integration, is the ultimate goal of developmental biology.

Alteration of Meristem Fate

Beyond the traditional questions that plant physiologists ask about the flower stimulus and its transmission, we need to know a great deal more about the details of evocation of the shoot apical meristem. Descriptively, what are the details of the cascade of changes in the cytology, biochemistry and molecular biology of the meristem in response to the arrival of the floral stimulus? Analytically, what are

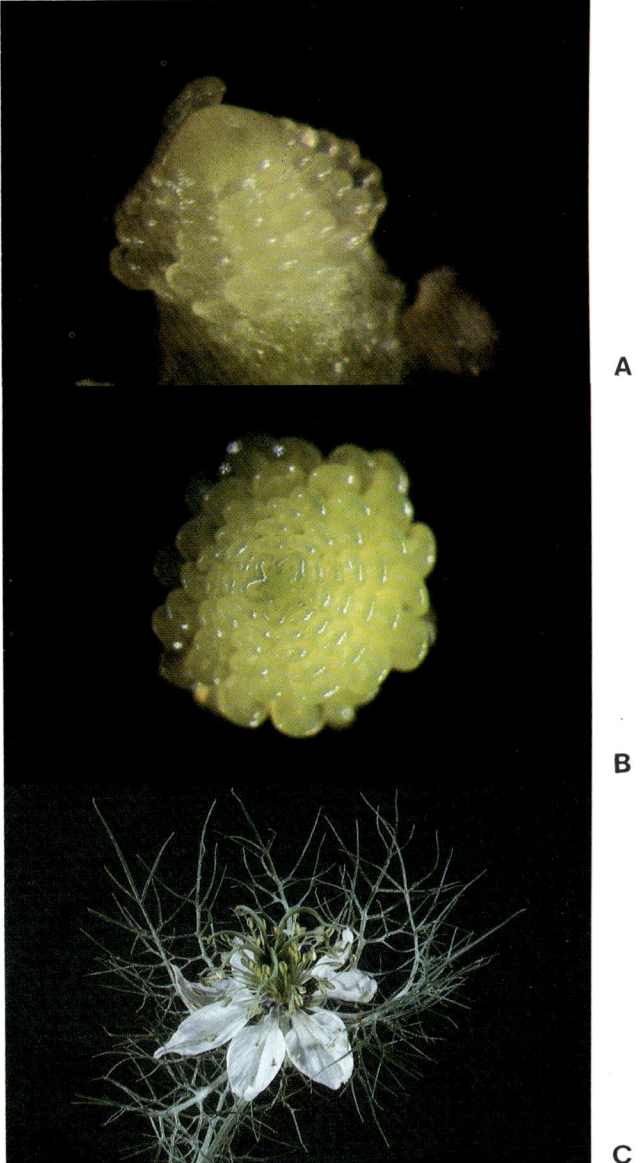

Fig. 1-1A–C. Three views of a developing flower meristem of *Nigella damascena* "single" as it grows and matures into a functional flower; time-lapse photography would be a superior medium. (A) Freshly dissected terminal flower meristem (bracts and sepals removed) at an early stage of organ initiation (approx. two weeks after initiation of LD regime). Most recently initiated (innermost) primordia are stamen primordia (approx. X60). (B) Freshly dissected (bracts and sepals removed) where all organs are initiated and organ differentiation is well advanced (three weeks after initiation of LD regime). The many stamen primordia block views of the nectariferous petals to the outside and surround the recently initiated carpels at the center (approx. X60). (C) A mature *Nigella* flower at anthesis possesses five kinds of floral organs: dissected leaf-like bracts, large showy sepals, nectariferous petals, stamens and a compound gynoecium (approx. X1/2).

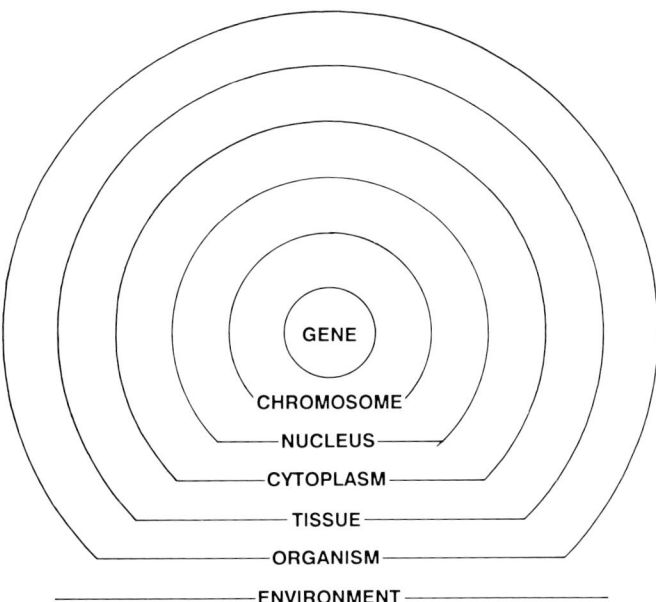

Fig. 1-2. A speculative view of the levels of organization (represented by the concentric shells) involved in biological development and, by implication, to the individual developing organs and to the whole flower. Some levels are known to interact with many other levels; other involvements are unidirectional. At most levels, especially the gene, additional levels of interaction are now recognized. Redrawn from Weiss (1963) with permission.

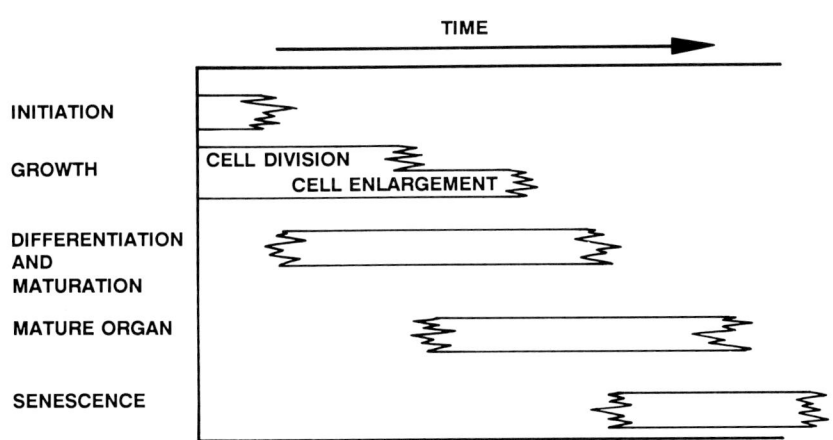

Fig. 1-3. A schematic generalization identifying the major cellular processes or stages in the development of a flower or its organs. During the development of some flowers, the time from initiation to senescence is many months. For others, development is completed in a few days.

the significant developmental events in the process? Are there stage-specific expression of inflorescence and flower genes? How are the properties of competence, commitment and determination explored in the evoked meristems?

Organ Initiation

As indicated in Fig. 1-3, the period recognized as "organ initiation" is brief and, in most cases, involves a small number of cells. It is assumed, though essentially unstudied, that this stage is a critical, perhaps determinative event in the developmental history of the organ. Questions appropriate to this stage include:

What is the cellular status, size and physiology of the evoked shoot apical meristem at the time of initiation of each organ type? How many cells are involved initially? What are the first cells of the meristem to be involved in organ initiation?

What details can be ascertained regarding the siting of organ initiation? What factors influence the pattern of subsequent initiation sites?

What are the cell dynamics at the site of initiation?

What physiological and molecular events trigger the differential activity at the site of initiation?

Can we uncover the genetic and molecular basis by which organ primordia, initially similar, differentiate through development?

Are primordia at the time of initiation essentially undifferentiated as to organ type, or are there organ-specific features operating from the outset? This question, regarding the timing, degree and nature of **determination** (Wareing 1978; McDaniel 1984b) is discussed in more detail in subsequent chapters.

What are the molecular concomitants associated with the developmental decisions relating to the siting and determination of organ kind?

Organ Growth

It is a truism that plant growth is the product of the combination of cell division and cell expansion (Fig. 1-3). It should also be added that important aspects of these cellular processes are their rates, duration and distribution within the organ and the orientation of the processes involved. With regard to the growth of flower organs, the following questions are appropriate:

What are the patterns and frequencies of cell division in the organ as it matures?

What are the patterns, rates and localizations of cell elongation as the organ grows and matures?

What are the physiological and biochemical concomitants, and in some cases, the causes of these events?

What is the cellular basis of organ shape and symmetry? What factors regulate and influence these phenomena?

Cell Differentiation Within the Organ

The process of cell differentiation is continuous and overlaps the growth stage (Fig. 1-3). The degree to which this is true in any single case will differ with the tissue and the cell type. The number of details about cellular differentiation in the different organs which could be recorded is potentially overwhelming. The discussion here is limited to the following points:

What different tissues and cell types have been identified in the different organs at maturity?

What are the details, timing and coordination mechanisms within the different tissues of the organ as it matures?

Can we describe the patterns of differentiation along organs?

Can we identify waves of activity along organs?

What factors are involved in cell differentiation?

To what extent does cell differentiation result from a response to fields of diffusible molecules, or is differentiation less regulative and more the temporal expression of early clonal differentiation and cell autonomy?

The Mature Organ

This topic is obviously more closely related to discussions beyond development. It will, however, be useful to consider some details of the mature physiology and biochemistry at this stage:

Can the physiological and cellular features of flower opening (aestivation) be unravelled?

What is the identity of any chemical products of the mature organ? Pigments? Nectar? Fragrance?

What are the cellular products of the organ—the male and female gametophytes?

What are the developmental details involved in the formation of sperm and eggs?

Organ Senescence

In the context of the plant life cycle, flower senescence can represent the onset of fruit, seed and embryo development. As such it is beyond the scope of this volume. However, some aspects are germane to flower development:

To what extent can senescence be interpreted as gene-regulated cell and organ death?

What physiological and environmental cues trigger senescence?

What plant growth substances are involved in the processes?

Can we identfy any molecular biology and genetic concomitants of senescence?

Unfortunately, even with this severely restricted set of questions answered in detail, the developmental biology of flowers will still be in its infancy. Analogous to the current status of our knowledge of animal zygotes, embryos and cell differentiation, many details of the cellular, physiological, biochemical and molecular aspects of flower development will remain unexplored. Our objective in preparing this text, then, is to present a summary discussion of flower development so that some of these studies into flower development are more easily planned and executed.

Traditional Approaches to Flower Structure

Developmental biologists rarely spend much effort justifying the philosophical basis for their discipline. The reductional, experimental approach is so much the common working environment of biochemists, geneticists and embryologists that such an exercise is thought to be unnecessary and might even be considered unscientific by many excellent developmentalists. Because, however, flower structure has not often been handled within this mechanistic tradition, it is appropriate to document as accurately as possible how traditional comparative flower morphology differs from modern developmental biology and to provide an overall explanation for the point of view taken in this text.

In the traditional literature of flower structure, two different intellectual objectives, quite distinct from ontogeny, can often be identified. These are: (a) **comparative morphology** (sometimes referred to as "classical morphology"); and (b) **phylogeny**.

Comparative morphology, in the strict sense, can be traced to a theme permeating eighteenth and nineteenth century pre-evolutionary biology, which attempted to make intellectual sense out of the structural diversity within the plant and animal kingdoms. One such attempt to provide a unifying view of biological diversity was given by the German philosopher Goethe (1790) [see Arber (1946) for translation and interpretation]. Perhaps more as a poet than as a scientist, he suggested that the similarity between floral organs and leaves was based upon metamorphosis. The goal of many flower morphologists since then has been to document this notion and to identify the ways in which floral organs are homologous with this presumed fundamental organ type—the leaf.

While much of the data collected within this intellectual tradition provided information on development, the developmental biology of the flower and its organs was a secondary objective for many researchers. The major objective was to uncover the mysteries of the hidden homologies within the plant.

Another, and not totally separate, objective of flower morphology which developed in the last half of the nineteenth century, attempted to understand the phylogeny of the angiosperms and the flower. It is convenient and appropriate to identify this approach with Charles Darwin. For, besides his contributions to the general theory of evolution and the descent of man, he was especially intrigued by the variability of flower structure and function (Darwin 1897).

In the minds of classical morphologists, these two themes, homology and phylogeny, came to represent the essence of flower morphology. The study of ontogeny became evidence in its support. It is not our intent to comment on the correctness or value of this scientific tradition for which there is a very full and argumentative literature: it is sufficient to acknowledge its existence. Identifying the tradition does explain, however, why flower physiologists and flower morphologists have typically had little to discuss when they met and why developmentalists find that traditional flower literature can be a difficult basis for their investigations.

Understanding the origins and history of the angiosperms and discovering homologies between different organs are obviously important research objectives in biology, but there is another significant scientific question about flower biology that requires definition and dedicated attention. We need to learn a great deal more about the cellular and biochemical properties of the transformed floral apex. We need to know when and what new genetic programs have been set into action in an apex committed to floral development. We need to learn how the florally determined apex produces the different organs at the different sites. Further, we must pursue more diligently than heretofore the different processes interacting with the environment that bring the flower to functional maturity. In other words, we have to treat the processes of flower development that are inferred in Fig. 1-1 with the same analytical zeal that motivated the animal embryologists over the last century.

It would be incorrect to suggest that botanists have ignored flower ontogeny completely in favor of the classical tradition. Eyde (1975) points to two early botanists, Schleiden and Hoffmeister, as examples of those who rejected the classical interpretation and its logic. In this century, Goebel, Wardlaw and Sinnott, in their individual ways, reflected a developmental approach to morphology. Some of these differences between traditional approaches to the flower and develpmental studies are also discused in Meyerowitz, Smyth and Bowman (1989).

It is true, however, that flower ontogeny has rarely received the concerted attention of botanists who researched from an experimental, developmentalist, reductionist tradition. These types of researchers have been attracted to embryos, tissue culture and plant physiology. The reasons appear to result from both scientific and philosophical choices made over the last century.

Scientifically, the intellectual choices that have contributed to the present situation include:

(a) The distinction between the studies of "form" and of "function" has been maintained more rigidly by botanists than by other biologists. The maintenance of this distinction between morphology and physiology would appear to be a serious impediment to an overall understanding of functional organ systems, and the implications of this schism for overall academic communication and pedagogy should be obvious.

(b) Botanists, both morphologists and physiologists, have been much slower than compatriot zoologists to accept Gehring's assumption (opening quotation to Chapter 9) that problems of development have a strong, in some cases overriding, genetic component of control. The consequences of this hesitation include: (i) the tendency to investigate developmental questions which can be phrased in terms of chemical integration, Plant Growth Substances (PGS) and other epigenetic factors; (ii) a general lack of awareness for genetically based variability in the material being investigated; and (iii) the consequent low priority given to identification, collection and maintenance of genetic variability of flower development. Recent attempts to rectify this situation include Marx (1983), Sheridan (1988) and Haughn and Somerville (1988).

(c) For many of those who have studied flowers, questions of phylogeny have been paramount: questions about ontogeny received less attention.

At a philosophical level, two antithetic positions have traditionally been expressed. On the one hand, it is argued that the complexities of development will be uncovered if the process is reduced to its myriad chemical, physical and cellular details: these details would represent the partial or intermediate causes of metaphysical scholars. This reductionist, inductive approach assumes that fundamental understanding of the whole will eventually accrue from an analysis of the parts. Alternative views, argued more tenaciously in botanical morphology than in most other biological disciplines, hold that this non-holistic strategy fails to provide depth of understanding regarding the broad regulatory features (final causes) which are intuitively anticipated. This controversy, like many other antitheses of biology (Woodger 1967) seems endless and unresolvable. As argued by Wardlaw (1968), these older holistic views appeal to fewer present-day scientists. A modern challenge to an overly simplistic cellular-reductionist model of plant organogenesis is argued by Kaplan and Hagemann (1991).

The power of the reductionist method for unravelling the mysteries of the natural, and in particular the biological, world is highly touted (Reiner 1968), though its utility is limited when attempts are made to interpret the interactions and complexities that arise between different levels of organization (Alberch 1980). Tennant (1986) provides a more formal analysis of the distinctions between reductionist and holistic approaches. Despite these reservations, for the foreseeable future, at least, the study of flower morphogenesis should benefit significantly from a thoroughly rigorous reductionist approach.

Ontogeny is not, however, the only unanswered question about flowers. Many different but valid questions are presently being explored more diligently than in the past, and these are variously described as flower biology, reproductive or pollination biology or flower ecology. Even phylogeny and taxonomy should benefit from a more rigorous analysis of ontogeny, especially at molecular levels. One should expect that all the traditional arguments of taxonomy and morphology will be phrased more sharply and with greater insight as the data on flower development, especially at the molecular level, accumulate.

Neither should we deceive ourselves into thinking that all questions about flowers will be solved by a single panacea. It is not necessary to pursue reductionism to the total exclusion of various synthetic and holistic approaches to biological organization and existence. Its purpose is solely to recognize, with precision, the nature of the questions of flower development that remain unanswered and to pursue them with a crisp conceptual grasp. The significance of our knowledge to other areas of the biosphere and how we use it, especially now that we are able to modify the genetic basis of development and apply this knowledge to agricultural situations, will require a wisdom and intellectual humility which is unfortunately not normally part of contemporary training.

Scientific exploration, with little regard for the consequences of knowledge, is increasingly questioned (Crouch 1990; Smith 1990). Let us hope that a combination of enlightened self-interest, good sense and genuine respect for earth's floral heritage will accompany our studies. For despite the prevalent pessimism about the negative consequences of manipulative and invasive research strategies, ignorance also has its down-sides.

Preview of the Next Chapters

Flower development is taken as the integration of all the events and effects that are involved as a vegetative meristem is transformed to produce a flower meristem and ultimately a functional flower. The point of view of this presentation is that this is to a great extent a function of cellular and tissue activities and that our questions about the flower reflect this assumption.

In Chapter 2, the basic terminology associated with flower structure is presented, and this is followed by a summary of the typical life cycle of an angiosperm. Chapter 3 outlines the subject that is frequently called flower physiology—the nature of the process by which the plant is induced to flower and the subsequent response of the meristem to those signals. Chapters 4, 5 and 6 present a summary of the initiation, growth and maturation of the perianth, the androecium and the gynoecium. Wherever possible the presentation attempts to answer the questions raised in this chapter about the cellular basis of organs. Chapter 7 broaches some developmental questions about the way flowers are aggregated into inflorescences. Chapter 8 focuses on the flowers and inflorescences of grasses. A synthesis of these ideas and information is attempted in Chapter 9.

Throughout, the objective is to present a generalization of flower development which is based upon the observations of cells and tissues and is thoroughly consistent with current interpretations of molecular biology, cell physiology and the dynamics of cell populations in higher plants. It is hoped that this can serve as a resource and a basis for further explorations into the molecular mechanisms that underlie these cellular activities.

Suggestions for Further Reading

In addition to the citations given earlier, the following references will provide an introduction to the data and controversy of traditional morphological literature: Mason (1957); Carlquist (1969); Kaplan (1971); Greyson (1973); Cusset (1982).

Recognition should be given here to the significant attempts of Rolf Sattler and his colleagues to place comparative morphology on a more objective plane. Beginning with concerns about traditional typology and concepts of organography (Sattler 1966), this group has progressed from examination of organs to investigation of the cellular dynamics of their ontogeny (Dubuc-Lebreux and Sattler 1984, 1985). For those who feel equally uncomfortable with the traditional speculation of "form" [see Arber (1970)] and with the unrestrained optimism sometimes attributed to the potential contributions of reductionism to our understandings of structure, Sattler's interjections represent a thoughtful alternative.

Discussions of the ways in which phylogeny and ontogeny can be related include Stebbins (1974, 1992), Gould (1977) and Endress (1990).

Current introductions to the subjects of plant and flower structure and physiology, which are the basis of discussions here, are available in Bell (1991); Esau (1977); Fahn (1982); Raven, Evert and Eichorn (1986); Stern (1991); Gifford and Foster (1989); Steeves and Sussex (1989); and Lyndon (1990) and the three volumes of Bernier and colleagues (Bernier, Kinet and Sachs 1981*a*, 1981*b;* Kinet, Sachs and Bernier 1985).

2

The Angiosperm Flower and Life Cycle

The Angiosperms: The Dominant Land Flora

The flowering plants [Division Magnoliophyta (Gifford and Foster 1989)] are recognized as the current, dominant land flora. As documented in the fossil record, this life form arose at the beginnings of the Cretaceous Period and now exploits an extraordinarily wide range of habitats on land and water and forms a significant component of the world's biosphere.

The biological success of the angiosperms has included extensive speciation such that approximately 270,000 species (Taylor 1988) are identified and arranged into 300–400 families (Hickey and King 1988) and approximately 80 orders (Stebbins 1974). While considerable variation exists in vegetative traits, the greatest diversification has taken place in the structure and arrangement of the reproductive structures and in their manner of function. Thus, the taxonomy of the group is based largely on a consideration of flower traits, though current taxonomic treatments attempt to incorporate a multi-variate input. A commonly accepted taxonomic treatment of the angiosperms is Cronquist (1981), which is used in Mabberley's (1989) portable dictionary of higher plants.

The term **flower** is reserved in this text for the characteristic aggregation of sterile and fertile reproductive organs of the angiosperms. While other authors occasionally apply the term to the cones and strobili of other seed plants and despite evidence from fossil or extant organisms of transitional forms (Doyle and Donoghue 1986), the limitation retains its utility.

Flowers occur individually (**solitary**) or are arranged in clusters (**inflorescences**). The descriptive morphology of inflorescences and a brief discussion as to how these structures might be studied by developmentalists is presented in Chapter 7.

Complications with this definition of a flower could also arise from evidence which shows that the angiosperms include groups with separate origins (polyphylesis) (Lam 1950; Meeuse 1965). While monophylesis is the traditional interpretation given to angiosperm origins, there are enough anomalies and difficulties with this assumption that polyphylesis is becoming increasingly considered. Some of these complexities are noted in subsequent discussions: the decision to devote a separate chapter to the flowers of grasses (Chapter 8) illustrates one aspect of the problem.

The terminology used to describe the flower is fraught with difficulties. How many different classes of organs exist? How many classes should be recognized? Can our terminology acknowledge historical relationships (phylogeny)? How are transitional situations interpreted? What is the significance of teratomas? What terms should be used to discuss flowers that differ significantly in their structure? The list of complications and problems seems endless and has represented a serious intellectual challenge to taxonomists, morphologists and phylogeneticists for two centuries. The literature is replete with redundant and archaic terminology.

Fortunately, for developmental studies, it is not necessary that definitions and terminology possess more meaning or connotation than that they are convenient and practical terms to categorize diversity and variability. Nor do we need to infer, as is required in traditional comparative studies, that organs with the same name possess a common ancestry or are homologous. It is obviously of interest if these claims regarding ancestry and homology can be made, but it is not essential to our discussion. All that is assumed when morphological structures are included under the same term or category is that they share, to some degree, similar attributes such as structure, ontogeny and function. In developmental studies the common baseline upon which discussion and investigation rest is the known or assumed genetic relationship of the line of organisms under investigation.

It may ultimately be possible to demonstrate similarity in the genetic and molecular sequences of phenotypic expression (developmental homology) between organs of the same category from different species, genera and families and thus contribute an insight into the historical relationships of organs between different groups. This work will, for the most part, be carried out by taxonomists and evolutionary botanists (see Roth 1988). For now, our lexicon of structural terminology, derived from traditional sources, is assumed to be one of convenience, largely based on overall similarity.

One further warning should be added. The concept of the flower used here is one which has been developed by botanists whose points of view matured in the midst of North Temperate floras and who, of necessity, approached the total world flora with preconceived views about the first flowers and the major evolutionary trends during the last 100 million years. The potential for an unnatural, arbitrary interpretation would seem significant.

It is possible, however, that the general picture of flower variability and history presented by current synthesizers is consistent with angiosperm history and that only the details are in question. Notwithstanding this perhaps overly optimistic interpretation, the dominance and diversity of the angiosperm populations and the relatively recent attempts at serious study, there will no doubt be surprises. We should be prepared to generate new insights and vocabulary, when necessary.

The Flower

As stated above, the term **flower** refers to the characteristic aggregation of spore-bearing and sterile appendages found in angiosperms. Despite the potential criti-

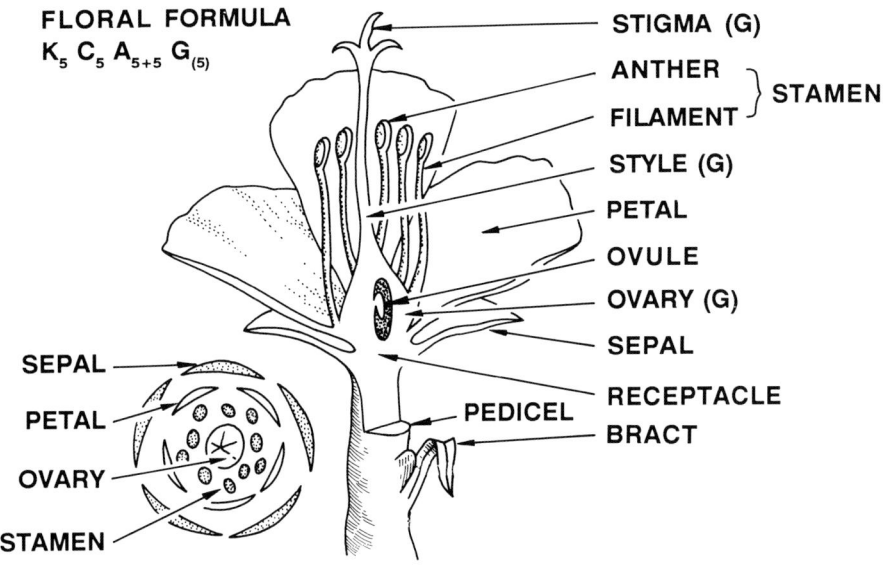

Fig. 2-1. A generalized solitary half-flower of a typical dicotyledon to illustrate the organs and their arrangement on the receptacle. The floral diagram and formula of such a flower indicate that the sepals, petals and stamens are individual and free. The gynoecium is interpreted as arising from the coalescence of five carpels: this interpretation is based, in part, upon the number of locules, septa and stigmas. Based on Hickey and King (1988), Benson (1957) and Sattler (1973a).

cism of circular logic, this distinction retains its original value through the assemblage of traits that are usually recognized as defining the angiosperms. Though perhaps with more uncertainty than heretofore, the ovule, with its integument(s) and its enclosure by the tissues of the gynoecium, the embryosac and double fertilization and its resultant endosperm still represent impressive evidence for the retention of the angiosperms as a distinct taxon and the flower as the reproductive organ of the reproductive phase.

An introduction to the general organography and terminology of mature flowers is provided in Fig. 2-1. The idealized flower can be interpreted as consisting of a variable number of organ classes arranged in helices or whorls at the tip (**receptacle**) of a floral shoot. The internode below the receptacle is called the **pedicel**, which may bear one or more (or no) **bracteoles** and be subtended by a **bract**. The sterile appendages (collectively the **perianth**) are frequently of two types: (a) an outer set, often green and more leaf-like, called **sepals**, collectively the **calyx**; and (b) the often colored and showy **petals**, collectively the **corolla**. When the perianth is undifferentiated, the individual members are termed **tepals** or perianth members. Internal to the perianth are the **stamens**, collectively the **androecium**, the **anther**-bearing appendages. The **ovule**-bearing organs at the center of the flower are collectively termed the **gynoecium,** though considerable variation in both the structure and its interpretation has led to a complex terminology.

Floral formulae and floral diagrams (Fig. 2-1) are useful conventions to summarize flower structure and provide a convenient short-hand description of flowers (Sattler 1973a; Weberling 1989).

Flowers possessing all four sets of organs—sepals, petals, stamens and gynoecium—are described as **complete,** while those missing all or part of the perianth or stamens or a gynoecium are **incomplete.** Flowers possessing both stamens and a gynoecium are described as **perfect, bisexual** or **monoclinous.** Perfect flowers are also sometimes described as being hermaphroditic. Flowers that lack either an androecium or a gynoecium are described as **imperfect, unisexual** or **diclinous.** While technically incorrect, the terms male and female are frequently used for the flowers in these situations. More appropriate terminology would be **staminate** (male) and **carpellate** or **gynoecial** (female) flowers. Individual species may exhibit a number of combinations of these flower features (see box below).

The individual organs of the flowers can be separate from each other **(free),** or they can be united or coalesced to varying degrees. Coalescence of organs of the same class (i.e., all the petals) is described as **cohesion.** Coalescence of organs of a different class (i.e., sepals and petals) is **adnation.** The cellular mechanisms by which these unions proceed is discussed later. In some cases fusion of organs is responsible. In other cases, intercalary activity of different types is apparent. A flower exhibiting a corolla of coherent petals is described as **sympetalous,** while one composed of free petals is **choripetalous** (Fig. 2-2).

Terminology Used to Describe Plants That Contain Different Combinations of Perfect, Staminate and Gynoecious Flowers

---

Monomorphic: Plants of the population of one type only.

Andromonoecious: Hermaphrodite and staminate flowers on the same plant.
Gynomonoecious: Hermaphrodite and carpellate flowers on the same plant.
Hermaphrodite: Flower contains both androecium and gynoecium.
Monoecious: Androecia and gynoecia in separate flowers on the same plant.
Polygamous: Hermaphrodite, staminate and carpellate flowers borne on the same plant.

Polymorphic: More than one type of plant in the population.

Androdioecious: Hermaphrodite and staminate flowers borne by different plants in the population.
Dioecious: Staminate and carpellate flowers borne by different plants in the population.
Gynodioecious: Hermaphrodite and carpellate flowers borne by different plants in the population.
Trioecious: Hermaphrodite, staminate and carpellate flowers borne by different plants in the population.

Source: From Heslop-Harrison (1972) with permission of Academic Press.

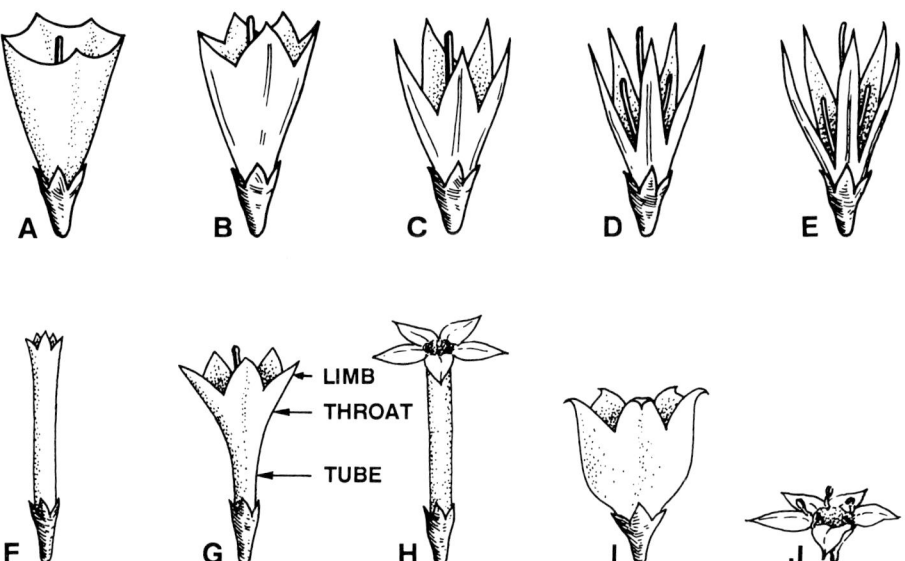

Fig. 2-2A–J. Examples of sympetalous corollas illustrating different degrees of lobing (A–E) and different shapes (F–J). (A) Toothed. (B) Lobed. (C) Cleft. (D) Parted. (E) Divided. (F) Tubular. (G) Funnelform. (H) Salverform. (I) Campanulate. (J) Rotate. Redrawn from Benson (1957) with permission.

The **stamens,** composed of **filament** and **anther,** are frequently free, as illustrated (Fig. 2-3), though many variations in the construction of the androecium are identified. In some families, various degrees of union are observed. If all filaments are joined to form a single tube, the stamens are described as **monadelphous;** stamens united to form two groups are **diadelphous;** and when they are

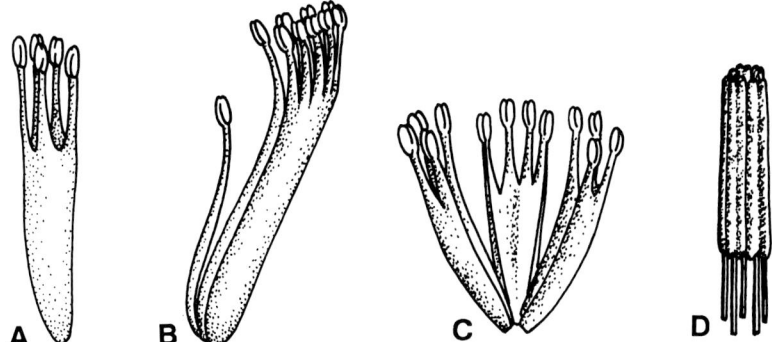

Fig. 2-3A–D. Examples of coalescence (coherence) of stamens. (A) Filaments cohere into a single tube (monadelphous). (B) Two sets of stamens of unequal length. Nine filaments are coherent plus a single free stamen (diadelphous). (C) Coherence of filaments into more than two groups (polyadelphous). (D) Coherent anthers (common in the Compositae). Redrawn from Benson (1957) with permission.

united into more than two **fascicles,** they are **polyadelphous.** In the Compositae, **coalescent** anthers (Fig. 2-3) are common. The androecium often includes sterile stamens **(staminodia).** Other anther variants are described by Weberling (1989).

Perhaps the most prominent example of organ cohesion is evident in the formation of the gynoecium. In some flowers it consists of one or more free **carpels** (i.e., apocarpous), but more frequently the gynoecium may be interpreted as being composed of joined carpels **(syncarpous** or **compound).** This compound gynoecium is frequently called a **pistil,** with three regions—**stigma, style** and **ovary** recognized (Fig. 2-1). The stigma is the pollen-receptive tissue and is frequently composed of hairs and secretory cells. It can be terminal or lateral and marginal and can be localized or widely dispersed over a number of separate elements. The style connects the stigma with the ovary and is the conduit through/on which the pollen tube grows towards the **ovules.**

The ovules form on the internal surface **(placenta)** of carpels or ovary. Common patterns of placentation are described as **axile, marginal, parietal, free-central** and **basal** (Fig. 2-4). The ovule is the site of megaspore formation and the development of the female gametophyte **(embryo sac)** and fertilization. Following fertilization, the ovule containing the **embryo** and nutritive tissue matures into the **seed.** The details of the formation of the gynoecium and its function are presented in Chapter 6.

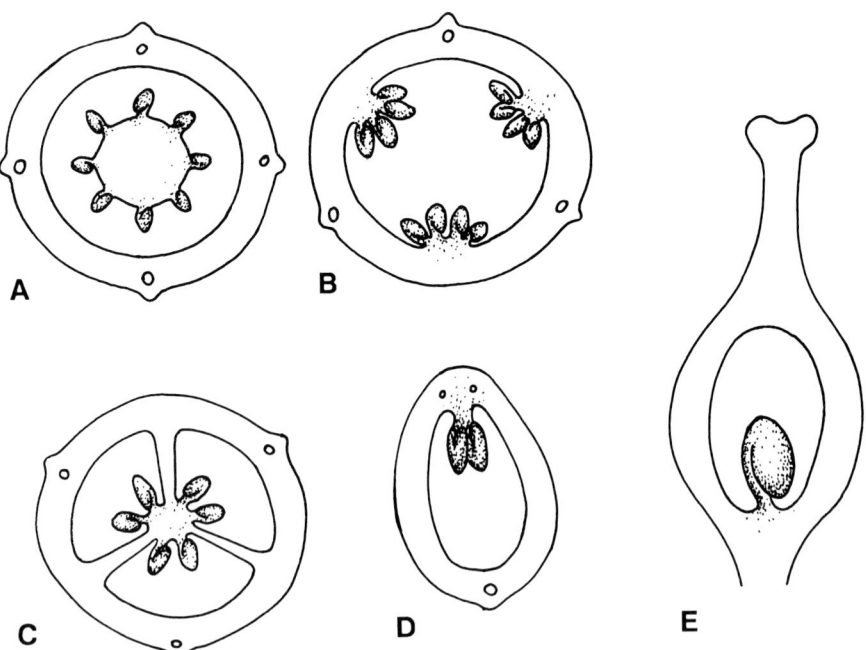

Fig. 2-4A–E. Sketches of cross sections of compound gynoecia to illustrate some types of placentation. (A) Free-central. (B) Parietal. (C) Axile. (D) Marginal. (E) Basal. Redrawn from Hickey and King (1988) with permission.

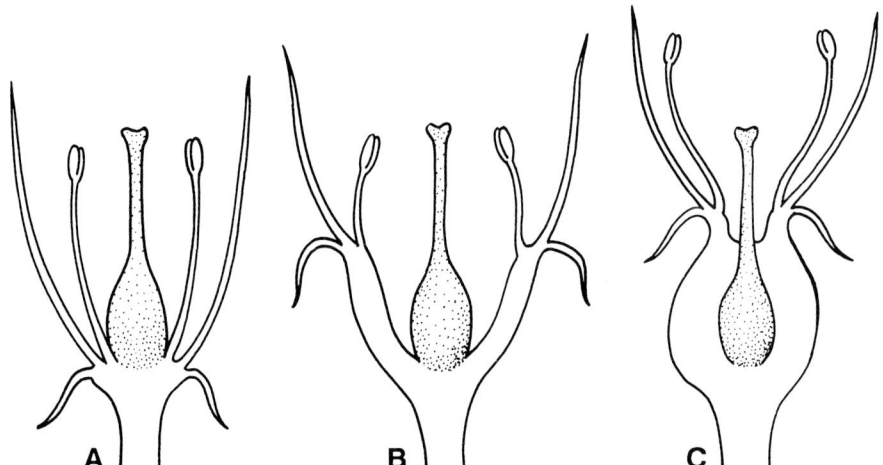

Fig. 2-5A–C. Outlines of flowers to illustrate different positions of the ovary relative to the attachment of the other organs. (A) Superior ovary; hypogynous flower. (B) Superior ovary; perigynous flower. (C) Inferior ovary; epigynous flower. Redrawn from Hickey and King (1988) with permission.

Depending upon the degree of adnation of the perianth with the androecium and gynoecium, three different flower forms are recognized (Fig. 2-5). **Hypogynous** flowers are those in which the sepals, petals and stamens are distinct—the gynoecium may or may not be compound. **Perigynous** flowers are those in which the joining of sepals, petals and stamens forms a cup or tube around the gynoecium. The ovary in these two situations is described as **superior,** for the attachment of the other organs to the receptacle is proximal to the attachment of the gynoecium. **Epigynous** flowers are those in which the joining of the sepals, petals and stamens are adnate to the gynoecium. Because the sepals, petals and stamens appear to arise from a distal portion of the gynoecium, the gynoecium is described as **inferior.** More details are given in later chapters on the processes by which these events take place. The traditional terminology used in the preceding paragraphs arose within the typological framework of comparative morphology. Closer examination of these structures within a developmental context may invalidate some distinctions and necessitate some revisions.

The Angiosperm Life Cycle

The flower, despite its importance and the attention it receives, is but a single and ephemeral structure in the sporophytic phase (generation) of the individual angiosperm organism. Figure 2-6 documents the essential features of this alternation between the spore-producing **(sporophyte)** generation and the gamete-producing **(gametophyte)** generation. In the angiosperm, life not only is a succession of mature adult individuals, as we frequently view human life cycles, but in a very

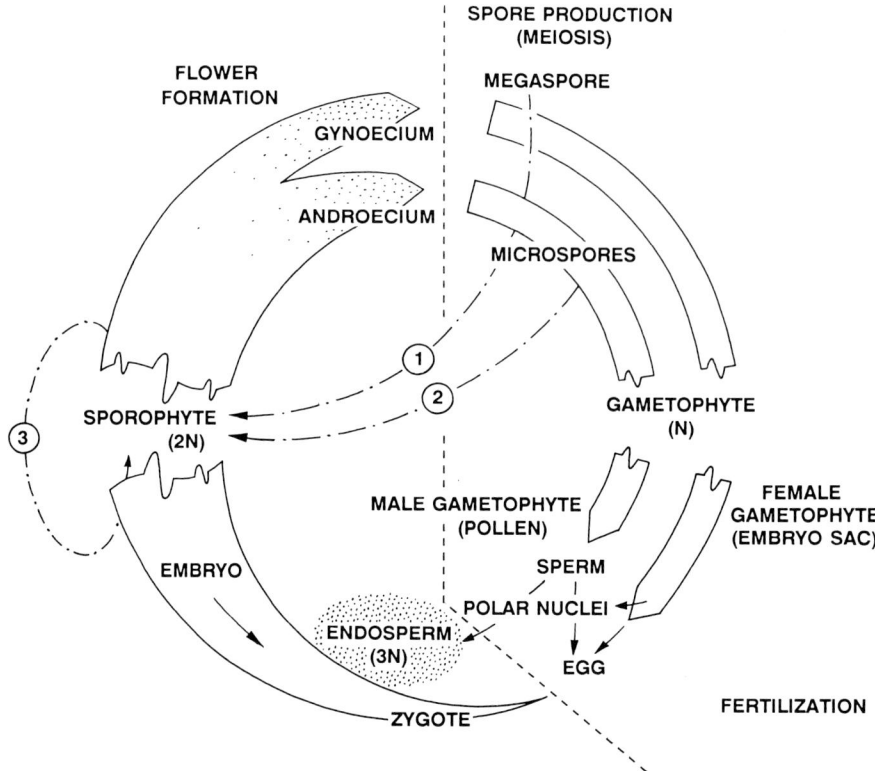

Fig. 2-6. A generalized life cycle schema for angiosperms. The essential features include: the alternation between a diploid (2N) sporophytic phase and a haploid (N) gametophytic phase; sporogenesis occurring in two distinct organs and giving rise to two different kinds of spores (heterospory) which develop as two different gametophytes; double fertilization, involving the syngamy of sperm and egg (to form a zygote) as well as the sperm with the polar nuclei to form the primary endosperm nucleus. Natural and/or artificial propagation is indicated at (1), (2) and (3): (1) indicates the possibility, primarily *in vitro*, of the production of haploid or diploid plants from the megaspore or other cells of the ovule; (2) acknowledges the production of haploid and diploid sporophytes through anther culture; (3) acknowledges the variety of *in vitro* and other procedures by which the sporophytes of many angiosperms can be propagated.

real sense is a process with the sporophyte representing only one portion of the two-phase cycle.

A schematic such as Fig. 2-6 unfortunately oversimplifies and obscures some of the realities involved. There are, for example, many chemical and cellular cycles that may have different starting and ending points and different periods than the organism cycle (Bonner 1974). Further, mitochondria and plastids, formed in the egg or earlier, represent the starting population for the zygote and embryo of the next generation. In many cases it is assumed that there is no effective mitochondrial and plastid contribution from the male parent, and it is known that this is the basis of some maternal inheritance patterns (Chapman 1986). Other cycles which may be out of phase with the chromosome cycle include mRNA synthesis and cytoplasmic organelles as discussed by Dickinson (1987).

Another negative consequence of such a schematic is that it overshadows considerable ignorance. First, our knowledge of the crude details of the life cycle of higher plants was worked out relatively recently, between 1840 and 1900. Because some aspects of the angiosperm life cycle are abbreviated, are hidden and involve one or only a few cells, the full understanding of the life cycle would be even more obscure if it were not possible to compare it with life cycles from the lower vascular plants (see Foster and Gifford 1974). Further, some details of fertilization have only become clear with the advent of transmission electron microscopy (TEM) and are based on a very small sample of the one-quarter million extant angiosperms.

A further oversimplification is the omission of reference to the environmental and biological context on which the successful completion of the life cycle depends. Little attempt is made in this text to remedy this defect.

Beyond a brief discussion of the two phases and sporogenesis (below and Figs. 2-6 and 2-7), further details of the features of these alternative expressions are provided in subsequent chapters.

The Sporophytic Phase

Because of the successes and advances made in exploring the descriptive features of the sporophytic phase of angiosperms through the sub-disciplines of cytology, anatomy, physiology, biochemistry, etc., we are tempted to assume that developmental questions have been answered. Unfortunately, insights about the regulation of various activities at different levels of organization remain vague and incomplete: much remains to be uncovered at all levels of organization from the zygote to mature functional organs.

Developmental analysis of the zygote as it becomes an embryo and seedling is a fascinating challenge for developmentalists. Beyond the traditional descriptive summaries, considerable opportunity now exists for in-depth exploration. These include: careful TEM analysis; analyses of protein differentiation (Sanchez-Martinez, Puigdomench and Pages 1986; Boothe and Walden 1990); studies with lethal mutants (Meinke 1985; Sheridan and Neuffer 1982); identification of embryo-specific genes (Crouch and Sussex 1981; Mayer et al. 1991; Jurgens 1992) and *in situ* localization of selected gene products. Frequently, research strategies will utilize a number of these techniques. In some cases, somatic embryogenesis from a variety of *in vitro* sources might be used to explore certain embryological problems. The basic findings of descriptive and experimental embryology are summarized in Raghavan (1976, 1986) and Johri (1982, 1984).

Sporogenesis

The maturation of the angiosperm sporophyte leads to the production of the flower, which produces two kind of spores. Angiosperms are thus described as being

heterosporous. Microspores arise in the microsporangia of the anthers, and **megaspores** form in the megasporangium (within the ovule). These events are discussed further in Chapters 5 and 6.

Meiosis, by which the chromosome number is halved, is the common cytogenetic component of both microsporogenesis (Chapter 5) and megasporogenesis (Chapter 6). Because of its critical and formative effects on development, it is appropriate to outline it briefly here, but more serious treatments should be consulted for details (Dickinson 1987).

Meiosis (Fig. 2-7), both in the anther and in the ovule, accomplishes two fundamental functions in the life cycle of angiosperms. First, it provides for the orderly reduction from the sporophytic diploid number of chromosomes to the haploid gametophytic complement. Second, through the mechanism of **crossing-over,** exchanges are made between non-sister chromatids. Further, through **independent assortment** of chromosomes, meiosis provides the genetic variability which can become the raw material of evolution.

In addition, it is suggested that in some organisms, meiosis may also provide mechanisms for DNA repair from damage and the consequences of ageing (Holliday 1984). The duration of meiosis in the anther, from 18h in *Petunia* to 400h in *Fritillaria*, is generally positively correlated with increasing DNA content and polyploidy (Bennett 1977). Another role which has been claimed for meiosis is the reversion of epigenetic traits such as the habituation of tobacco tissues to cytokinin independence (Meins and Binns 1979) or some somoclonal traits produced from tissue culture.

These events during microsporogenesis and megasporogenesis are discussed in more detail in Chapters 5 and 6. It is sufficient at this point to record that the products of the two processes are different. Microsporogenesis generates four viable microspores from each microsporocyte. Megasporogenesis, while initially producing four products of meiosis, results, in the most common case, in the production of a single functional haploid megaspore. Selective cell death consumes the other three products. Variations in these events are presented in more detail by Willemse and van Went (1984).

The Gametophytic Phase

The single-celled microspore and megaspore represent the beginning of the gametophytic phase. Because the microspore becomes the free-living pollen, it represents a very useful object for developmental studies. The alternative paths to pollen or embryos via anther culture are one of the experimental opportunities available (see Chapter 5). The megaspore, because it is enclosed (normally one per ovule) is a much more unyielding research material, though the questions it generates are no fewer.

Fertilization

While the reality of **double fertilization** (the fusion of sperm nuclei with egg nucleus and polar nuclei) has been appreciated since first described independently

The Angiosperm Flower and Life Cycle

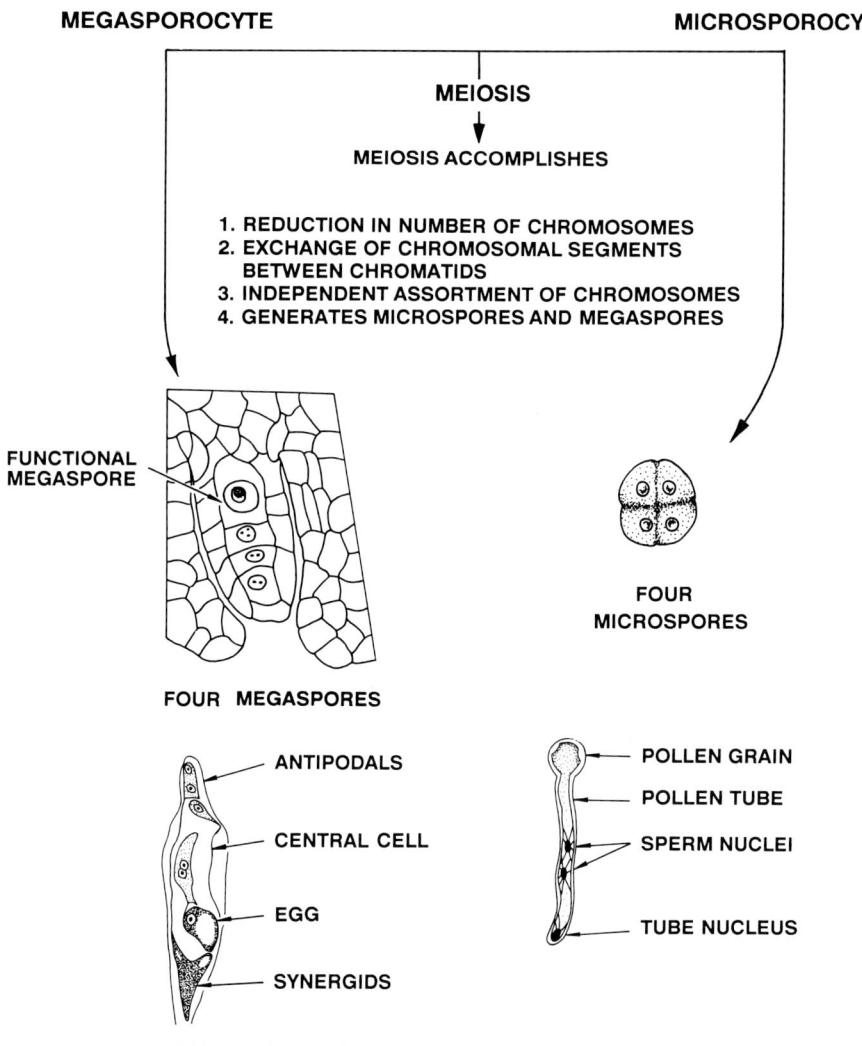

Fig. 2-7. A summary of the cytological contributions of microsporogenesis, megasporogenesis and gametogenesis in the life cycle of angiosperms.

by Nawaschin and Guignard in 1898–99 (Maheshwari 1950), the fine details of the process have been uncovered only recently and are currently being expanded (see review by van Went and Willemse (1984)). The usual consequence of double fertilization is the formation of the diploid **zygote** and the triploid **primary endosperm nucleus** from which the **endosperm** is produced. This description fits the situation in the majority of angiosperms, but other patterns are known.

Speculation about the significance of double fertilization in the success of the angiosperms is ongoing. In essence, it would appear that this mechanism provides an opportunity for the differential segregation of those traits associated with the production of nutritive tissue (endosperm) for the seed from those traits associated with successful embryo development. With the initiation of each tissue (zygote and endosperm) from separate fertilizations, the opportunity exists for the optimal, independent specialization of each. In gymnosperms, where the nutritive tissue is derived from the female gametophyte (as is the egg) this potential for specialization is more remote.

In concluding this section it is worthwhile to consider how the angiosperm life cycle might be viewed developmentally, in contrast to earlier, primarily phylogenetic or comparative interpretations. The changes in cell expression and changes in growth and physiology associated with different cell types and organs that occur from zygote to sporophyte maturity can be viewed, in current developmental theory, as involving changes in genetic expression and interaction with a variety of environmental cues. The changes associated with the alteration between the sporophytic and gametophytic phases at sporogenesis and between the gametophytic and sporophytic phases at fertilization can also be viewed within this conceptual framework. The clarification of the details of these interactions will be the responsibility and opportunity of future developmentalists.

Suggestions for Further Reading

The morphology of flowers from selected species of 100 families is illustrated with drawings and described by Hickey and King (1988). Macrophotographs of portions of fifty different flowers are presented in Sattler (1973a), many using the epi-illumination technique (Sattler 1968). Many views of the early stages of initiation are presented. Other discussions of the variations in flower structure are found in Benson (1957), Weberling (1989), Davis and Cullen (1989) and Bell (1991). The first five chapters of Cronquist (1988) provide an interesting introduction to the complexity and the challenge provided by studies of the taxonomy and assumed history of the angiosperms. Comparative observations on the life cycles of angiosperms and other land plants are provided in texts such as Gifford and Foster (1989).

3

Photoperiodic Induction, Evocation and Floral Initiation

CARL N. McDANIEL

Our primary objective in this book is to cast the formation of the flower within a developmental perspective. In the first part of this chapter, however, it seems appropriate to make a diversion from the shoot apical meristem and briefly consider inputs from other parts of the plant that are intimately associated with the developmental behavior of the shoot apex. In some plants the control of these inputs is tightly linked to environmental cues such as light and temperature, while in others the control appears to reside, in large part, with endogenously regulated cues. It is well established that the leaves can provide stimulatory as well as inhibitory signals (Lang 1987; Murfet 1989), while in some plants the roots have been reported to have an inhibitory influence on flowering (McDaniel 1984a). In some plants the influences from the leaves and roots appear to be integrated by the shoot system so as to ensure a period of vegetative growth prior to reproductive growth (Schwabe and Al-Doori 1973; Miginiac 1978; Gebhardt and McDaniel 1991).

Photoperiodic Induction

Photoperiodic induction refers to the photoperiod-influenced processes that occur in the leaves resulting in the transmission of an as yet uncharacterized signal(s) whose primary influence appears to be manifest at the shoot apex (Evans 1969a; Lang 1987). The importance of the photoperiod in the flowering of some plants was elucidated in the early part of this century when Garner and Allard (1920) firmly established the photoperiod as a major factor controlling flowering in soybean and tobacco and subsequently in many plants.

We have come to classify plants as a function of their flowering response to the photoperiod: short-day plants (SDP), long-day plants (LDP), and day-neutral plants (DNP) (Table 3-1). Some plants require a sequence of photoperiods such as short days followed by long days (see Lang (1965) and Bernier, Kinet and

Table 3-1 Photoperiod and Flowering: Three Basic Developmental Patterns

Photoperiod (24-hr day)	Long-day Plant	Short-day Plant	Day-neutral Plant
Critical Day Length			
A. Day length less than species-specific critical length	vegetative	floral	floral
[Light break a]	floral	vegetative	floral
B. Day length greater than species-specific critical length	floral	vegetative	floral

[a] Light break in the dark period leads to the indicated response in some long-day plants and in many short-day plants.

Sachs (1981a) for detailed consideration of flowering types). In some cases the photoperiod merely accelerates the time to flowering, a facultative response, while in others flowering will not be initiated without the appropriate photoperiod, an obligatory response. It must be realized, however, that other environmental factors (e.g., temperature, irradiance, nutrition) can markedly influence and in some cases eliminate a photoperiodic requirement (Bernier, Kinet and Sachs 1981a).

Extensive research on the relative importance of the light and dark periods has firmly demonstrated that the importance of each depends not only upon the response type but also upon the variety being characterized. In general, however, the night period has been considered to be of major importance because for SDP light breaks during an otherwise inductive night can prevent flowering while light breaks in an otherwise inhibitory night can allow some LDP to flower. Elegant physiological studies have demonstrated that there are endogenous rhythms of sensitivity to light breaks, and these rhythms can be complex (Vince-Prue, Thomas and Cockshull 1984; Deitzer 1989).

Temperature can also be a controlling factor in the flowering response. In some plants a period of cold, for vernalization, may accelerate or be required for flowering. Whereas daylength is perceived by the leaves, low temperatures act on the meristem; however, other cells/tissues can be influenced (Purvis 1961; Lang 1965; Bernier, Kinet and Sachs 1981a). A less extreme response to temperature is observed in plants like tobacco where the day and night temperatures can substantially alter the number of leaves produced by a meristem before forming a terminal flower (Thomas et al. 1975). In this instance it is unclear whether the temperature is acting on leaf or meristem processes.

By employing grafting procedures (e.g., shoot tips, leaves, approach grafts) and by exposing various parts of a plant to different photoperiods, it has been established that the leaves are the primary recipient of the photoperiodic signal and that leaves are, in photoperiodic plants as well as in DNP, the source of stimulatory and, in some cases, inhibitory signals that influence the developmental expression of shoot apical meristems (Zeevaart 1958; Lang 1965, 1987; Murfet 1989). Phytochrome appears to be intimately involved in the photoperiod-sensing mechanism but the details of how photoperiodic time is measured and of the explicit biochemical changes that are responsible for the inductive activity of leaves have eluded researchers (Vince-Prue, Thomas and Cockshull 1984; Deitzer 1989).

The acquisition of competence by a leaf to respond to an inductive photoperiod, as well as the maintenance of inductive capacity by the leaves, varies among plants. Almost every imaginable situation has been observed (Evans 1969a, Halevy 1985): the cotyledons are sensitive and respond to a single inductive photoperiod in *Pharbitis nil* cv. Violet (morning glory) and *Brassica campestris*; the first leaves produced are insensitive or at least far less sensitive than ontogenetically older leaves in *Xanthium strumarium* (cocklebur) and *Nicotiana tabacum* cv. Maryland Mammoth (tobacco); leaves must be exposed to at least several inductive photoperiods in *Silene armeria* and *N. tabacum* cv. Maryland Mammoth; once induced, leaves maintain the capacity to induce flowering until they senesce in red-leaved *Perilla crispa* and *X. strumarium*. The above discussion indicates that photoperiodic induction refers to leaf processes that are most likely not the same in all plants. In general, photoperiodic induction leads to changes in the export of signals from the leaves that influence the flowering process at the apex; in different plants, however, the upstream control mechanisms will most likely exhibit diversity.

It is important to recognize that photoperiodic induction is not an inductive process as defined by developmental biologists. As a generic developmental term, induction refers to the action of a developmental signal on a cell/tissue that leads to a change in or restriction of the developmental fate such that the cell/tissue is determined for a particular fate. In photoperiodic induction, the photoperiod acts on a leaf to change its physiology. The important developmental event in leaf formation, as far as photoperiodic induction is concerned, appears to be the commitment of a leaf to develop competence to respond to the inductive photoperiod.

There is also an interplay among temperature, photoperiod and irradiance such that a given plant may be stimulated to flower under several quite different sets of conditions (Bernier, Kinet and Sachs 1981*a*, 1981*b;* Bernier 1986). All of this complexity has confounded attempts to provide general models applicable to photoperiodic induction in a wide variety of plants (Bernier 1988; McDaniel 1984*a;* McDaniel, Singer and Smith 1992). From evolutionary and developmental perspectives this immense complexity is not unexpected, as will be discussed in the last section of this chapter.

It is clear that the leaves are a source of signals that influence the developmental patterns expressed by shoot apical meristems. The hypothetical floral stimulus, or the slowest moving component if the signal has more than one component, appears to move from the leaves to the apex with rates of translocation that vary from several cm/h (*Lolium temulentum*, Evans and Wardlaw (1966)) to almost 40 cm/h (*Pharbitis nil*, Takeba and Takimoto (1966); King, Evans and Wardlaw (1968)).

To date there has been no accepted identification of a signal from the leaves that acts on the shoot apex to cause or to inhibit floral initiation. Numerous attempts have been made to identify these signals, and the literature contains many promising reports which have not stood the test of time. The complexity of the situation can be seen by the response of *L. temulentum* strain Ceres to gibberellins (GA). As has been known for over thirty years, a number of long-day plants, including *L. temulentum*, will flower in non-inductive photoperiods if GA is applied to the leaves (Evans 1969c; Lang 1965). Recently Evans et al. (1990) reported on the structural features of the GA molecule that give it high floral-inducing activity in *L. temulentum* when the GA is applied to the leaves. These data implicate GA as influencing flowering and might even lead one to consider a specific GA to be the floral stimulus in *L. temulentum*. When apices are explanted from non-induced plants and placed on a GA-containing medium, however, they do not flower as would be expected if GA were the inductive signal from the leaves (McDaniel, King and Evans 1991). Since apices from induced plants initiate almost normal inflorescences more consistently on media with GA than without it, GA appears to play a role in floral morphogenesis if not initiation. Although GAs as well as other PGS have been shown to influence the flowering behaviour of many plants, most flowering physiologists have yet to be convinced that they function as the inductive signals from the leaves.

Although grafting experiments clearly indicate communication between the leaves and the apex, the lack of chemical identification of the compounds involved has led some to question the hormonal basis of this communication. It has been proposed that the apex is nutritionally deprived and that flowering results from a reallocation of nutrients to the apex (Sachs 1977; Sachs and Hackett 1977). Even if this proved to be the case, the control of nutrient allocation by the leaves must then be explained. It has also been proposed that hormones may be involved not as a function of a concentration change but rather as a phase of a pulsed signal (Khait 1986). Other types of signals such as membrane permeability changes are also possible, but few data have accumulated on alternative explanations.

Current dogma coupled with success in explaining some developmental phenomena via changes in gene expression has led many researchers to expect inductively active leaves to express different genes than leaves that are not inductively active. Identification and characterization of new gene products that are associated with the inductive capacity of leaves have not been accomplished although differences have been observed (Warm 1984; Lay-Yee, Sachs and Reid 1987). New techniques and approaches will be required before we will be able to understand the mechanistic bases for the various leaf processes associated with the capacity of leaves to influence floral morphogenesis.

Evocation

The origin of the word **evocation** in the flowering literature came from Evans (1969b) who wanted to remove the confusion about what was meant by "photoperiodic induction." Usage of this term had evolved to make no sharp distinction between leaf processes and apex processes. Evans wrote:

> On the grounds of priority, the term photoperiodic induction should be used for the processes occurring in the leaf. Those at the shoot apex would then need to be qualified as apex induction, or given a new term. One suggested long ago by Waddington for comparable events in embryology seems appropriate: evocation. In the sense of calling forth memories and summoning to new energies and a higher fate, evocation is an apt term to replace shoot apex induction. It has the further advantage that Waddington has repeatedly emphasized its separation from the succeeding processes of differentiation (Waddington 1966a). (From Evans (1969b).

Evocation is a developmental term equivalent to induction but specifically referring to processes in the apex associated with the fating of cells for reproductive development. Photoperiodic induction results in the movement of a signal(s) from the leaves to the apex where the signal(s) acts on the apex by an inductive process called evocation to commit the apex for a pattern of floral morphogenesis. Thus, evocation results in a developmental state change in the apex such that the apex becomes florally determined.

It is clear that both stimulatory and inhibitory signals can influence the response of the shoot apex (Lang 1987; Murfet 1989). It is possible that flowering may ensue when the inhibitor level decreases. In practice, since neither the floral inhibitor(s) nor the floral stimulator(s) have been identified, a decrease in inhibitor level would be indistinguishable from an increase in stimulator concentration. In this chapter we will focus on the floral stimulus from the leaves; however, it is likely in some cases to be a simplification of the interactions among leaf signals that are occurring during floral initiation.

Numerous patterns of morphological change follow evocation, leading to a multiplicity of inflorescence and flower types. Morphologists and anatomists have exquisitely described these numerous patterns at the light microscope (LM), TEM and SEM levels. It appears that there are many ways to initiate an inflorescence or a flower. The developmental importance of these patterns will be considered in later chapters (see Chapters 4, 5, 6). Of interest within the context of floral initiation is the observation that, during that transition from vegetative growth to floral development, the patterns of cell division and the organization of the meristem exhibit marked changes such that cells which had long cell cycles begin to divide more rapidly (Figs. 3-1 and 3-2).

These observations led researchers in the 1950s to propose that a group of meristem cells are passively carried until the time for reproductive development (Buvat 1952). The morphological aspects of floral initiation that follow evocation would then involve the activation of these cells which subsequently form the re-

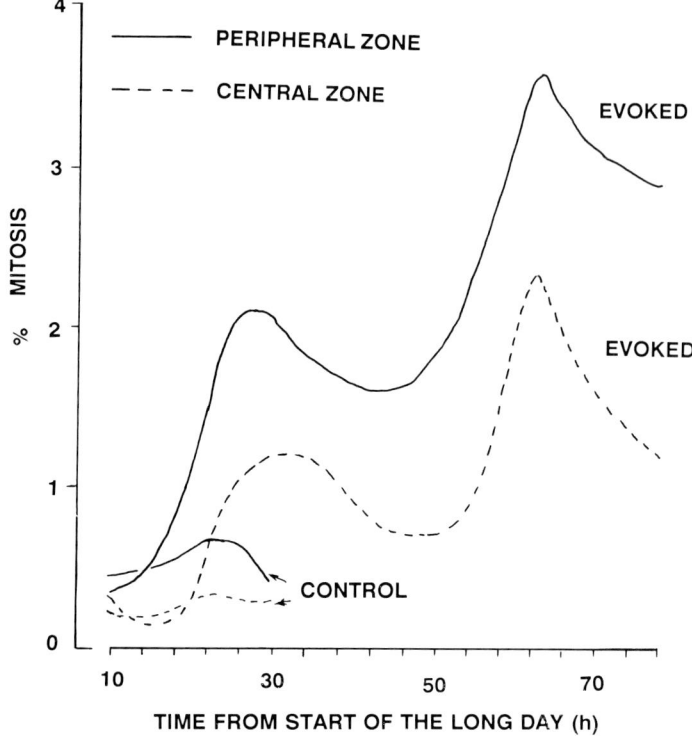

Fig. 3-1. Mitotic index in the peripheral and central zones of the meristem of *Sinapis alba* as a function of time after start of the inductive LD. Vegetative meristems (control) and evoked meristems as indicated. From Bernier, Kinet and Bronchart, *Physiol. Veg.* 5:311, (1967). Reproduced by permission of Gauthier-Villars, ed., Paris.

productive structures. Clonal analysis studies in maize (McDaniel and Poethig 1988; Fig. 3-3A) and sunflower (Jegla and Sussex 1989; Fig. 3-3B) have established that descendents of the same meristem cell can participate in the formation of both vegetative and reproductive structures, thereby providing direct evidence against this hypothesis. At this time there is no reason to postulate that there are germline cells as found in many animals because, in plants, embryonic cell lineages have not been linked exclusively to gamete production (Johri and Coe 1983; Dawe and Freeling 1990).

Floral Determination

Evocation results in changes in an apex such that it becomes florally determined. It is the expression of this developmental state that leads to the initiation of a floral apex and ultimately to the formation of flowers. At this time we do not understand the mechanism nor the molecular changes involved in the commitment

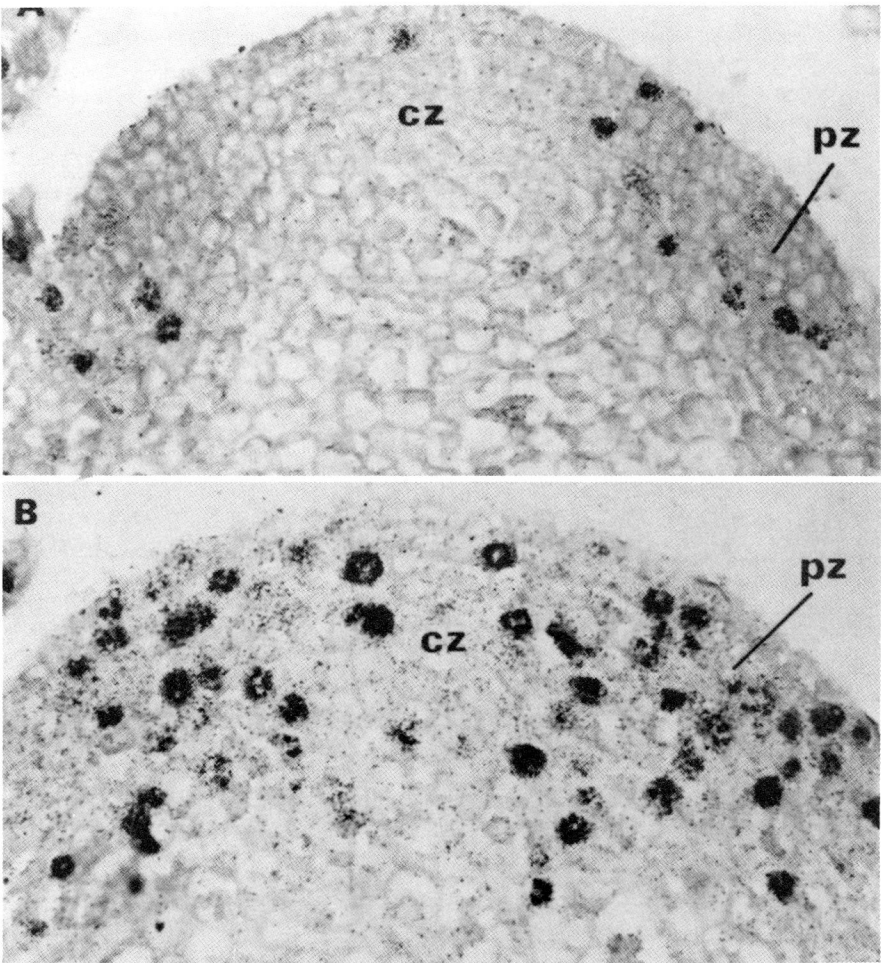

Fig. 3-2A–B. Autoradiographs of meristems of the LDP *Sinapis alba* labeled with [^{3}H]thymidine for 4 hr, showing sites of synthesis of DNA. (A) An intermediate meristem of a 2-month-old plant growth in SD. (B) A pre-floral meristem collected 38 hr after start of induction by a single LD. Note the marked increase in the number of labeled nuclei in both the peripheral (pz) and the central zone (cz). Reprinted with permission from Bernier, Kinet and Sachs (1981*b*), *The Physiology of Flowering,* vol. 2. Copyright CRC Press Inc. Boca Raton, FL.

of an apex to a developmental fate of floral morphogenesis. As a result of this lack of knowledge, floral determination is an operationally defined concept. States of determination such as leaf determination or floral determination are evaluated by assessing developmental fates in different environments (Table 3-2). The classical procedure is to contrast the developmental fate *in situ* with that in another environment in which the developmental fate would normally be different (Slack

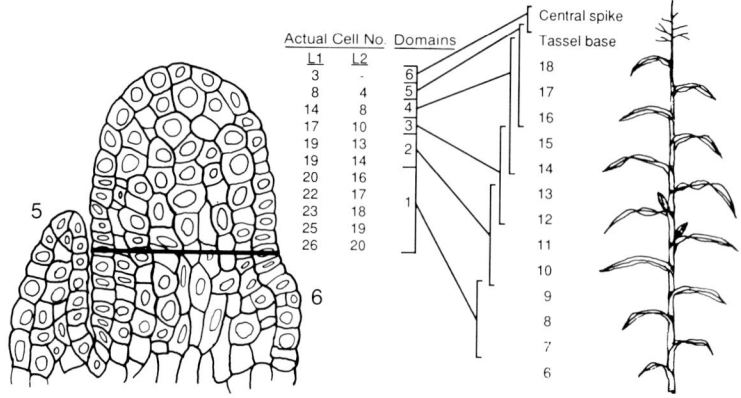

Fig. 3-3A. Fate map of the shoot apical meristem of corn at the time of germination. A camera lucida drawing of a median, longitudinal section represents the meristem of a germinating embryo. The horizontal line delineates the meristem proper, which produced shoot structures above the sixth leaf. The numbers 5 and 6 indicate primordia which produced leaves 5 and 6. At the time of germination the height of the meristem proper was about 90μm. The number of cells in the LI and LII layers was estimated by counting cells in serial cross sections and in median longitudinal sections of three meristems. The portion of the mature corn plant which was produced (except leaf 6) by the meristem proper is shown diagrammatically on the right and was about 2.2m tall. Fates for domains of cells in the meristem were assigned to lineage units of the adult corn plant. From McDaniel and Poethig (1988) with permission.

1983). In the case of floral determination, if an apex flowers with the same pattern in the new environment as it would have in its original location, it is considered to be florally determined (McDaniel 1984*b*).

One might ask at what organizational level is floral determination programmed: whole plant, shoot apex (shoot apical meristem in association with several leaves and leaf primordia), shoot apical meristem (cells distal to leaf primordia), or cell level? The classical whole plant shifts of photoperiodic plants from inductive to non-inductive conditions establish that the plant, as a whole, in inductive conditions gains the capacity to flower even when returned to non-inductive conditions. The kinetics of whole plant floral determination are diverse. Some plants like *X. strumarium* or *L. temulentum* require only a single inductive photoperiod, while others like *S. armeria* and *N. tabacum* cv. Maryland Mammoth require a week or more of inductive photoperiods. What accounts for these differences? For the most part we do not know; however, leaf sensitivity, signal strength, signal stability, and apex sensitivity are good candidates. We must realize that the mechanism and the point of developmental control in one species will not necessarily be predictive for other species even if they have similar physiologies (e.g., not all LDP employ the same control processes to influence the time of flowering).

The timing of floral determination has also been assessed in the context of the whole plant by the application of metabolic inhibitors or growth regulators (Evans 1964, 1966; Kinet et al. 1971). These treatments, in effect, modify the environ-

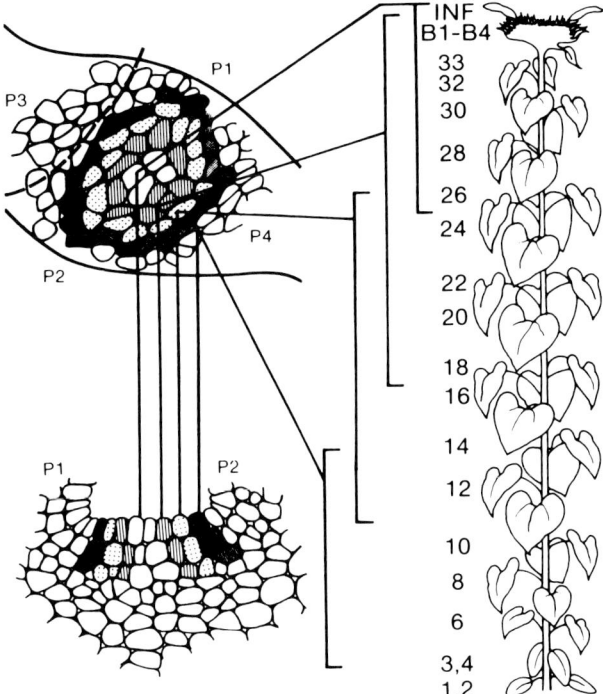

Fig. 3-3B. Fate map of the sunflower shoot meristem in the dry-seed embryo. Camera lucida drawings show a surface view of a meristem (above) and a median longitudinal section of a meristem (below). P1 to P4 label the positions of the leaf primordia present on the meristem. Cells are differentially shaded to show how the meristem could accommodate four circumferential populations of cells from its periphery to its center. Comparison of the two sections shows that the number of cells in the LI, LII and LIII layers of the meristem is similar. Brackets indicate the proposed contribution of the different populations of cells to the mature sunflower shoot diagrammed on the right. Numbers of the diagram label nodes: B1–B4 are the involucral bracts; INF is the inflorescence (florets and floret bracts). The involucre and inflorescence are shown in longitudinal section. The fate of cell populations is indicated. The embryo meristem is approximately 50μm in diameter, the shoot is as tall as 2m at maturity and the inflorescence is approximately 20cm in diameter. From Jegla and Sussex (1989) with permission.

ment of the apex. This approach is only practical in those plants where it is possible to know with some precision when the leaf-derived floral stimulus arrives at the apex. Although it is reasonable to assume that the metabolic inhibitors or growth regulators are acting on the cells at the shoot apex where evocation is occurring, data from these types of assays only confirm that floral determination can be expressed at the whole plant level.

Apex culture with numerous species has established that, once floral initiation has begun, there is a strong tendency for flower development to continue when the culture medium is appropriate (Dickens and van Staden 1988). In a small number of cases apices were cultured prior to flower initiation. For example, by

Table 3-2 Assessment of Developmental Determination

Starting Material	Procedure	Outcome: Developmental Fate	Inferred Developmental State of Starting Material
Leaf primordium	Stays intact on apex	Forms leaf	—
	Removed and cultured	Forms shoot apex	Not determined for leaf formation
		Forms leaf	Determined for leaf formation
Shoot apical meristem which has produced 28 leaf primordia	Stays intact on plant	Produces 2 leaves and then a flower	—
	Removed and cultured	Produces 30 leaves and then a flower	Not determined for floral development
		Produces 2 leaves and then a flower	Determined for floral development

culturing apices of *P. nil* cv. Violet (SD dicot), first Bhar (1970) and then Matsushima et al. (1974) established that isolated apices gained the capacity for floral morphogenesis within several hours (Matsushima et al. 1974) to 36h (Bhar 1970) after the end of the inductive night. Larkin, Felsheim and Das (1990), employing a bud grafting assay, provided data indicating that the bud of a six-day-old seedling (meristem plus six leaf primordia with the largest being 2–3 mm in length) becomes florally determined within hours of the end of the inductive night. The tight correlation between whole plant and apex floral determination has also been demonstrated for *N. tabacum* cv. Maryland Mammoth (Gebhardt and McDaniel 1987; Smith and McDaniel 1992) and *L. temulentum* (McDaniel, King and Evans 1991).

In 1955 Skoog reported that isolated stem pieces of *N. tabacum* cv. Wisconsin 38 would regenerate floral shoots in culture (Skoog 1955). This phenomenon of *in vitro* regeneration of a meristem which would form a flower directly or a floral shoot with several leaves has been observed in other species but has been most extensively characterized in tobacco (Tran Than Van 1981; Fig. 3-4). Rajeevan and Lang (1987, personal communication) have established that explants of epidermal layers of pedicels from all photoperiodic types in the genus *Nicotiana* have the capacity to regenerate flower buds in culture. The capacity of explants to produce floral buds is correlated with the source of the explant and with the developmental age of the source plant. In day-neutral tobacco, explants from young vegetative plants produce very few floral shoots while those from plants at anthesis produce many (McDaniel, Sangrey and Jegla 1989). Explants from flowering tobacco plants produce floral shoots almost exclusively when taken from the inflorescence, with the frequency decreasing to almost zero when taken from the base of the plant (Aghion-Prat 1965; McDaniel, Sangrey and Jegla 1989). These observations indicate that cells throughout the tobacco plant can organize two

Fig. 3-4. Flowers regenerated from *Nicotiana tabacum* thin layer. Original Courtesy of Stefan Eberhard.

different types of shoot apical meristems: (1) those that form several leaves and then a flower, or a flower directly; and (2) those that form vegetative shoots. Does this mean that the floral stimulus from the leaves can act on cells throughout the plant via the evocation process which normally occurs at the apex? Is floral determination programmed at the level of cells?

Cells, organized as a shoot apical meristem, form lateral organs in specific spatial patterns and temporal sequences. A vegetative meristem expresses one pattern and sequence, while a floral meristem expresses a different pattern and sequence. It is difficult to know exactly what type of information is responsible for such morphogenetic patterns and how the expression of such information is programmed. It is not possible for a single cell to form a flower, so how can it be possible for a cell to be florally determined, i.e., programmed to form a flower? Clearly a cell can only be programmed to participate in the formation of a flower as a part of a group of cells that have been organized as a meristem. The fact that explants from several sources in the day-neutral tobacco plant have the capacity to organize developmentally different types of meristems, including those which form a flower directly, indicates that an early state of floral determination may be induced in cells, perhaps individual cells. This early state of floral determination appears to be related to the communication among cells within a meristem to allow for the expression of the spatial and temporal patterns of lateral organ initiation associated with flower formation. There are no compelling reasons to believe that the induction process leading to floral determination in cells outside of a shoot apical meristem is different from evocation that occurs in cells of the meristem. Thus, although Evans (1969*b*) considered floral evocation to be a process(es) uniquely associated with the shoot apex, it may be more generally applicable.

Studies of vernalization have also shown that, although cold temperature acts at the shoot apex to induce a developmental state change, other cells not organized into shoot apical meristems can respond. For example, petioles from leaves of *Lunaria annua* (honesty) and *Thlaspi arvense* (pennycress) can regenerate shoots when isolated. If the leaves are vernalized, then the regenerated plants flower; while if they are not vernalized, only vegetative shoots are regenerated and these require vernalization before flowering (Wellensiek 1961; Metzger 1988). Thus, vernalization brings about a developmental state change which appears to be imprinted at the cellular level.

Evocation Period

In two plants it has been possible to measure with some degree of precision the duration of the evocation period. Such a measurement requires two things. First, it must be possible to estimate the arrival time of the floral stimulus at the apex. Second, an assay must be available for assessing when the apex has become committed to a pattern of floral morphogenesis. The evocation period would then be the time interval between arrival of the floral stimulus at the apex and floral determination of the apex. The duration of the evocation period has been estimated in *P. nil* cv. Violet (short-day dicot) and *L. temulentum* cv. Ceres (long-day monocot).

In *P. nil* seedlings the cotyledons respond to the photoperiod and export a floral stimulus which travels at a rate of almost 40 cm/h (Takeba and Takimoto 1966; King, Evans and Wardlaw 1968). In six-day-old seedlings the distance from the base of the cotyledons to the apex is only several cm; thus, the signal should arrive at the apex within less than an hour of its export from the cotyledons. Studies of Bhar (1970) and Matsushima et al. (1974) showed that a state of floral determination, as assessed by an apex culture assay, was established in the apex somewhere between 5 and 36 hours after the end of the inductive night. Employing a grafting assay, Larkin, Felsheim and Das (1990) demonstrated that axillary buds 3 and 4 were florally determined within an hour of the arrival of the floral stimulus, while floral determination of the terminal bud took several more hours (Fig. 3-5).

Although *L. temulentum* is a long-day plant and a monocot, the kinetics of floral determination are strikingly similar to those of *P. nil*. The floral stimulus moves more slowly in *L. temulentum* than in *P. nil* and should reach the apex about 4h after it exits from the leaf blades (Evans and Wardlaw 1966). Using an apex culture assay it was established that floral determination occurred in some apices at about the time that the floral stimulus was estimated to arrive at the apex and in all apices within about 9 hours of probable stimulus arrival (Fig. 3-6).

If *P. nil* and *L. temulentum* are typical of plants where the time of flowering is primarily influenced by the output of floral stimulus from the leaves, then the

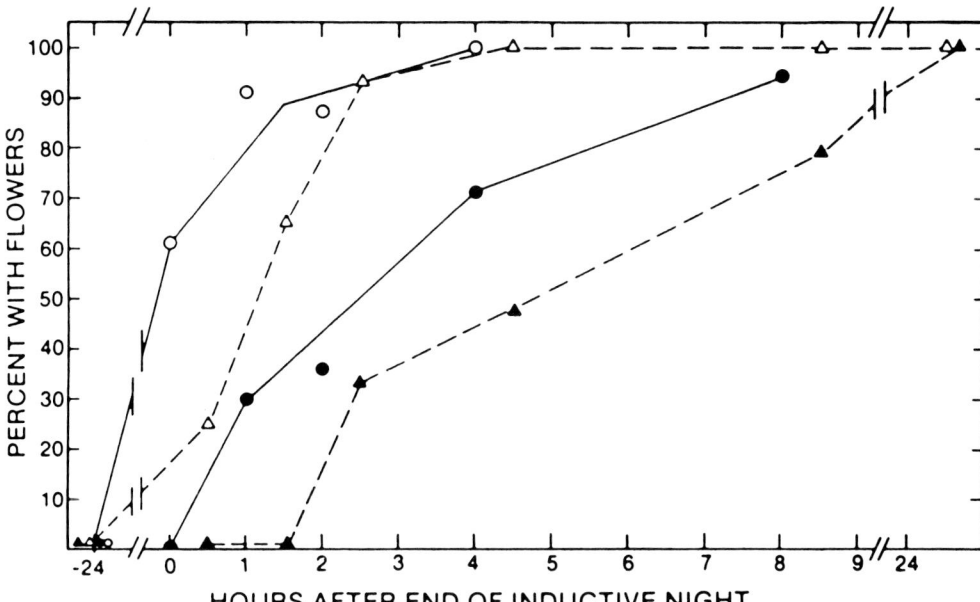

Fig. 3-5. Flowering response of *Pharbitis nil* Chois cv. Violet cotyledonless plants and grafted terminal buds. Percentage with axillary flowers (plants ○, buds △) and terminal flower (plants ●, buds ▲) plotted as a function of time of cotyledon removal or bud grafting. Cotyledons were removed and terminal buds grafted onto uninduced plants before and at various times after the inductive night. Plants were subsequently grown in noninductive conditions and scored for flowering. Data from Larkin, Felsheim and Das (1990) with permission.

evocation period in these developmental types is only several hours long under ideal conditions. Other ways of regulating the time of flowering have evolved and in some of these the evocation period may be of minimal consequence. For example, in day-neutral *N. tabacum* cv. Wisconsin 38 there is evidence that the floral stimulus is being constantly produced by the leaves but the apex does not become florally determined until the plant has attained some minimal number of nodes (McDaniel 1980; Singer and McDaniel 1986; Gebhardt and McDaniel 1991). Thus floral stimulus appears always to be present and one can only assess when the apex becomes florally determined. By employing a terminal bud rooting assay, it was established that considerable vegetative growth preceded floral determination but that only a several-day period was required for a specific bud to go from a vegetative to a florally determined state. These and other observations indicate that in day-neutral *N. tabacum* the time of flowering is not primarily controlled by photoperiodically regulated leaf processes but rather by an interaction among leaf signal(s), roots, nodes-internodes, and meristem competence (Gebhardt and McDaniel 1991; McDaniel 1992; McDaniel, Singer and Smith 1992; Singer, Hannon and Huber 1992).

The data from the assays employed to ascertain a florally determined state in

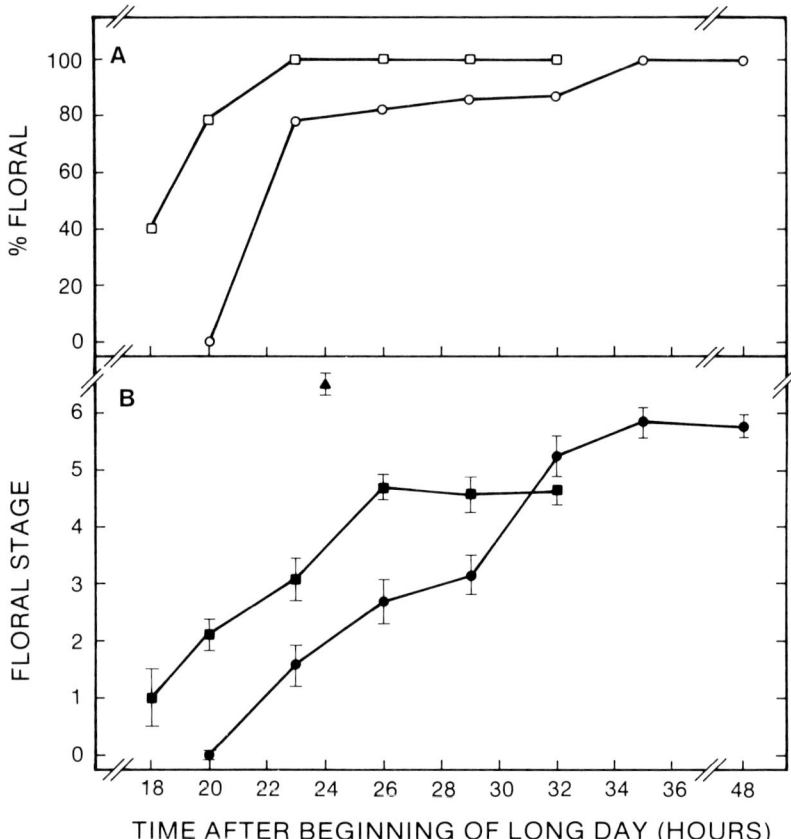

Fig. 3-6A–B. Flowering response three weeks after the removal of leaf blades or excision and culture of shoot apices from plants of *Lolium temulentum*, strain Ceres, exposed to one long day when 48 days old. (A) Percent exhibiting floral determination; i.e., percent of plants (□) or of cultured apices (○) reaching at least the double ridges stage of floral development. (B) Average floral stage of plants (■) or culture apices (●) based on the scale: 0, vegetative; 1, shoot apex elongated; 2, double ridges; 3, advanced double ridges; 4, glume primordia; 5, lemma primordia; 6, floret primordia. All apices excised from control plants grown in short days remained vegetative during the three-week culture period (data not plotted). The flowering stage of intact plants given one long day is indicated for comparison (▲). The vertical bars represent 2 × standard error, with 18 replicates per treatment. Data from McDaniel, King and Evans (1991) with permission.

a shoot apex or meristem are grouped in a way that implies an all-or-none change in meristem fate. As illustrated for *P. nil* (King and Evans 1969; Larkin, Felsheim and Das 1990) and for *L. temulentum* (Evans 1960), however, varying the photoperiodic-induction treatment can modify the rate as well as the extent of floral morphogenesis. Even the imposition of an additional inductive night after the initiation of floral morphogenesis in *L. temulentum* can accelerate inflorescence development. Additionally, there are many documented cases of reversion

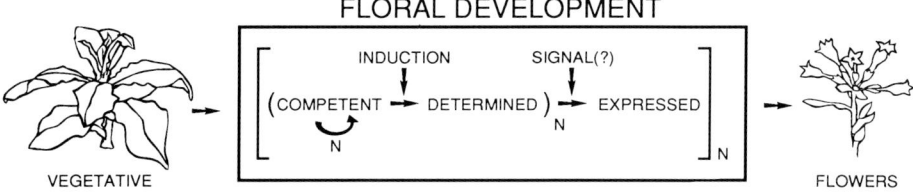

Fig. 3-7. Some of the possible events which may occur during the ontogeny of an individual plant. Arrows indicate the normal direction of events but do not imply irreversibility. Competent cells/tissues are those which have the ability to respond to a developmental signal in a specific way. Cells and tissues can acquire different competences via either endogenous or exogenous means. Induction occurs when a developmental signal acting on competent cells/tissues to determine them for a particular developmental fate. The determined state is subsequently expressed. The "N" indicates that the bracketed sequence may be reiterated. From McDaniel (1989) with permission.

from floral development to vegetative growth (Battey and Lyndon 1990) (see also Chapter 6). Flowering, like much of plant development, exhibits the capacity to be influenced by environmental inputs almost up to the formation of the terminal phenotype. This fact clearly illustrates that fating in plants is not an irreversible process, nor is it a single event (Fig. 3-7). Floral determination as assessed in *P. nil, L. temulentum* and *N. tabacum* appears to represent a stable developmental fating that permits, under suitable conditions, a continuation of the developmental state changes associated with the formation of an inflorescence and flowers.

Considerable data have been collected on the molecular and cellular events that were thought to be associated with evocation, in an attempt to identify those involved in the commitment of a meristem to floral morphogenesis (Bernier, Kinet and Sachs 1981*a;* Bernier 1988). As discussed above, the evocation period has only recently been shown to be of short duration, perhaps less than several hours. Since fating for floral morphogenesis appears to occur quickly and is immediately followed by floral morphogenesis in photoperiodic plants like *P. nil* and *L. temulentum* it will be difficult to separate events involved in fating from those associated with the expression of the florally determined state. As indicated at the end of this chapter and in Chapter 9, developmental genetics may be the preferred approach for gaining an understanding of the molecular and cellular events that influence fating.

Flowering as a Developmental Process

Flowering plants came into existence more than 100 million years ago (Cronquist 1988). These millions of years have permitted an enormous number of changes to be made in the original developmental sequence(s) which led to flowering in the first angiosperm(s). These changes have enabled species to be reproductively successful in essentially all terrestrial and many aquatic environments.

The fossil record gives us some impression of what the earliest flowers might have looked like, but we can only guess about their physiology, their morphogenesis and the developmental processes which influenced their formation. Of course, what we do examine is the physiology, morphology and development of extant phenotypes, trying to decipher fundamental patterns from tremendous diversity. Each disciplinary approach makes unique contributions.

The morphologist describes at various organizational levels, from the organismal to the cellular, the shape changes which produce the diversity of inflorescence and flower types. This not only has told us what has evolved but also has instructed us as to how these morphologies are produced. The physiological approach enables us to understand the critical processes that are employed for the regulation of flowering in a specific plant. This approach led to the discovery of photoperiodically controlled leaf processes, the identification of phytochrome as the major light receptor pigment, vernalization and other processes employed to regulate reproductive development. Although elegant and highly sophisticated physiological studies have given us the bulk of our knowledge on how flowering is regulated in an enormous variety of plants (Halevy 1985), a coherent understanding of the flowering process itself has not emerged (Bernier 1988). Probably, it is unreasonable to expect that a physiological approach by itself would be able to provide such a coherent understanding since, by their nature, physiological analyses tend to focus on the differences which have evolved.

The shoot apical meristem formed during embryogenesis produces leaves and only later forms a flower either directly or indirectly. This must be the ancient pattern because making a flower prior to any vegetative growth is not a pattern consistent with the evolutionary history of angiosperms (Cronquist 1988). Thus, simultaneous with the evolution of the flower would be the evolution of control processes which permitted the transition to reproductive growth after some period of vegetative growth. Central to all of development is the instructing of cells as to their developmental fate, followed by or coincident with the expression of the acquired fate. In transiting from vegetative to floral morphogenesis, we are concerned with cells acquiring and expressing new fates. An emphasis on fate and the developmental states associated with the acquisition, maintenance and expression of the developmental fates characterizes the developmental approach to flowering. This approach makes it possible to identify in a specific plant, under a specified set of environmental conditions, the developmental process which has the greatest influence on floral initiation (McDaniel 1992; McDaniel, Singer and Smith 1992). As a consequence of this type of analysis it may be possible to establish those processes which are common to flowering in perhaps all angiosperms (McDaniel 1984a).

If one focuses on the initiation of a floral meristem, two developmental states are acquired and expressed in order for a meristem to initiate floral morphogenesis. First, the meristem cells become competent to respond to the developmental signal which evokes them into a florally determined state. Second, they are evoked into a florally determined state and then express that state. It should be possible to organize plants into groups as a function of the primary developmental state or process that is employed to establish when flowers will be formed as well as to establish any associated secondary process(es).

It is difficult to demonstrate that a meristem is incompetent to respond to a developmental signal, because negative results are not easily interpreted. This is especially true if the developmental signal has not been identified and characterized, as is the case for the floral stimulus(i) and inhibitor(s). Since many woody plants have a juvenile period measured in years during which they do not flower, meristem incompetence would be a good candidate for the developmental process that is regulated. Unfortunately, the literature is not clear on this, and competence can only be indicated as a possible control point (Hackett 1985). Recent work by Singer, Hannon and Huber (1992) indicates that seedling meristems of day-neutral *N. tabacum* cv. Wisconsin 38 and long-day *N. silvestris* are less sensitive to floral stimulus than ontogenetically older meristems. Thus meristem competence appears to be regulated in shoot apical meristems of some *Nicotiana* species. Perhaps the best evidence that competence is a unique developmental state will eventually come from analyses of flowering mutants with phenotypes like *veg* in *Pisum sativum* (pea) (Reid and Murfet 1984) and *floricaula* in *Antirrhinum majus* (Coen 1991), where the floral stimulus appears to be present but flowers are not formed.

The acquisition of a florally determined state by a meristem has been established to be the point of regulation in a number of species. For example, in single-cycle photoperiodic plants like *P. nil* or *L. temulentum*, floral initiation is controlled primarily by regulating when the apex becomes evoked. Once sufficient floral stimulus reaches the apex, evocation occurs quickly and inflorescence differentiation commences almost immediately (Larkin, Felsheim and Das 1990; McDaniel, King and Evans 1991). In these plants and others like them, floral stimulus release is photoperiodically controlled in the leaves. Floral initiation in day-neutral tobacco is also a function of the time of floral determination in the apex; however, it is not dependent upon a level of floral stimulus that is primarily influenced by the photoperiod. Rather it is a complex interaction among meristem competence, root inhibitory influence, floral stimulus from the leaves and the number of node-internode units separating the apex from the roots (Table 3-3) (Gebhardt and McDaniel 1991; McDaniel 1992; McDaniel, Singer and Smith 1992; Singer, Hannon and Huber 1992).

Expression of the florally determined state is the process regulated in another group of plants. Sunflower is a day-neutral plant like tobacco that forms a uniform number of nodes in a given environment before initiation of reproductive development. Unlike tobacco, however, grafting and rooting of terminal buds has established that the number of leaves produced by a terminal meristem cannot be increased but only decreased (Habermann and Sekulow 1972). When seedling apices and half apices were cultured, the apical meristem and the new meristems that formed initiated reproductive development after producing just a few leaves (Paterson 1984). The simplest interpretation of the data from these grafting and isolation experiments is that the sunflower terminal meristem, as well as cells not in an organized shoot apical meristem, is florally determined soon after germination and that, when released from whole plant influences, reproductive development is commenced.

Bougainvillea "San Diego Red" forms a meristem in the axil of each leaf and under inductive SD conditions develops into an inflorescence (Hackett and Sachs 1985). Under non-inductive long-day conditions the meristem forms bracts, brac-

Table 3-3 Proposed Sequence of Developmental State Changes Associated with Flowering in *Nicotiana tabacum* cv. Wisconsin 38[a]

Pattern of Growth	Vegetative Growth	Vegetative Growth	Vegetative Growth	Floral Morphogenesis
Number of leaf primordia initiated	5–10	10–25	25–32	32–35
Signals:[b]				
Leaf signal strength at apex	+	+ +	+ +	+ +
Root signal strength at apex	+ +	+	+/0	0
Developmental State:[c]				
Meristem competent	−	+/−	+	+
Meristem florally determined	−	−	−	+

[a] The above illustrates the sequence of processes and events which are proposed to occur in a plant from a population of *N. tabacum* cv. Wisconsin 38 plants that produce, on average, 35 nodes below their terminal flowers.

[b] Proposed signal strength changes at the apex are based upon results of defoliation, grafting, and rooting experiments, as well as the interpretation that the number of nodes/internodes separating roots from the shoot apical meristem is important in establishing when flowering occurs. Symbols: 0, no signal; +, some signal; + +, more signal.

[c] Proposed developmental state changes are based upon rooting and grafting experiments. Symbols: −, not in state; +, in state. The factors influencing competence are not known. Floral determination is apparently influenced by the interaction between root and leaf signals with the leaf signal eventually evoking a competent meristem into a florally determined state.

Source: From McDaniel, Singer and Smith (1992).

teoles and a thorn but not florets. When axillary buds are excised from non-induced vegetative plants and grown in culture, they develop into inflorescences (Steffen, Sachs and Hackett 1988*a;* 1988*b*). Floral initiation appears to result from a lack of appropriate nutrients, and the role of the inductive photoperiod is to permit nutrient reallocation to the axillary buds. As with sunflower the simplest explanation of these observations is that flowering in *Bougainvillea* "San Diego Red" is controlled at a point after floral determination. In this plant the floral stimulus from the leaves appears to permit the expression of a florally determined state rather than evocation of a florally determined state.

The above examples illustrate how flowering plants can be developmentally grouped into a small number of classes. In general we observe that cells, meristems and organ primordia acquire competences to respond to signals, inductive signals are transmitted and fates are changed as new states of determination are imprinted and expressed (Fig. 3-7). Thus, the initiation of floral development is regulated by an interrelated network of sequence(s) of developmental state changes.

Despite all of the evolutionary modifications, the developmental perspective should also allow for the identification of processes which are common to flowering in all angiosperms. Competence to be evoked into a florally determined state and the acquisition of a florally determined state appear to be common. For different plants, when evocation occurs and when the florally determined state is expressed appear to be influenced in various ways. The fact that more than one set of environmental conditions can permit floral initiation is not inconsistent with there being common developmental states and processes in all angiosperms, since it should be possible to arrive at a specific developmental state by more than one pathway.

In flower organ formation we are beginning to understand how fate is assigned (Coen and Meyerowitz 1991). Homeotic mutations have allowed for the identification of gene products which appear to be transcriptional regulatory elements that directly or indirectly instruct the cells of an organ primordium as to what organ (i.e., sepal, petal, stamen, gynoecium) to form (see Chapter 9). How the inductive signal from the leaves leads to this pattern of gene expression is unknown. Mutations like *veg* and *floricaula* which block, in some way, the formation of flowers may permit one to relate flower initiation to flower formation.

We have a far clearer understanding of the regulation of the flowering process now than in 1920, when the photoperiod was identified as a major controlling factor in some plants. The list of major questions, however, is substantial. How does the photoperiod cause a leaf to become inductively active? What are the leaf and root signals? Is signal translocation regulated? How and where do stimulatory and inhibitory signals interact? How is competence regulated? What role do the known PGS play? How are developmental fates programmed? How are developmental fates remembered and expressed? Clearly the opportunities for expanding and enhancing our understanding of floral initiation abound.

4

The Perianth

Initiation and Early Growth of Leaves

The similarities of leaves to sepals and petals are so striking and fundamental that it is appropriate to begin the discussion of perianth development with a brief summary of what is known about leaf initiation, growth and development in leaves. Fortunately, a great deal of information is available on many aspects of leaf development (Dale and Milthorpe 1983).

Important questions—e.g., Why do leaves arise where they do? Why do they arise from the meristem at all? What are the descriptive and experimental bases for understanding the cell dynamics of leaf growth?—have been studied from a variety of points of view and in many different species. In fact, there is at present a very good descriptive basis for future exploration into the molecular aspects of leaf initiation and differentiation (Freeling 1992).

With few exceptions (for example, some organogenic tissue cultures), leaves arise on the flanks of shoot apical meristems. In theory, a primordium could arise from increased rates of cell enlargement or cell division or an alteration of the planes of these activities. The sequence of cellular events in particular meristems—probably some combination of the three—is difficult to determine, so few precise analyses are available. In *Pisum,* the formation of a primordium apparently results from changes in the direction of growth rather than from changes in the rates of growth (Lyndon 1970, 1990).

The first cellular indication of leaf initiation is frequently recognized as one or more periclinal divisions. In some cases, cell elongation perpendicular to the surface may be a prior event (Foarde 1971). In dicotyledons, the first periclinal divisions (cell plates parallel to the epidermal surface) are frequently observed in the hypodermal cell layer (2nd Tunica, T2). In grasses, periclinal divisions are observed in both the outer layer (T1) and the hypodermal layer (T2).

It is generally acknowledged, however, that periclinal divisions are only one event in a complex web of causality. Besides the chain of cellular events (assumed to begin with novel gene products) which lead to altered cell behavior, certain physical, environmental and chemical cues must be integrated into any morphogenetic hypothesis. Note that development is a continuous process involving many interactions between epigenetic and genetic components. Further discussion of the complexities of morphogenesis is left for Chapter 9.

More difficult to identify than the cellular dynamics associated with initiation are the factors responsible for site selection (Halperin 1978). These have been the subject of a great deal of speculation by different researchers. Physical pressure (Adler 1974), a field of chemical gradients within the meristem (Wardlaw 1965*b*), influences from the lower, older tissues (Plantefol 1948; Larson 1983) and available space on the meristem (Snow and Snow 1931) are among the most popular possibilities. Jean (1984) has provided a serious examination of these possibilities in the context of his mathematical analysis of phyllotaxis.

Of these, the most commonly accepted and perhaps the best studied prospect assumes that primordia arise at locations on the growing meristem when signals from overlapping inhibitory gradients permit or trigger the cellular events described above. At least for those meristems on which leaves are initiated singly in a helical (frequently called spiral) sequence [see data on leaf initiation in *Linum* (Williams 1975)], such **field hypotheses** are the most provocative (Steeves and Sussex 1989). Such hypotheses are also open to successful computer modelling (Mitchison 1977; Veen and Lindenmayer 1977). At present, however, there are no solid data on the chemical and molecular nature of these morphogenetic gradients. The direction of the helical sequence of initiation on any plant or stem is apparently randomly set, so both clockwise or anti-clockwise patterns exist in most populations, though factors are known which modify or regulate this trait.

Less well studied and still unresolved are models and hypotheses which might explain the siting of primordia in those cases which are genetically predisposed to initiate two (opposite and decussate leaf arrangement) or more (whorled) primordia simultaneously. Despite our earlier claim that leaf initiation and growth are among the best studied in plant developmental research, the actual number of carefully studied cases where all aspects of the problem of leaf initiation are covered is limited. Therefore, many details about this important process remain unexplored (Lyndon 1990).

Attempts at the experimental manipulation of leaf initiation sites have demonstrated that some movement in the location of initiation sites is possible. The results of carefully timed surgical damage, exogenous application of PGS and *in vitro* meristem culture all tend to support, though not unequivocally, the "field" interpretation of the meristem (Steeves and Sussex 1989).

Green (1980, 1984*a*) and colleagues (Selker, Steucek and Green 1992) have focused on the events taking place in the cell wall during growth and differentiation and have revealed a series of interrelated events and conditions that suggest the importance of biophysical factors. Their work suggests that cell plate formation and the antecedent alignment of the microtubules is based upon the strain derived from the stress within the tissues. Presumably, altered orientation of microtubules and therefore of cell plates and cellulose microfibrils is brought about by localized stretching, frequently of the protoderm. The source of tension is assumed to result from some combination of cell extension and wall reaction. Green, without minimizing the importance of molecular-genetic components, differs from many current developmental researchers by emphasizing the importance of biophysical factors in the wall as providing the cues for the initiation of organ formation.

Initiation, the first event in the origin of the leaf, involving one or very few hypodermal cells, is followed by corresponding division and enlargement of surrounding cells and anticlinal division patterns in the protoderm so that an obvious protuberance is formed.

These highly localized initial events are difficult to quantify, but the ability to make predictions about their location and timing leads directly to a number of penetrating developmental experiments. Surgery (Sussex 1955) and *in vitro* culture (Steeves 1966), combined with biochemical analysis, can reveal when the primordium is determined as a leaf. Data suggest that, after an early undetermined condition when manipulation to alternative expressions is possible, the developmental options become restricted early in development (Steeves 1966). The extension of these experiments to other meristems in conjunction with modern methods of molecular analysis should provide many fresh insights.

Leaves and Axillary Buds

The common association of shoot primordia (adaxial, axillary buds) with leaves has stimulated considerable speculation with regard to organ relationships within the flower. If similarities between leaf and floral organ are to be drawn, is there any reason to consider a similar two-organ relationship within the organization of the flower?

Fortunately, there are enough examples of non-correspondence to this leaf–bud relationship among the flowering plants to discredit any simple and universal extension of this concept. Leaves exist without axillary buds and stems develop in the sites of buds without leaves. In *Utricularia,* the evidence for the development of a relativistic morphology and "process morphology" rather than arbitrary organ classes is well supported (Sattler and Rutishauser 1990). In *Nymphaea,* leaves, vegetative buds and flower buds arise from equivalent sites (Cutter 1957*a*), and the sepals are equivalent to the bract: organs therefore are not axillary (Cutter 1957*b*). By way of contrast, Lacroix and Sattler (1988) discuss some of the complications that arise when stamens are adaxial to and superposed on perianth parts. Thus, despite the use of terms such as **antisepalous** and **antipetalous** (See Chapter 5) to describe the positional relationship of stamens to the perianth members, no obligatory developmental relationship is implied, at least in this presentation.

The Shift from Leaf to Perianth Organ Initiation

The sequence of initiation of the organs of the perianth on the transformed meristem can continue directly from the leaf-forming pattern through the transitional stage to the initiation of the perianth. The sequence of initiation in *Ranunculus* (Meicenheimer 1979), which has traditionally been assumed to be of this type, is shown on closer inspection—through careful measurements of angular divergences and radial dimensions from SEM micrographs—to be more complex (Fig. 4-1). Leaf, bract and sepal primordia arise in a helical sequence, whereas petal primor-

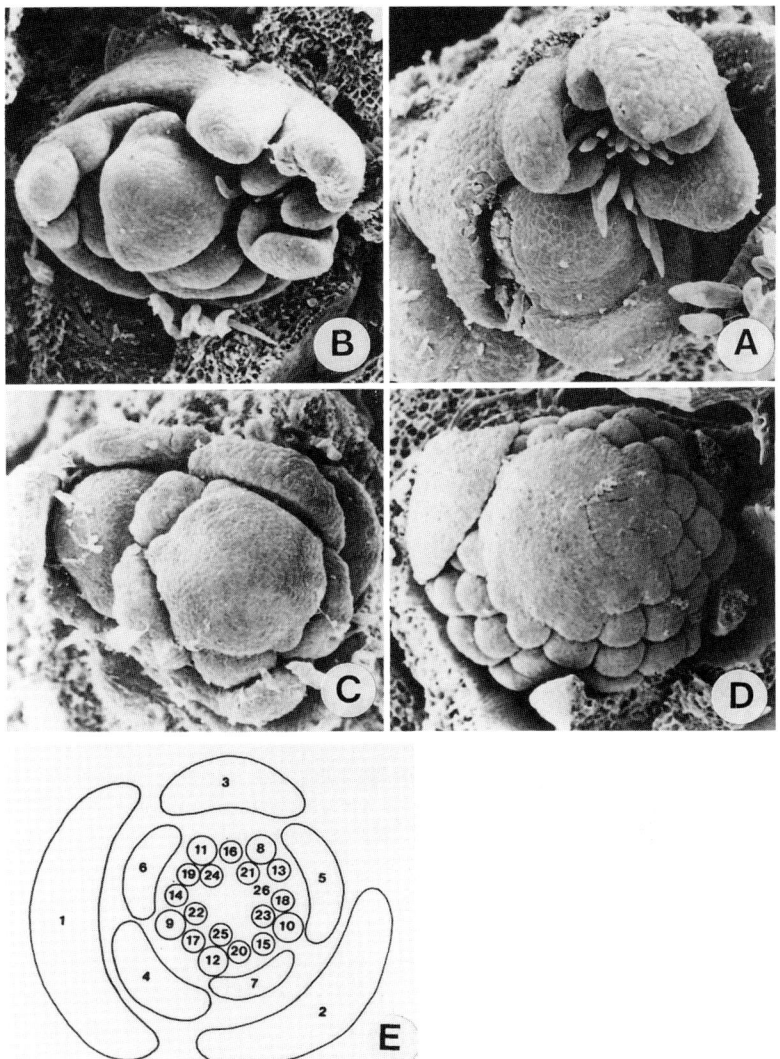

Fig. 4-1A–E. SEM micrographs of the sequence of initiation of vegetative and perianth organs on the shoot apex of *Ranunculus repens* (A–D) and an interpretive summary (E). (A) SEM micrograph of a meristem at an early transitional stage, with N9–N11 (primordia number) arranged along a left-handed generative helix (X100). (B) Later phase in transitional stage. N9–N14 are arranged along a right-handed helix (X100). (C) Petal initiation stage. Bract with axillary bud, bracteole, five sepal, five petal and two stamen primordia arranged on a right-handed generative helix (X100). (D) Stamen stage. Left-handed generative helix (X109). (E) Summary diagram of the sequence of organ initiation of organs on a *Ranunculus* meristem at an early stamen stage, illustrating the calculated relative positions of the primordia. Primordia 1–2, floral bracts; 3–7 (contact parastichies (1, 2)), sepals; 8–12 (contact parastichies (2, 3)), petals equidistant from the center and evenly spaced; 13–26, stamens. Original photographs presented as stereo pairs. From Meicenheimer (1979) with permission of *American Journal of Botany*.

dia appear to arise as a whorl of five equally spaced primordia equidistant from the apex center. Thus, the generation of **parastichies** of organs on the mature meristem may not, in all cases, indicate that a strictly helical sequence of organ initiation has been followed.

Pattern of Sepal and Petal Initiation

More frequently the initiation of sepals begins a new or altered phyllotactic pattern. In *Silene* (Lyndon 1978), leaves are initiated in pairs (Fig. 4-2) but the sepals are initiated singly, leading to a generative helix of five sepals. In Fig. 4-2A two sepal primordia and the site of the third are indicated. The pattern of initiation in this case, as viewed from above, is anti-clockwise. The locations of primordia No. 4 and No. 5 can be predicted. At a later stage (Fig. 4-2B), with five sepals initiated and developed, other primordia are also initiated. Petal primordia initiate simultaneously, or in quick succession, as small bulges low on the meristem between the sepal primordia and do so when the first few stamens are being initiated. The petal primordia are therefore all approximately the same size.

The rate at which primordia are initiated on the meristem has rarely been recorded, but we can infer, though the answers will be highly variable, that it will fall between a common rate for the initiation of leaves (one leaf every one to three days) and the rate calculated for the initiation of a flower on a sunflower inflorescence (1 flower/ 10 min) (Palmer and Steer 1985). One should expect this process to proceed, therefore, with the initiation of a sepal every few hours. Obviously, these values will vary considerably with the organ, the species and the environmental conditions during initiation.

In *Brassica* (Polowick and Sawhney 1986), flower primordia arise helically on an inflorescence apex (Fig. 4-3A and B). The adaxial sepal on each flower is initiated first, followed by the two lateral primordia. The fourth (abaxial) sepal primordium appears last. The sepals first appear as broad ridges which become continuous, at least initially. Individual sepals subsequently arise from these connected primordia (Fig. 4-3C and D).

Petal initiation, in *Brassica,* does not occur until all six stamens have appeared. They arise simultaneously at the edge of the short stamens in an antisepalous position. Initiation of the perianth involves a non-helical, zygomorphically symmetric pattern of sepal initiation and the centripetal initiation of petals. It appears that variability in the details of these events, between species in a genus and to a greater degree within a family, should be anticipated. For example, this description for *Brassica* does not describe sepal initiation for all members of the Cruciferae. In *Cheiranthus,* sepals are initiated simultaneously (Sattler 1973a).

To describe the pattern of initiation accurately and with precision is often difficult. Traditionally, casual inspection of freshly dissected meristems coupled with drawings or photographic records was acceptable. Some information about initiation patterns can often be obtained from carefully oriented transverse microscopic sections. Pattern may also be studied from killed and fixed specimens examined with epi-illumination microscopy (Sattler 1968; Posluszny, Scott and Sat-

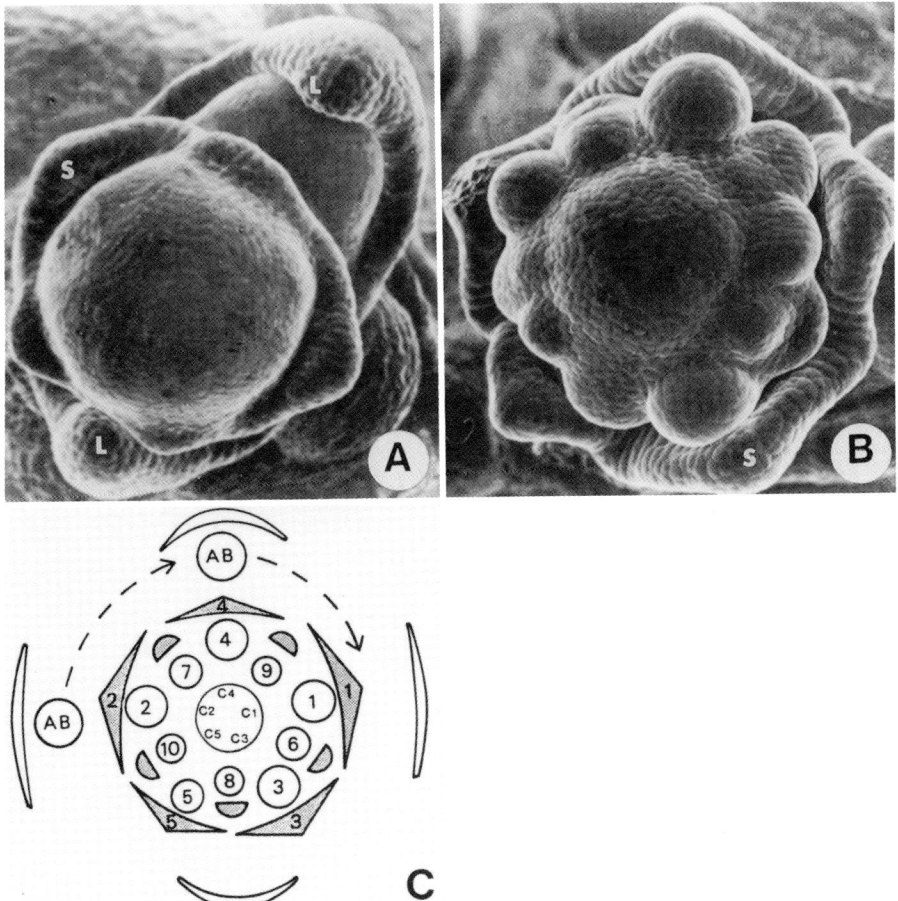

Fig. 4-2A–C. The initiation of primordia from vegetative and floral meristems of *Silene coeli-rosa*. (A) SEM micrograph (plan view) of an early floral apex with an opposite pair of leaf primordia (L), five sepal primordia (S) and perhaps some stamen primordia initiating (approx. X200). (B) SEM micrograph (plan view) of a floral apex with five sepals (S), five petals, five antesepalous stamens (1–5) and stamens 6 and 7 present; stamens 8 and 9 are just being initiated, and stamen 10 is not yet visible. All leaves have been removed (approx. X200). (C) Summary diagram illustrating the sequence of initiation of primordia. The sepals (shaded) and stamens (circles) are numbered according to the sequence of initiation in each whorl. Stamen 1 is initiated after sepal 5 and carpel 1 after stamen 10. Petals are indicated by shaded half circles. The acropetal helix of unequal axillary buds (AB) is indicated by the dashed line. The penultimate pair of leaves is shown at the sides and the last pair above and below the apex. From Lyndon (1978) with permission of *Annals of Botany*.

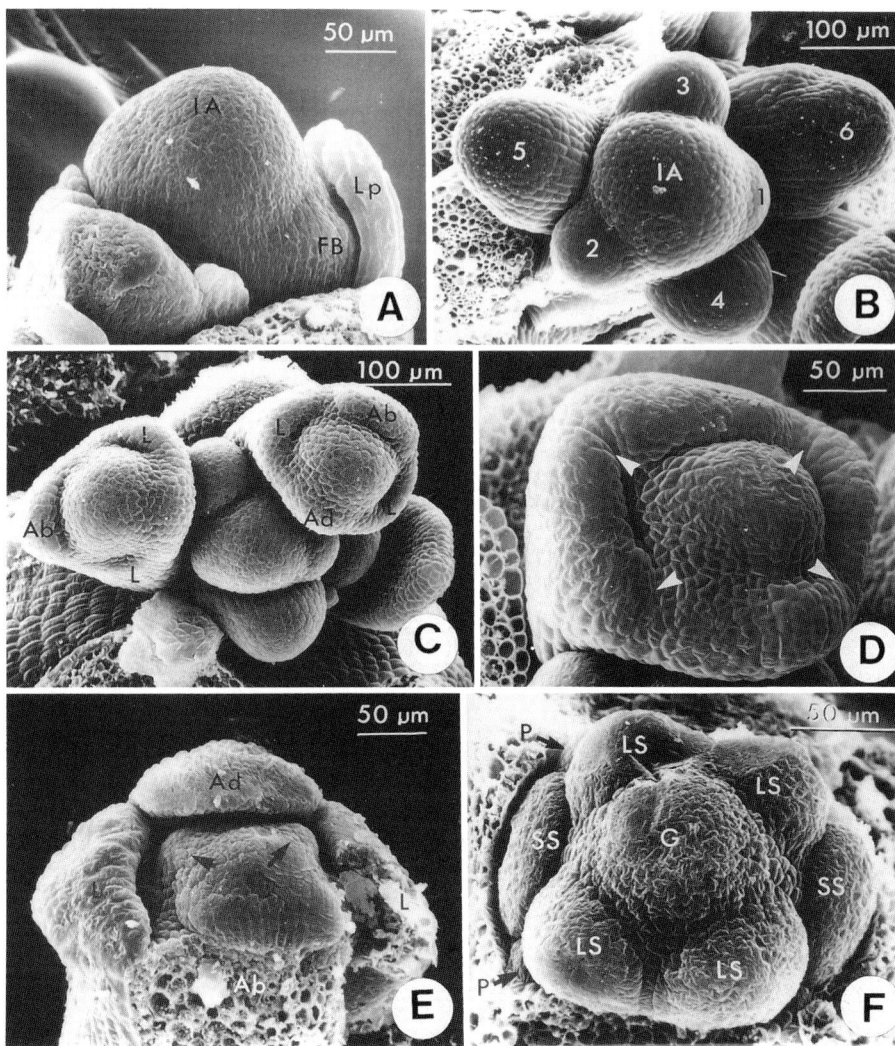

Fig. 4-3A–F. Sepal and petal initiation in *Brassica napus*. (A) A young inflorescence apex (IA) with one floral bud (FB) enclosed by the youngest leaf primordium (Lp). (B) A young developing inflorescence showing the helical pattern of flower bud initiation (1–6) from the inflorescence apex (1A). (C) An immature inflorescence showing the initiation of sepal primordia: abaxial (Ab), adaxial (Ad), and lateral (L) sepal primordia. (D) A young flower bud showing connections (arrows) between the four sepal primordia. (E) A young flower bud with abaxial sepal removed to show the primordia of the four long stamens (arrows). Ad = adaxial sepal primordium; L = lateral sepal primordium. (F) A floral bud with sepal primordia removed to show the long (LS) and short (SS) stamen primordia. Petal primordia (P) are located on either side of the short stamens; only those visible are indicated by arrows. G = region which will develop into the gynoecium. Reproduced from Polowick and Sawhney (1986) with permission of *American Journal of Botany*.

tler 1980). More recently, because of its greater depth of field and clarity of image, SEM is the technique of choice for many specimens (Tucker 1988*b;* Bowman et al. 1992). Meicenheimer (1979) has demonstrated how these images can yield quantitative information about the pattern of organs on the meristem. For many situations, subtleties will only be detected if the possibilities for varietal and environmentally induced variability are controlled in observations and experiments.

In those situations in which non-helical patterns of initiation are uncovered, it might be necessary to postulate different inductive mechanisms than those so far proposed. Certainly, the unidirectional and overlapping patterns described below (Tucker 1984*b*, 1989) challenge any simplistic view of apical meristem dynamics based upon helical initiation patterns. While biological models may be difficult to formulate and investigate, it is encouraging to note that mathematical models and computer simulations of whorled patterns are available (Meinhardt 1982).

Variations from the two most common patterns of perianth organ initiation, helical and whorled, are common. Because present techniques at identifying the timed events are relatively crude and indirect, future studies will probably detect many more variations.

One variant, frequently described, involves the initiation of a common primordium from which petal and stamen primordia arise. Sattler (1967) describes some of the different possibilities based upon different relative positions and timing of organ initiation. *Cyclamen* (Fig. 4-4) illustrates one of these situations in which petal and stamen primordia arise from a common primordium (Sundberg 1982*a*).

The meristem of the pea *(Pisum)* flower, in addition to producing four common primordia from which petals and stamens arise, exhibits a number of additional variations in organ initiation (Tucker 1989). These include: (a) unidirectional initiation within each whorl, beginning at the abaxial side; (b) overlapping initiation of organ formation between whorls; (c) centrifugal inception of the inner

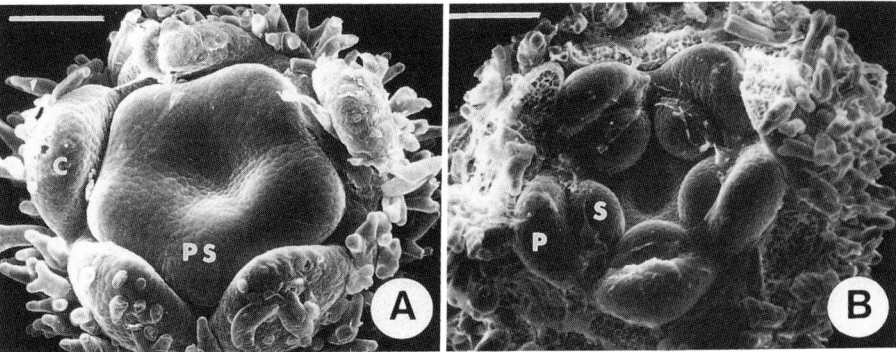

Fig. 4-4A–B. (A) Flower meristem of *Cyclamen orbiculatum* which has initiated sepal (C) and common petal-stamen primordia (PS). (B) Flower meristem at a later stage where distinct petal (P) and stamen (S) primordia have developed. Bar = 100μm. From Sundberg (1982*b*) with permission of *American Journal of Botany.*

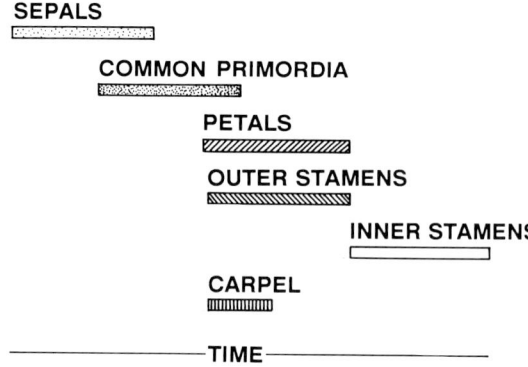

Fig. 4-5. Relative overlap of initiation time of the organs on the flowers of *Pisum sativum*. The length of each line is proportional to the total duration of initiation of all members in the whorl. From Tucker (1989) with permission of *American Journal of Botany*.

stamens; (d) initiation of fusions of organs of different types; (e) beginnings of zygomorphic symmetry. These findings are reviewed by Tucker (1984*a*, 1984*b*, 1987, 1989).

Each of these is illustrated in Tucker's studies of *Pisum* and other members of the legume family. In *P. sativum* var. "Sugar snap," the first sepal initiates in the abaxial position. Subsequently, sepals initiate individually, alternating from side to side on the meristem. This pattern is described as **unidirectional** (Tucker (1984*b*, 1989). This represents the first visible manifestation of zygomorphic symmetry in *Pisum*.

Although the phenomenon of overlapping organ initiation applies more to the initiation of androecium and gynoecium, variations also occur, to different degrees, in the initiation sequences of the perianth organs of legumes. An example is illustrated in Fig. 4-5 (Tucker 1989), which is an interpretation of the relative time of initiation of different organ patterns. In *Pisum*, petals, outer stamens and carpels initiate simultaneously and are followed by the initiation of the inner stamens. Other examples of centrifugal initiation are discussed in Chapters 5 and 6.

Another pattern variation, also illustrated in *Pisum*, is the initiation of a common primordium from which petal and stamen primordia arise. This is illustrated, as viewed from SEM micrographs, in Fig. 4-6, where two petals and one stamen are seen to arise from the adaxial, common primordium.

Undoubtedly, considerable scope for fresh developmental insight is possible in this area through a thoughtful combination of quantitative analysis of the cell dynamics of carefully selected examples and the extension of molecular techniques to the primordia involved. For some developmental traits and for some species, differences might also be recognized at the varietal level and within the same plant. It is important, therefore, to be careful and specific about the material being described and the growth conditions employed. It should also be obvious that hypotheses about the space-regulating influences will be different from those that are postulated to explain helical patterns.

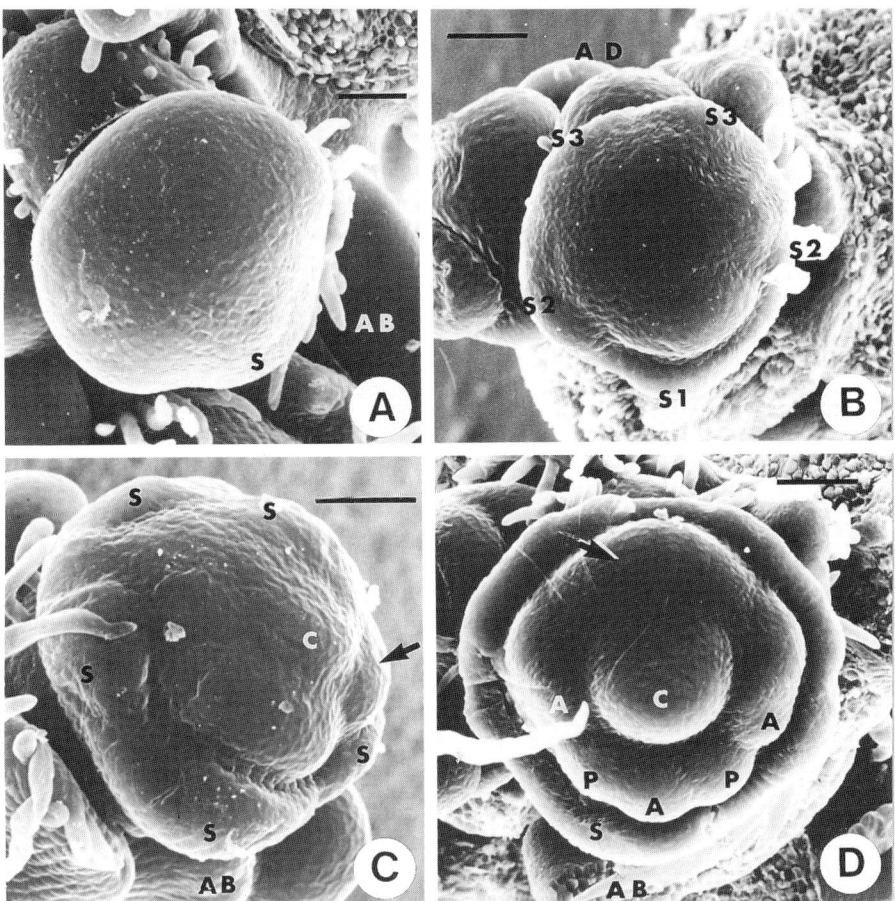

Fig. 4-6A–D. SEM micrographs of floral meristems of *Pisum sativum* during sepal and petal initiation. (A) Plan view of a floral apex with one sepal primordium (S) in the abaxial (Ab) position (scale bar = 50μm). (B) Plan view of meristem after the initiation of five sepals: S1 in the abaxial position; S2 on either side of S1; S3 sepals on either side of the adaxial position. (C) Oblique view of flower with all five sepals and a common primordium (large arrow). From the common primordium arise two petal primordia and a single median abaxial stamen. The beginning of the carpel (C) and one of the two lateral common petal-stamen primordia (small arrow) are also visible. (D) Plan view of flower in which each of the lateral common petal-stamen primordia has produced a petal primordium and an antisepalous stamen primordium (A). The adaxial common petal-stamen primordium is at the arrow. From Tucker (1989) with permission of *American Journal of Botany*.

The Cellular Basis of Perianth Organ Initiation

The ideal conditions for a critical study of cellular features of perianth organ initiation would include: (a) the selection of a sensitive, inducible flowering species; (b) the timed sampling of meristems following induction for fixation; (c) the preparation of replicated, longitudinal and cross-sectional views from microscopic sections through the sites of organ initiation (perhaps confocal microscopy will also prove useful for this task); (d) a quantitative analysis of the cellular changes taking place at different times; (e) a parallel study of the changes in the frequency and orientation of mitotic activity at the site of initiation; (f) the opportunity to explore the relative contributions to the mature organs of the different layers of the meristem through the examination of periclinal chimeras (Blakeslee, Satina and Avery 1940).

Since no study is presently available which meets all these requirements, a composite impression must be generated from a number of different studies. For sepal initiation, the most widely accepted view is that the cellular events closely resemble those taking place during the initiation of a leaf primordium. Thus, periclinal divisions take place in the hypodermal (sub-protoderm) layer and are commonly recognized as the first visible evidence of initiation at the site. Further periclinal divisions and concomitant anticlinal divisions and cell enlargement produce a localized protuberance. In this interpretation, the protodermal layer contributes, through anticlinal divisions, to the epidermis of the developing organ. Whether, in fact, this is true in all cases is not known and can only be resolved definitively by quantitative studies, as were performed on leaves (Maksymowych 1973), or by an analysis of carefully constructed periclinal chimeras.

In *Datura,* Satina and Blakeslee (1941) found that the floral apex has three independent germ layers (Fig. 4-7A). The first layer (L-1) forms the epidermis. Cells beneath the third layer (L-III) form the central core and are derived of this layer. The initiation and development of the leaf, sepal and petal are similar and depend primarily on the activity of the second germ layer (L-II) (Fig. 4-7B–D).

Many of these generalizations are illustrated in the study of sepal and petal initiation in *Nicotiana* (tobacco) prepared by Hicks (1973) (Fig. 4-8 and Fig. 4-9). Both sepals and petals arise through an integrated series of cell divisions in the outer tunica layer (T1), the hypodermal tunica layer (T2) and the corpus cells at the site of initiation. In *Nicotiana,* T1 contributes only to the epidermis, while the majority of the internal cells of the primordium are derived from T2. As described later in this chapter, the sepal primordia may ultimately coalesce, as do the petal primordia, so that the calyx and corolla may be tubular.

Of necessity, this and other such interpretations (Kaplan 1968; Dengler 1972) about these dynamic processes in the floral meristem are qualitative and are based on a relatively small number of observations from any single stage. As biological phenomena are noted for their variability, it is possible that many fresh insights are yet possible about these early stages if more care is given to experimental design and sampling procedures. Future research, combining a variety of microscope technologies, *in situ* localization of developmentally significant enzymes

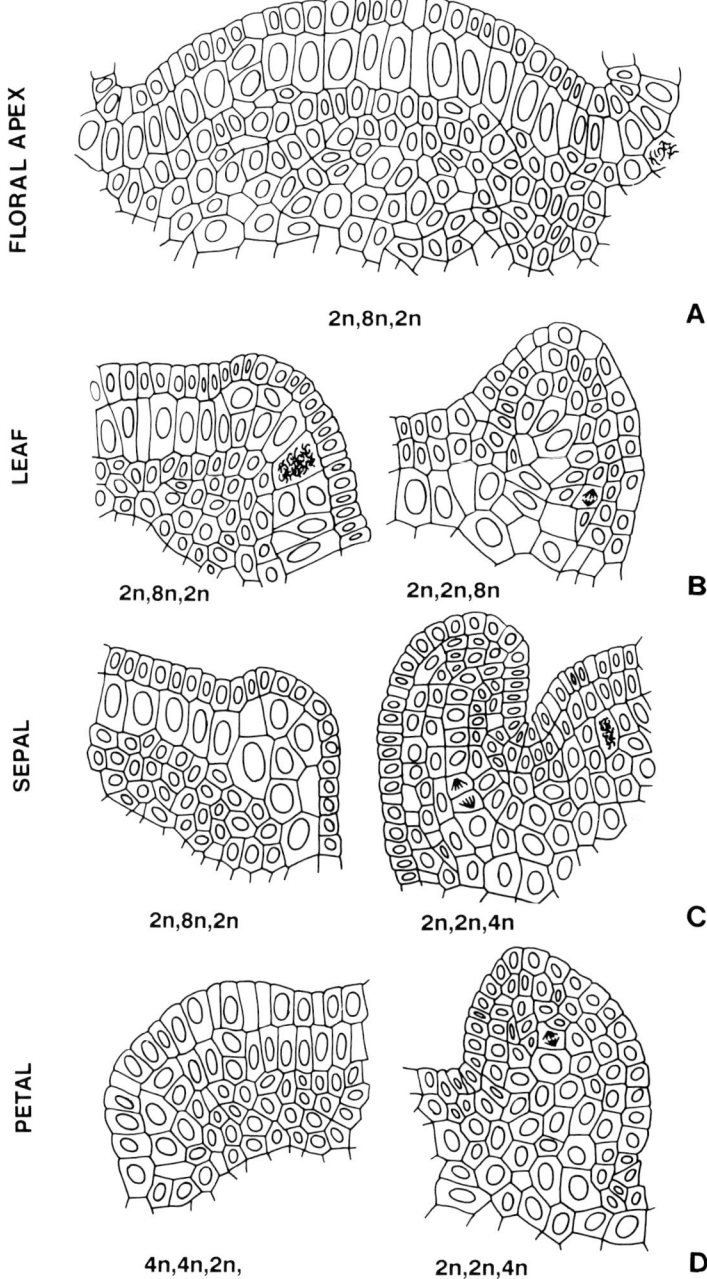

Fig. 4-7A–D. Median L/S views of the shoot apical meristem and primordia which document different periclinal chimeras of the germ layers of *Datura stramonium*. (A) Floral apex. (B) Leaf primordia. (C) Sepal primordia. (D) Petal primordia. Ploidy levels of different layers indicated. From Satina and Blakeslee (1941) with permission of *American Journal of Botany*.

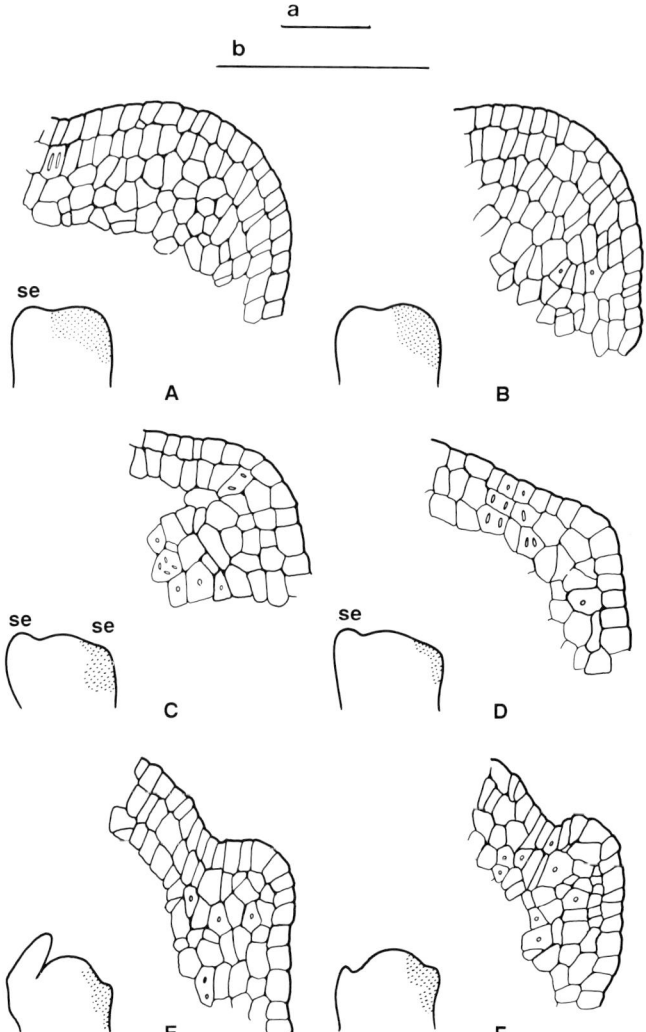

Fig. 4-8A–F. Sepal initiation and emergence in *Nicotiana tabacum*. (**GENERAL ORIENTATION:** Each figure letter is for a pair of drawings. The smaller of each pair is an outline of a longitudinally sectioned floral bud, with stippling indicating area from which larger drawing was taken. Scale "a" above designates 200µm for all smaller drawings. Scale "b" designates 100µm for the larger drawings. Symbols for the particular primordia: se = sepal; pe = petal. All buds sectioned longitudinally.) (A) Floral meristem before the initiation of a sepal opposite the first sepal. Note well-defined T1 and T2 layers. (B) Note periclinally divided cells in T2 down meristem flank. Primordium has not emerged. (C) Sepal emergence. T2 periclinal divisions more widespread at telophase figure and below. Corpus cells have divided recently. (D) An example of early sepal initiation, showing anticlinal division in T2. (E) and (F) Later views of sepal development. The contribution of T2 to the sepal can be inferred by following it from the left side of each drawing towards the sepal. Reproduced from Hicks (1973) with permission of the National Research Council of Canada from the *Canadian Journal of Botany* 51:1611–1617.

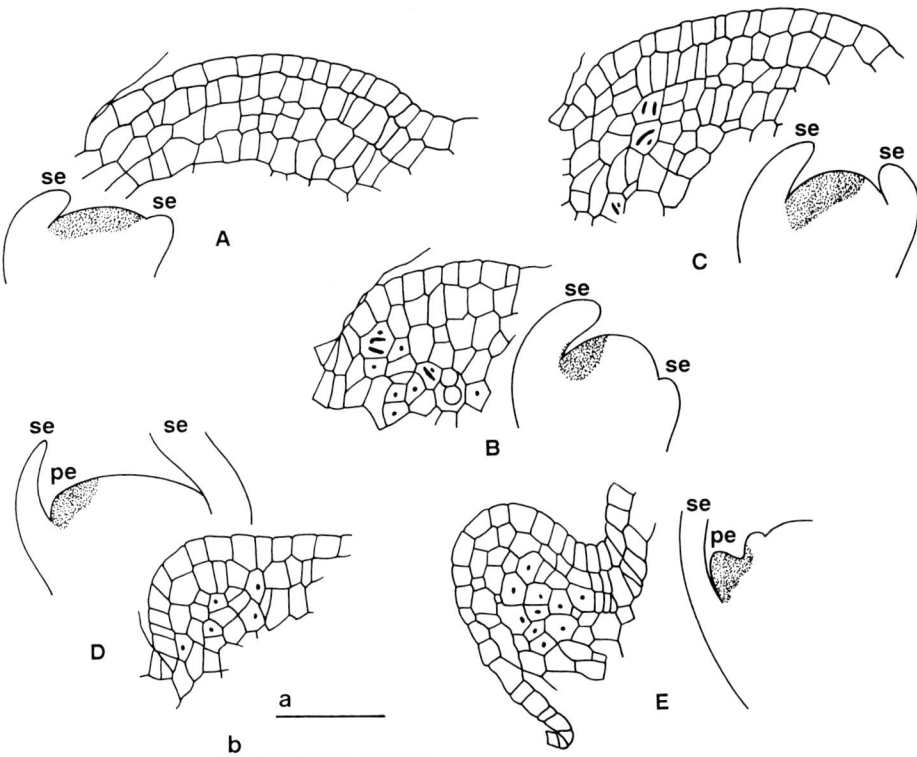

Fig. 4-9A–E. Petal initiation and emergence in *Nicotiana tabacum*. See Fig. 4-8 for general orientation and symbols. (A) Floral meristem just before petal initiation, showing two-layered tunica and underlying corpus cells. (B) and (C) Meristems before petal emergence. Detail near the base of the oldest sepal. Note the periclinal or oblique division of T2 cells on extreme left flank of meristem and division figures or recent cross walls in subjacent corpus cells. (D) At petal emergence. Periclinal T2 divisions are evident on abaxial flank (left) of rudiment. Corpus cells under T2 have probably divided recently at right angles to direction of outgrowth. (E) Later petal stage. By following T2 from meristem (right) to left one can see the numerous periclinal divisions on the adaxial side of the primordium. Reproduced from Hicks (1973) with permission of the National Research Council of Canada from the *Canadian Journal of Botany* 51:1611–1617.

and procedures from molecular biology, should enhance our insights about important developmental events in organ initiation and clarify the ways in which sepals and petals are developmentally similar to as well as different from each other and the degree to which they share similar developmental features with leaves.

The Growth of Sepals and Petals

Early Post-initiation Growth of Sepals and Petals

A perusal of representative longitudinal microscopic sections through immature sepals and petals document that cell division is the most significant contributor to overall growth at these early stages. Cell enlargement proceeds at a rate which maintains average cell size at approximately the initial size. Despite the lack of explicit quantitative detail, we can infer that restriction of the planes of cell division and subsequent enlargement and the localization of division rates in different regions are responsible for establishing the beginnings of bilateral symmetry of some sepals and petals and the differential development of the adaxial and abaxial sides and resultant curvature of the organs. The early extension of the margins form a lamina, and the early differential growth yields unique organ morphologies including tube and lobe formation.

The origin of the typical planar symmetry of many leaves begins at the surface of the meristem through the lateral extension of the site of initiation, but it is maintained through the extension of margins. It has been fashionable to infer the identification of **marginal meristems** from isolated images from cross sections of sepals and petals. These are assumed to be regions along the margins with relatively high mitotic activity and to contain the marginal and sub-marginal initials. These initials are assumed to represent the histogens from which the leaf cell layers arise. More recently, because of the greater awareness of the complexities involved in assessing the dynamics of cell populations, such characterizations are generally acknowledged to be uncritical and probably incorrect. Quantitative analyses of leaf development, such as that on *Xanthium* (Maksymowych 1973), fail to support the traditional view of cell lineage patterns from marginal and submarginal initials and meristems (Steeves and Sussex 1989). These ideas are also reviewed by Cusset (1986).

Data about the quantitative growth dynamics of sepals and petals are meagre, but we assume they would resemble those of leaves. In *Solanum dulcamara* (Dubuc-Lebreux and Sattler 1984) and *Nicotiana* (Dubuc-Lebreux and Sattler 1985), both members of the Solanaceae, mitotic activity was found to be dispersed across the petals such that the concept of marginal meristems was deemed inappropriate (Fig. 4-10). The generalization of these interpretations to other genera, however, has been questioned (Hagemann and Kaplan 1989). Because of the different morphologies achieved by different sepals and petals of other species, future research should be able to settle this controversy and perhaps uncover some novel patterns of cellular growth not previously identified. Detailed analysis of the cell dynamics, as above, and more detailed cell-cycle kinetics (Fig. 4-11) will necessarily accompany future investigations of the molecular aspects of organ initiation and growth.

While the degree to which mitotic activity is restricted to the marginal meristems is debated, so too is the concept of initials. The relevant question, then, is whether cell layers are restricted to a lineage from identifiable initials at the margins or whether the layers, while appearing to arise from initials, actually have a mixed origin. Two different tests can be used to distinguish between the possibilities: (i) identification of the frequency and location of periclinal divisions in the different layers of the developing leaf; and (ii) the use of periclinal chimeras.

In a descriptive comparison of corolla development in six species of the Apocynaceae, Nishino (1982) noted occasional periclinal divisions of the surface cells of the marginal meristem. In order to know whether this observation relates di-

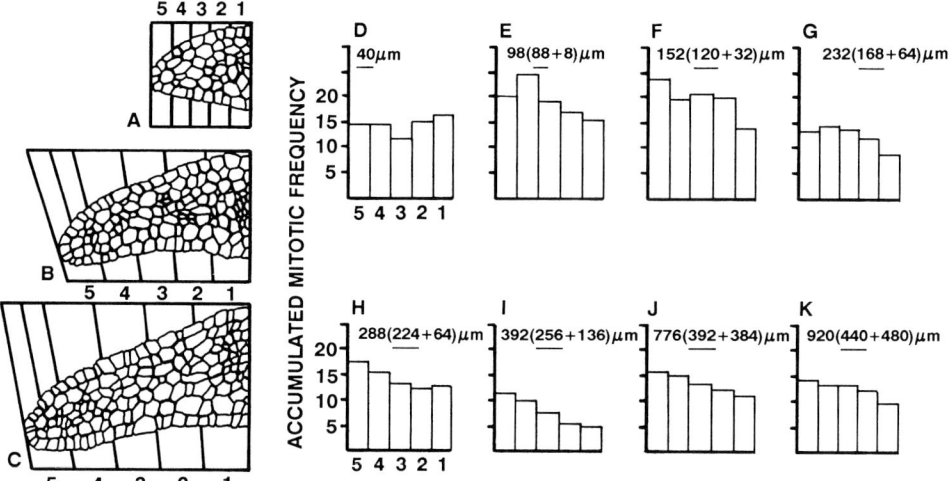

Fig. 4-10A–K. Summary of analysis of observations on the growth of the corolla of *Nicotiana tabacum*. (A)–(C) Representative outlines of cross sections of one-half petal primordia or petal lobes from corollas of different lengths. (D)–(K) Histograms of the relative mitotic activity at five zones across the mid-region of the petal (or lobes). These are also illustrated on Figs. A–C. Zone 1 is the median portion of the petal primordium (or lobe). Zone 5 represents the margin. The lengths of the corolla (μm) is indicated at the top of each histogram. The length of the petal lobe and tube are given in brackets. The value for the lobe is underlined. (***INTERPRETATION:*** These data demonstrate that the expansion of the lamina of the petal lobe is accompanied by mitotic activity throughout the breadth of the lobe. Additional statistical implications of the data are presented in original paper.) From Dubuc-Lebreux and Sattler (1985) with permission of *Phytomorphology*.

rectly to the point at issue, quantitative studies as were performed on leaves (Maksymowych 1973) will be required. At present we are unaware of quantitative data from periclinal chimeras that bear on the topic.

Later Growth of Sepals and Petals

Following the initiation of individual primordia, sepal and petal growth, as with leaves, is the product of cell enlargement accompanied by cell division. Average cell size and average cell number, therefore, rise to final values. The different shapes at maturity reflect the different patterns of relative rates, timing and loci of these cellular processes. Our discussion will use those insights gained from the few examples available. Considering the vast number of candidates available for investigation, it is probable that a number of novel patterns of these events are yet to be described.

As with other biological systems, measurements of sepal and petal growth generate sigmoid-like curves. This holds true for cell number per petal (Fig. 4-12A) and for any other set of measures such as sepal length or width. Some

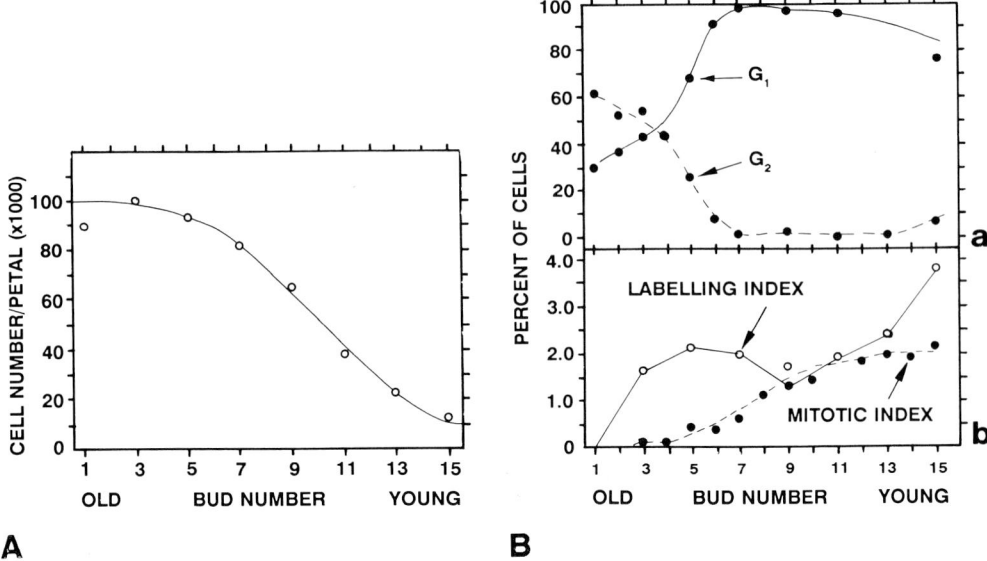

Fig. 4-11A–B. Observations of the cell kinetics of cells of the growing petals of *Tradescantia* (Clone 4430). (A) The cell number/petal as a function of bud number within an inflorescence. (B) The percent of cells in various stages of the cell cycle expressed as a function of bud number: (a) % of cells in G_1 and G_2 was measured cytophotometrically; (b) the labelling index was determined by autoradiography after a 1h incubation of the petals in [^{3}H]thymidine, and the mitotic index was measured on freshly fixed material. (*INTERPRETATION:* Despite the fact that "stage of bud" rather than "time" is used, growth in terms of cell number approaches a sigmoidal curve. Further, mitotic activity in petals of young buds is high and declines in older buds. Information on the distribution of mitoses in the petal as it grows is not available.) From Kudirka and Van't Hof (1980) with permission.

organs grow in synchrony with other organs or organ parts and, when their lengths are plotted, generate straight lines [Fig. 4-12B (Nos. 1 and 5)].

In other cases (Fig. 4-13), growth of individual organs is non-synchronous. These situations raise intriguing developmental questions. In the case of synchrony: Is synchrony based on communication between the parts, or are the parts or organs growing at similar rates but with no interaction? In the case of the non-synchronous patterns, such as the difference between the palea and the lemma in the floret of the wheat flower (Fig. 4-13), the obvious questions relate to the factors that initiate and maintain the growth differential.

Data from growth studies are often presented in allometric comparisons where the parameters of two organs, or parts of organs, are plotted logarithmically (Huxley 1932; Gould 1966). While the correlative information from this technique will rarely constitute compelling evidence for or against a proposition, allometry can suggest important developmental relationships that might otherwise go undetected. For example, Lord's (1982) study of **chasmogamous** and **cleistogamous** flowers of *Lamium* (Fig. 4-14A) documents the morphological similarity of the corollas

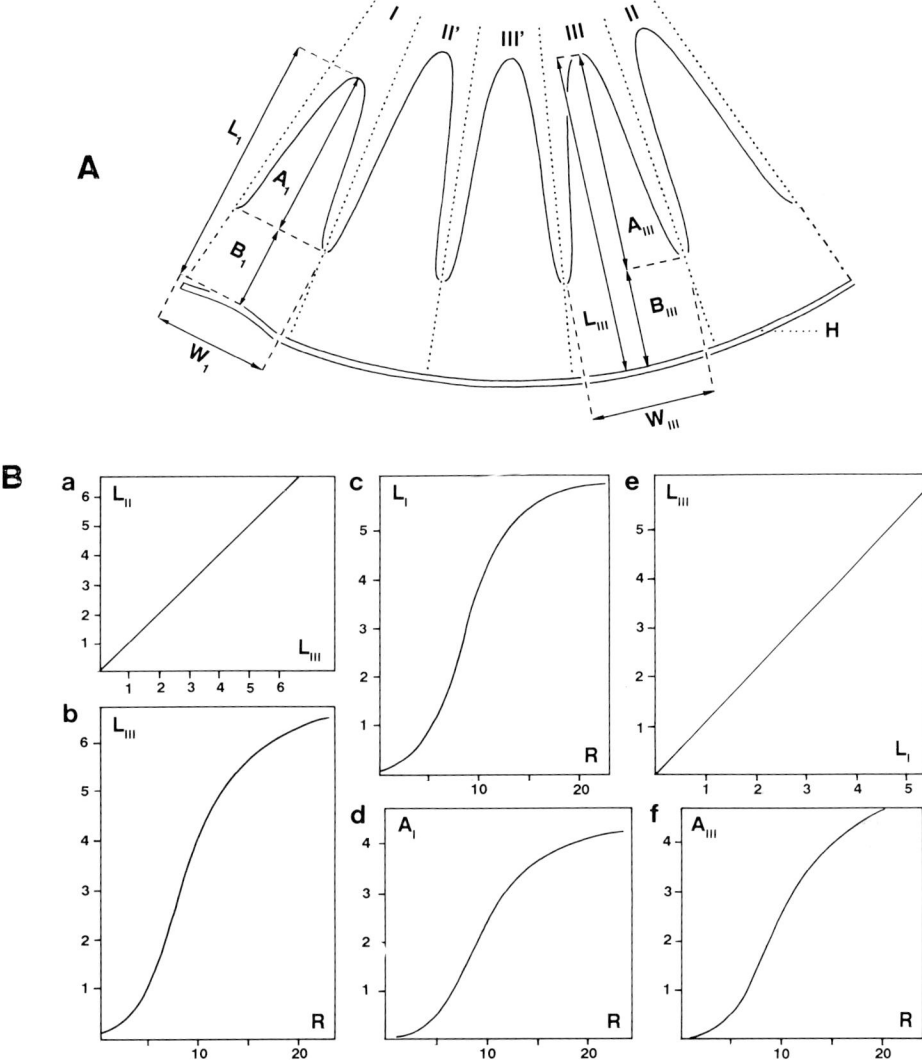

Fig. 4-12A–B. (A) Outline sketch of the calyx of *Fittonia vershaffeltii* to show the individual sepal lobes and tube of which it is composed. The sepals of this zygomorphic flower are numbered 1, 11, 111, 111' and 11'. At the base, the short attachment to the corolla forming the hypanthium (H) is indicated. The individual measurements made on these flowers of different sizes along the inflorescence were: A = length of sepal lobe; B = length of sepal tube; L = total length of sepal; W = width of sepal at the base of the blade. (B) Graphical display of measurements made from flowers of different stages of development: (a) L11 X L111; (b) L111 X stage; (c) L1 X stage (S); (d) A1 X stage; (e) an allometric-like comparison of L111 X L1; (f) A111 X stage. Y-axis units "mm" in all cases. X-axis units are "flower number from the youngest to the oldest." Reproduced from Ferrand and Cusset (1983) with permission of the National Research Council of Canada from the *Canadian Journal of Botany* 61:2694–2701.

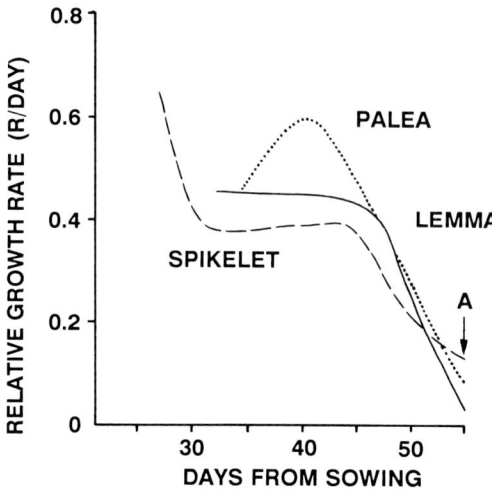

Fig. 4-13. Relative growth rates for median spikelets, their lemmas and paleae from the basal florets of *Triticum* inflorescences. "A" refers to anthesis. (***INTERPRETATION:*** The relative growth rates during the period before anthesis are not uniform but, in general, decline. The rates of the individual organs of the wheat flower are quite different. The growth of the palea, independent of the spikelet and the lemma, undergoes a surge of growth at approximately 40 days after sowing. From Williams (1966*b*) with permission of *Australian Journal of Biological Science.*

of these two different flowers produced on the same plant until the corollas are 0.7–0.9 mm long. At this point the slopes of the regression lines diverge significantly. Another plot (Fig. 4-14B), however, suggests that some morphological differences can be detected early in development. Direct studies on the onset of molecular and other differences between organs with divergent growth patterns, such as these, should eventually settle questions about the timing of developmental events.

Tucker's (1989) observation of "equalization," a common feature of legume flower development in which the members of a whorl initiate successively but become equal in size and grow synchronously, means that in growth study comparisons care has to taken to identify the organs individually, as well as by class.

Corolla Growth and Plant Growth Substances (PGS)

It is a popular assumption amongst plant physiologists that a spectrum of PGS is a natural component of the control of cell division, cell enlargement and cell differentiation. This is not always obvious or demonstrable. In some species, however, during the later stages of corolla growth, primarily during the cell elongation phase, GA involvement is easily demonstrated. The corolla is a particularly useful organ to study: (i) in many flowers, after initiation and an early slow enlargement, it grows substantially and rapidly in primarily one dimension: and (ii) it is readily accessible for analysis and experimentation. The role of PGS in corolla matura-

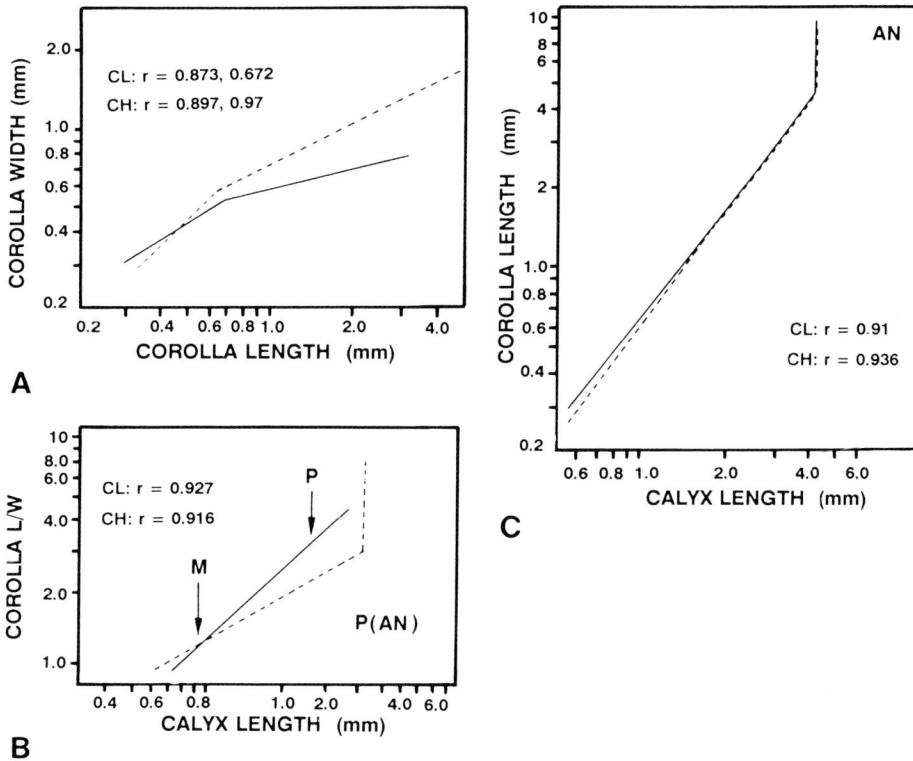

Fig. 4-14A–C. Allometric growth comparisons from the cleistogamous (CL)(solid) and chasmogamous (CH) (dotted) flower organs of *Lamium amplexicaule*. (A) Comparative allometric plots of corolla length versus width from CL and CH flowers. (B) Allometric growth plot of corolla length/width ratio in CH and CL flowers. (C) Allometric growth plot of corolla length versus calyx length in CL and CH flowers. P = pollination; AN = anthesis; M = meiosis. Solid lines represent CH flowers and dotted lines represent CL flowers. From Lord (1982) with permission of the University of Chicago Press.

tion, however, turns out to be a complex relationship among a number of substances which differ at different times during development and probably differ significantly between different species.

Traditionally there have been four ways to study PGS involvement in the growth of an organ; (i) extraction and estimation of endogenous levels; (ii) exogenous application through carriers or *in-vitro* culture; (iii) application of PGS inhibitors; and (iv) use of selected genetic variants. Research frequently combines two or more of these approaches. The recent ability to introduce selected genes which affect PGS level and function (Klee and Estelle 1991) represents a major contribution to this area.

An obvious and common approach is to extract and assay for the concentration of relevant PGS at different stages during organ growth. Murakami (1973) demonstrated that the petals of *Pharbitis nil* were relatively rich sources of GA-like

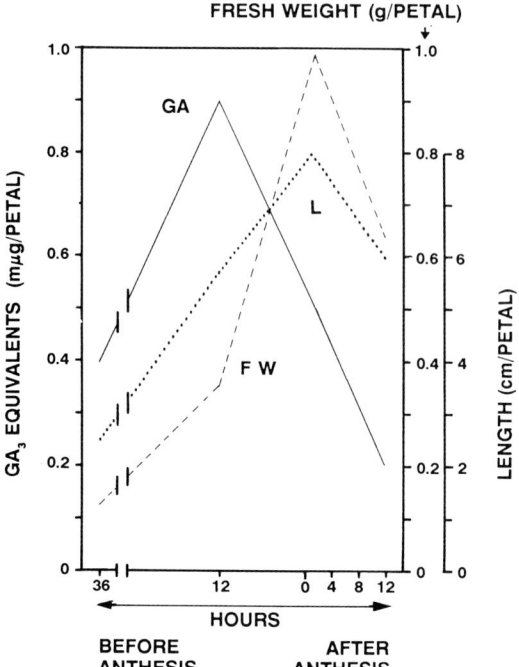

Fig. 4-15. Observations on petal growth (length (L) and fresh weight (FW)) in *Pharbitis nil* and the changes in the extractable levels of GA-like materials (GA) (expressed as GA_3–equivalents at different times during the growth of the petal). The observations and the assay results are for 36h before anthesis and 12h after anthesis. Extracts were separated with TLC using isopropyl ether/acetic acid (95:5, v:v) chromatography solvent, and the assay was the Tanginbozu dwarf rice bioassay. The changes in fresh weight and corolla length are also indicated. From Murakami (1973) with permission.

substances and that the highest concentrations were reached approximately 12h before anthesis and before the petals reached their final length and greatest fresh weight (Fig. 4-15). Comparable correlations were detected during the petal growth of the four-o'clock [*Mirabilis jalapa* (Nyctaginaceae)] (Murakami 1975). In this case, it was also shown that floral tube growth was strongly inhibited following early anther removal but was stimulated by exogenous application of gibberellic acid (GA_3). Thus, despite the difficulties and artifacts involved with extraction procedures, these and other observations (Jeffcoat and Cockshull 1972; Koning 1984) suggest that for some flowers the concentrations of GA-like substances correlate strongly with the extension phase of petal growth.

As Koning (Koning 1986; Raab and Koning 1987) points out, however, ethylene may also be involved at this pre-anthesis stage. These workers conclude, from experiments in which both stimulators and inhibitors of both GAs and ethylene were applied, that corolla expansion depends (in part) on compensatory shifts in the levels of the two PGS. After attaining full size, the *Pharbitis* corolla unfolds prior to pollination and then senesces. These regulatory events are thought

Table 4-1 Comparison Between Cell Number (CN) and Cell Size (CS) of Cleistogamous (CL) and Chasmogamous (CH) Corollas of *Lamium amplexicaule, in Vivo* and *in Vitro*

Corolla Area Measured	In Vivo		In Vitro	
	Control	GA_3-Treated	Control	GA_3-Treated
Corolla Base				
CL CN	37.6 ±5.5	41.1 ±4.3	44.9 ±4.8	48.7 ±6.4
CS (mm)	0.04±0.01	0.1 ±0.2	0.03±0.01	0.1 ±0.02
CH CN	36.9 ±2.33	44.9 ±4.1	51.2 ±8.5	44.3 ±6.0
CS (mm)	0.12±0.03	0.12±0.02	0.24±0.01	0.09±0.4
Upper Corolla				
CL CN	32.4 ±3.7	33.3 ±4.6	23.8 ±2.9	24.4 ±4.4
CH CN	44.9 ±4.3	35.4 ±5.7	36.0 ±5.4	32.2 ±3.6

INTERPRETATION: These data document the differential sensitivity of these corollas, and the cell elongation response, to applications of GA_3.

Source: From Lord and Mayers (1982) with permission.

to include phytochrome, abscisic acid (ABA) (Kaihara and Takimoto 1983) and ethylene (Beyer and Sundin 1978).

Another corolla which exhibits a pronounced sensitivity to GAs is that of the cleistogamous (CL) flower of *Lamium amplexicaule* (Lord and Mayers 1982). This sensitivity to GA3, applied either exogenously or by *in vitro* culture, suggests a normal physiological involvement of GAs with corolla growth. Data on the differential cellular responses are given in Table 4-1. As in *Pharbitis,* however, the web of causality in this and other cleistogamous examples may be complex, involving other regulators and metabolites in conjunction with sensitivities to environmental cues. Lord's review (Lord 1981) documents some of this potential and provides an excellent foundation for further studies.

Koning's (1984) study of the factors involved in the later stages of corolla growth of the ray flowers of *Gaillardia grandiflora* is instructive, for it details both the response of applied PGS and data on changes in endogenous content (Fig. 4-16). Further, it demonstrates that, in this flower, GAs are the main PGS involved in regulating corolla growth. Further discussion of the roles of PGS are included in ensuing chapters.

In vitro culture can also provide information on the PGS requirements of flower meristems and their organs during development. For example, the importance of GAs for sepal growth was illustrated in the culture of *Aquilegia* floral meristems (Karpoff 1974) (Fig. 4-17). Gibberellic acid was necessary for sepal growth at all phases of development. This growth stimulation involved both cell division and cell extension. Indole acetic acid (IAA), however, while stimulating cell extension slightly, had no effect on cell division. These data on sepal growth were obtained at the expense of overall bud development and, by contemporary culture criteria, are less than optimal. They do, however, point out an organ-specific sensitivity to GAs, even though GA_3 might not be one of the naturally occurring forms of GA in this plant. The identification of the naturally occurring forms and the biochemical pathways by which they are formed and metabolized, comparable to those

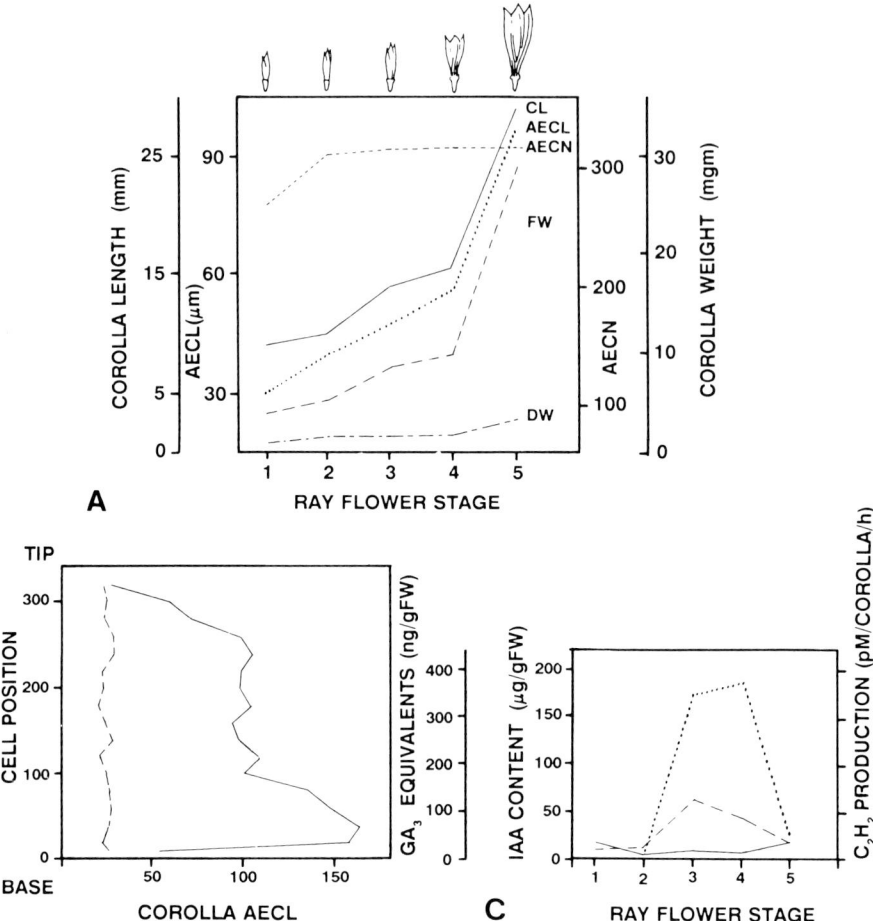

Fig. 4-16A–C. Observations from studies and experiments on the growth of the corolla of *Gaillardia grandiflora*. (A) Parameters of corolla growth. The mean corolla length (mm) (CL) during the five stages of ray flower opening is compared with the changes in the adaxial epidermal cell length (μm) (AECL), adaxial epidermal cell numbers (AECN), fresh weight (FW) and dry weight (DW) of the corollas. (B) Corolla AECL as a function of position in a median column of cells extending from the base of the corolla to the tip of the median lobe both before (dotted) and after (solid) elongation in ray flowers. (C) Levels of GA_3-equivalents (ng/g FW) (dotted), IAA (μg/g FW) (solid) and ethylene production (pM/corolla/h) (dashed) during the five stages of flower opening in *Gaillardia*. From Koning (1984) with permission of *American Journal of Botany*.

worked out for scarlet runner bean seeds and seedlings (Crozier 1981) or maize plants (Phinney and Spray 1982) remains to be documented.

By contrast, isolated petal primordia of *Nicotiana* (McHughen 1977) grew on a basal medium containing no PGS and achieved approximately 32% of their *in vivo* length but did possess many normal features. The triangular lobes of the cultured petals were, on average, larger than those on *in vivo*–grown corollas.

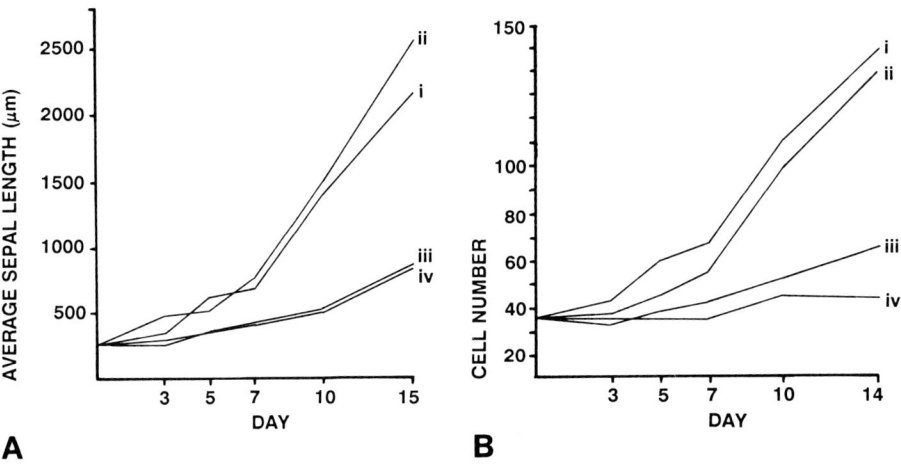

Fig. 4-17A–B. Observations on sepal growth on *in vitro* cultured flower meristems of *Aquilegia*. (A) Average length of sepal No. 2 from buds cultured for different times on media + and − GA$_3$ (2.0 mg/l) and + and − IAA (1.0 mg/l). (B) Average number of epidermal cells along the length of sepals from flower bud cultured for different numbers of days. (**Media used:** (i) +GA$_3$ −IAA; (ii) +GA$_3$ +IAA; (iii) −GA$_3$ −IAA; (iv) −GA$_3$ +IAA.) From Karpoff (1974) with permission of *American Journal of Botany*.

While there are no data from experiments which explore the sensitivity of tobacco petals to GAs and other PGS, this work highlights the vastly different species-specific requirements for, and responses to, PGS in culture. This work further demonstrates the level of autonomy that the petal has for its own development at this early stage.

Since Darwin's time (Darwin 1877; Muller 1883; Knuth 1906–9) botanists have been intrigued by the number of examples amongst gynomonoecious and gynodioecious species where the gynoecious flower is smaller than the hermaphrodite or staminate flower. Baker (1948) identified seventy-four species in sixteen dicotyledonous families where the gynoecious flower was smaller than the hermaphrodite flower. Only seven dioecious species with a similar flower-size relationship were listed. The developmental basis of these differences remains virtually unexplored.

The Cellular Basis of Adnation and Cohesion

A good deal of the variation in flower morphology involves the so-called joining together of different organs to varying degrees (Chapter 2). A substantial literature, reviewed by Cusick (1966) and Sattler (1978), attempts to identify trends and mechanisms and to delineate when and where the term "fusion" should be applied. At the risk of oversimplification, our discussion here is limited to two ways by which organs become united.

The first, which occurs during and after the initiation of the individual primor-

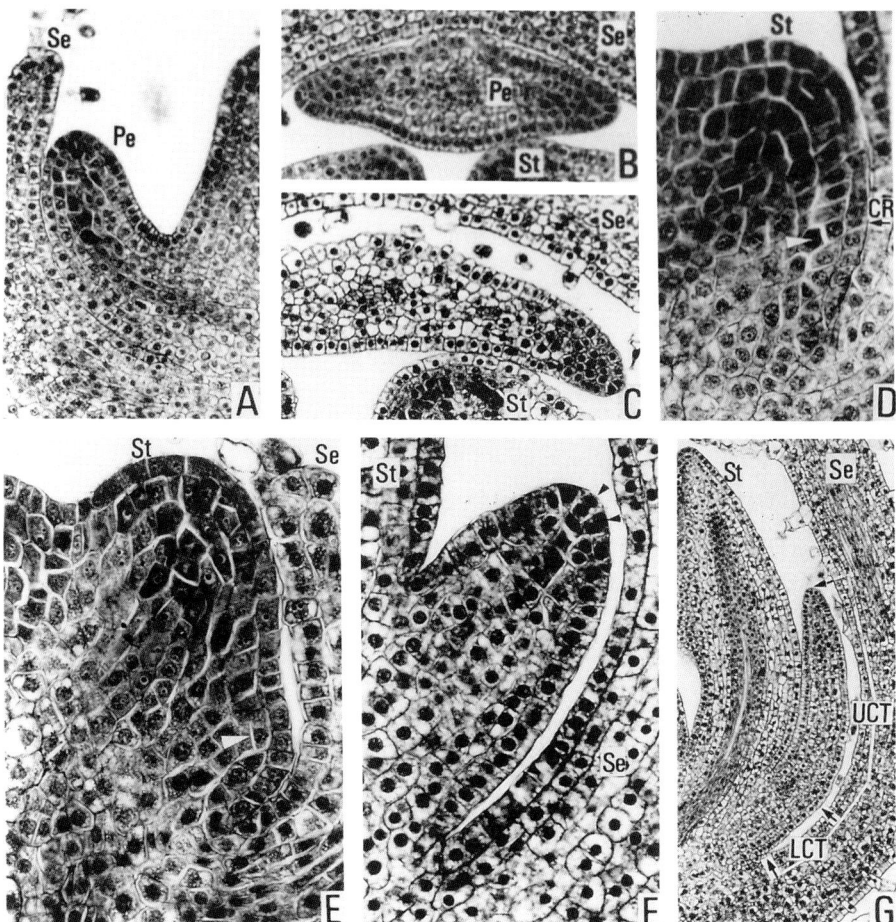

Fig. 4-18A–G. Corolla development in *Nicotiana*. (A) Median L/S of petal primordium (X150). (B) T/S of petal primordium of about 330μm in height cut at 110μm above the base. (C) T/S of petal primordium of about 640μm in height cut at 320μm above the base. (D)–(G) Radial L/S of the connected regions of petal primordia at the stages of about 5μm (D), 20μm (E), 400μm (F), and 700μm (G) in height. CR = connected region of the petal primordia; UCT = upper portion of corolla tube; LCT = lower portion of corolla tube; Pe = petal; Se = sepal; St = stamen. (A–C: X150; D–F: X300; G: X75.) From Nishino (1978) with permission of Botanical Magazine (Tokyo).

dia, involves the extension of the initiation process between the primordia until they join and a ring of initiation is complete. It is generally conceded (Daniel and Sattler 1978; Tucker 1984*a*), that the cellular events of this early stage that lead to a tubular calyx or corolla, as well as floral tube development, are not well described by the term "fusion." In *Nicotiana* (Nishino 1978), after petal primordia are initiated (Fig. 4-18), cell divisions in the floral meristem at the lateral edges of the primordia extend into the interprimordial region so that primordia become continuous, both among themselves and with the abaxial side of the stamen primordia. Subsequent intercalary growth in the different regions produced from these connections leads to the mature corolla morphology. It should also be apparent that the common primordia (Tucker 1989) described earlier will frequently lead to stamens attached to petals, and meristematic activity below and between corollas and calyxes combined with intercalary growth will lead to floral cups.

A second process, to which the term "fusion" can legitimately be applied, takes place later in development and involves the coming together and fusion of organs that are initially separate. In *Catharanthus,* Boke (1948) demonstrated that the margins of the petals fused and the epidermal layers at the point of fusion disappeared. Further work (Nishino 1982) showed that, in addition to this type of fusion, corolla tube formation involves interprimordial development (as above) with the extension to include stamen primordia, followed by intercalary growth of the different regions (Fig. 4-18). The fusion of the adjacent corolla lobes occurs relatively late in development.

Given the possibilities for variation, both within and between families, considerable opportunities exist for future descriptive study. Especially attractive is the potential for experimental exploration of the cellular mechanisms involved in organ fusion. Further discussion of experiments with fusion in the gynoecium of *Catharanthus* are presented in Chapter 6.

Variation in Perianth Traits

The appeal and intrigue of flowers is not only based on the number of species and their diversity but, as well, on the variability within populations. The importance of variable traits such as corolla size (Baker 1948; Bell 1985; Stanton and Preston 1988) and color (Stanton 1987a, 1987b) in the maintenance of populations is well studied. From the many different aspects of variation, the following topics have been selected for their developmental features.

Symmetry

A significant component of flower variability involves symmetry. Many flowers are radially symmetric **(actinomorphic, regular),** where organs of the same class of equal shapes and sizes, are arranged around the receptacle. Technically, all lines through the center of the flower divide it into two mirror-image halves (Fig. 4-19A). Other flowers are irregular. Some, such as *Dicentra,* exhibit two planes of symmetry **(bisymmetry** or **bilateral)** (Fig. 4-19B) (Weberling 1989). Others, with only a single plane of symmetry, are termed **zygomorphic** (Fig. 4-19C); a single line through the center of the flower will yield mirror-image halves. (Note: In some presentations "bilateral" is synonymous with "zygomorphic."). Variations in symmetry will arise in the extent of the modification and the degree to which organs other than the corolla are involved. Since the appraisal of flower traits is made most frequently on a casual, visual basis, one should expect that the

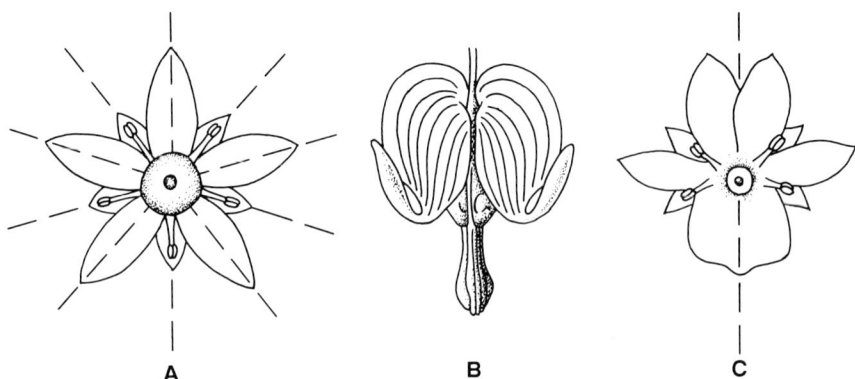

Fig. 4-19A–C. Symmetry in flowers. (A) Radially symmetric flower. Flower can be divided into mirror-image halves along a number of axes. (B) Bisymmetric flowers possess two planes of symmetry. (C) Zygomorphic flower. Division along one axis yields mirror-image halves.

use of more objective methods will identify many more cases of flower asymmetry.

The most obvious organs involved in variations of symmetry are the petals, but a wide range of variation has been recorded. Because many families possess species which exhibit both regular and irregular flowers, it is assumed that zygomorphic symmetry has evolved independently many times. From a developmental perspective one should therefore expect to reveal a variety of developmental mechanisms that underlie the development of different examples of irregular symmetry.

Different symmetrical arrangements raise many different developmental questions. At what point during flower formation is zygomorphic symmetry expressed? Is it based on differences in the patterns and rates of cell division and therefore cell numbers, or is differential cell extension responsible? Can physiological and histochemical differences be detected prior to morphological expression? To what extent can genetic and environmental factors be tied to the expression? What is the basis of developmental variability, both heteroblastic (exhibiting a strong genetic input) and plastic (responding to environmental cues), in the variable expression on the same plant?

There are many excellent materials in a variety of families for careful descriptive studies on symmetry. The mints (Labiatae), orchids (Orchidaceae), legumes (Fabaceae) and violets (Violaceae) include many excellent species for detailed study. Because the snapdragon (*Antirrhinum majus* L., Scrophulariaceae) is so well studied genetically, because various mutant forms of flower shape are available (Table 4-2) and because it is an easily grown annual, it is chosen here as the candidate for emphasis. The recent studies on the molecular biology of the gene chalcone synthase (CHS) and other steps in the synthesis of flower pigments (Martin et al. 1987) and the availability of transposon technology are further justifica-

Table 4-2 Selected List of Genes of *Antirrhinum* (2N = 16) Which Affect the Structure of Flowers

Gene Symbol	Gene Name	Description
Cyc	Cycloidea*	A series of multiple alleles transforming the symmetry of the flower from zygomorphic to radial.
Cyc^{neo}	neo-hemiradius	Flower not completely radial; number of anthers 4 or more.
Cyc^{rad}	radialis	Flower with radial symmetry.
Def	Deficiens*	A series of multiple alleles transforming the corolla, stepwise, into small greenish petals and reducing male fertility.
trans	transcendens	Malformation of the corolla and reduction in the number of anthers.

*Color illustration in Meyerowitz, Smyth and Bowman (1989).

Source: Harte (1980) with permission of Plenum Press and the author. A more extensive list has been prepared by Stubbe (1966).

tions for seriously considering *Antirrhinum* as a model system for developmental studies.

Studies on the inheritance of corolla shape date back to Darwin (1868), who crossed normal irregular-flowered snapdragons with regular (peloric) flowered plants. Current genetic studies document a number of genes which modify the corolla. For example the "Cycloidia" (Cyc) phenotype is expressed by a series of multiple alleles (Table 4-2) which induce a variety of radial flowers and other effects such as stamen number, male sterility and modifications of the petals (Harte 1980). Another gene, "transcendens" (trans) induces malformed corollas with a reduction in the number of anthers. As yet, there is no information on how the expression of these genes is connected to the growth patterns of the flower.

As discussed earlier, graphical depiction of organ growth in one dimension through time is relatively straightforward and common. So too, is the quantitative description of correlative growth relationships between two organs or two dimensions of an organ through time. These are frequently plotted as allometric relationships. More complex growth patterns are difficult to portray and have rarely been attempted.

One attempt is the novel analysis of the growth parameters of the corolla of *Calceolaria* (Scrophulariaceae) (Ritterbusch 1976, 1980a, 1980b, 1980c; English presentation in Ritterbusch and Wunderlin 1989). For this study, Ritterbusch has developed a number of original techniques, including a microtechnical method to prepare thick strips of flowers at different stages (Ritterbusch 1977), the development of a normalized plastochron scale for flower development and a pictorial portrayal of change utilizing Bildscharen and trajectories (Ritterbusch 1980c). The essentials of the procedure are illustrated in Fig. 4-20.

Attempts such as this, especially those which utilize computers for digitization, illustration and analysis of the data, should be pursued with other examples of complex corolla organization. Such spatio-temporal analyses provide a holistic view of organ growth and, in particular, highlight the differential growth between different parts. This descriptive approach can reveal interactions that would not

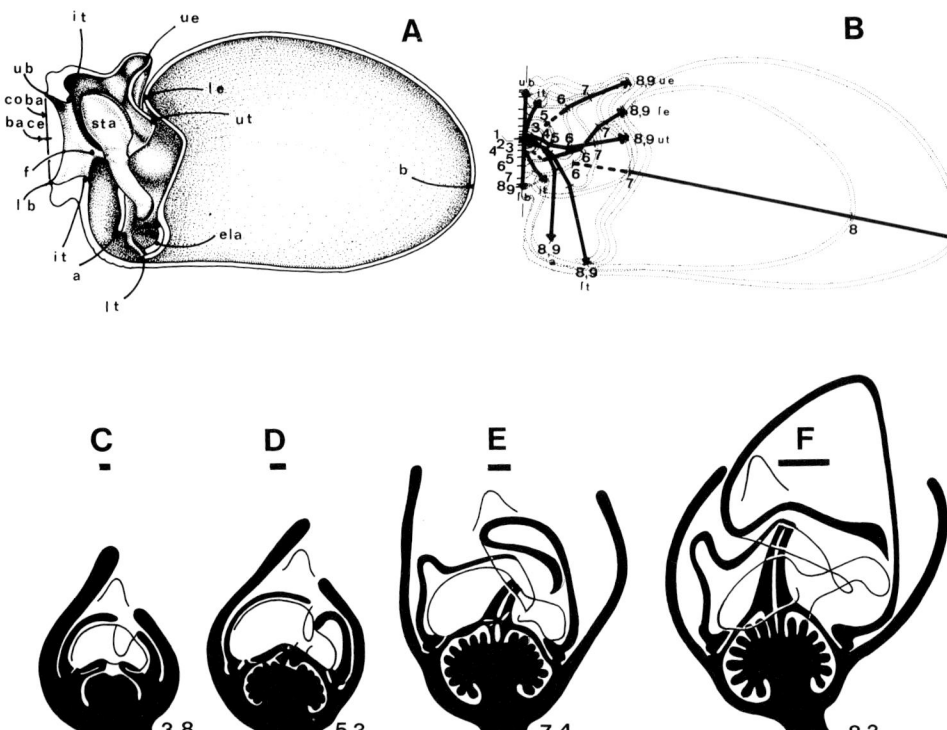

Fig. 4-20A–F. Analysis of the growth of the flower of *Calceolaria tripartita*. (A) Drawing of a median longitudinal view of a nearly mature corolla of *C. tripartita*. Localization of marks for the construction of trajectories are abbreviated as follows: coba = corolla base; bace = base center; ub and lb = upper and lower lip bases; it = inner torus (of upper and lower lip); ue and le = upper and lower lip edge; ut and lt = upper and lower lip tip; a = angle between the upper and lower lip; b = balloon tip; ela = elaiophore; sta = stamen; f = fusion of stamen corolla. The corolla base and the base center serve as reference loci. (B) Bildscharen (dotted lines) and trajectories (solid lines, dashed lines if prolongated backwards) of the non-manipulated linear transformed developmental sequence of the *Calceolaria* flower; identical numbers along the trajectories refer to the same stage; the age of any two consecutive stages differs by one relative time unit. (C–F) Four outlines of median longitudinal sections from a series of flowers studied at different stages of development. Stage is identified in Normalized Plastochrons (NP). Median parts are black; paramedian parts are in outline. All stages are oriented similarly. Sizes are adjusted by sliding scales in which the oldest stages are magnified the least and successively younger stages are magnified more. Scales in each case correspond to 0.2 mm. Reprinted with permission from Ritterbusch and Wunderlin (1989). Copyright (1989) Pergamon Press.

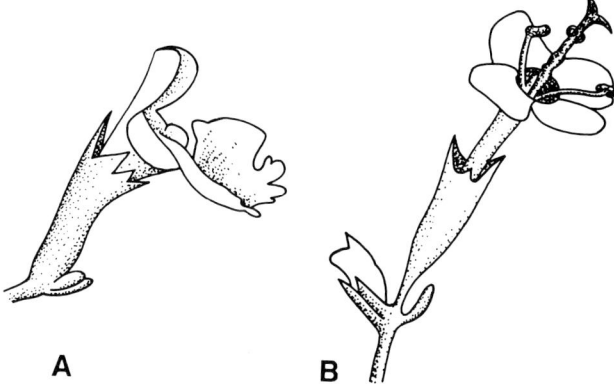

Fig. 4-21A–B. Variation of flower symmetry in *Nepeta mussimi*. (A) The normal zygomorphic symmetry of flowers along the stem. (B) A regular terminal flower. From Wardlaw (1965*a*) with permission of Longman Group UK. Original by E. G. Cutter.

normally be available through current molecular or reductionist analyses (Ritterbusch and Wunderlin 1989), though the eventual integration of genes with localized spatial and temporal expression will make a major contribution and is a logical goal of developmental studies (Stebbins 1992).

Symmetry need not be strikingly obvious nor so rigidly determined as it is in many of the Scrophulariaceae. Goebel (1900) cites a number of examples in which the lateral flowers of an inflorescence are zygomorphic but where the terminal flower is peloric. An example is provided in Fig. 4-21. While gravity appears to be an operative factor contributing to zygomorphic symmetry in some of these cases, a full developmental understanding of this condition seems as obscure today as it did in 1900 (Wardlaw 1965*a*). It seems quite possible that current experimental techniques might be able to distinguish better than heretofore among the heteroblastic (developmental), positional and environmental contributing factors. Meyerowitz, Smyth and Bowman (1989) provide some additional detail on the implications of pelory in the interpretations of flower development.

Even greater and perhaps more enigmatic plasticity was reported in the common garden weed, *Stellaria media* (Matzke 1932), where variable and subtle alterations of flower symmetry are complicated by variability of the number of organs. These data document, on the one hand, considerable plasticity in the number of different organs of each class but simultaneously a consistency as to where alterations occur. In the majority of cases studied, supernumerary petals and supernumerary petaloid stamens occurred in definite positions in the flower. These flowers could most often be described as zygomorphic rather than asymmetric.

Thus, a developmental understanding of the origin of symmetry of flower form has yet to be established. Many examples of appropriate research specimens are available, though many will require extensive preliminary descriptive preparation. The literature on the subject provides many clues. Many different mechanisms are to be expected: some will have a strong genetically determined basis, and others may be based in regulative mechanisms of which some may arise in response to environmental cues.

Meristic Variation

There are few aspects of plant development in which the question "Do plants keep track of numbers?" is as appropriate as in the formation of organs on the flower meristem. In many species, the precision and constancy in the number of organs of each type on the flower receptacle is impressive and is an important angiosperm taxonomic trait. The ability to generate floral formulae (Chapter 2) which are reliable in many of the wild species and are consistent within families, as well as simplistic "natural history" rules such as "parts in fours and fives" for the dicotyledons and "parts in threes" for monocotyledons, is further evidence of this phenomenon.

General consistency, however, does not mean inflexibility. Within many, perhaps most, populations variability in the number of petals, stamens, carpels or stylar lobes can be uncovered. Tucker (1991), under the term **merosity,** explores its extent in the legume *Gleditsia*. This variability represents a component of what floral morphologists have described as "chaotic" in some species and groups.

An example of the factors that operate in the maintenance of one system among wild populations is provided by Heuther's (1969) study of natural populations of *Linanthus*. He found from samples taken in 1964 that the normal corolla lobe number (five) was maintained throughout a number of populations, as is characteristic of the Polemoniaceae family, but that small percentages of plants, between 1% and 4%, had from four, six or seven petal lobes. In the following two years, environmental factors were found which disrupted the constancy, thereby causing an increase in the number of abnormal flowers. He further showed (Heuther 1968) that this variability was based upon a wide degree of genetic heterogeneity and limited by the canalization or developmental damper on expression which was sensitive to environmental cues such as heat, moisture and grazing by animals. Examples of natural and experimentally induced variability from this study are illustrated in Fig. 4-22A and B.

Another study of variation in flower organ number in wild populations is that of Ellstrand (1983), who discovered that nearly 33% of the plants in thirteen populations of *Ipomopsis aggregata* (Polemoniaceae) had at least one anomalous flower and almost 10% of the flowers displayed abnormal floral formulae. As with *Linanthus,* petal number varied approximately 3% but differed in that deviant flowers tended to have more petals than normal. Floral inconsistency varied significantly among populations, and the author concludes that both genetic and environmental differences may be in play.

More direct evidence of the environmental impact on the number of perianth parts was obtained from male plants of *Cannabis sativa* (Heslop-Harrison and Wood 1959) (Table 4-3 and Fig. 4-22C). In this experiment, plants induced to flower at a low temperature produced significantly more abnormal flowers than those induced at control temperatures. A variety of perianth arrangements includ-

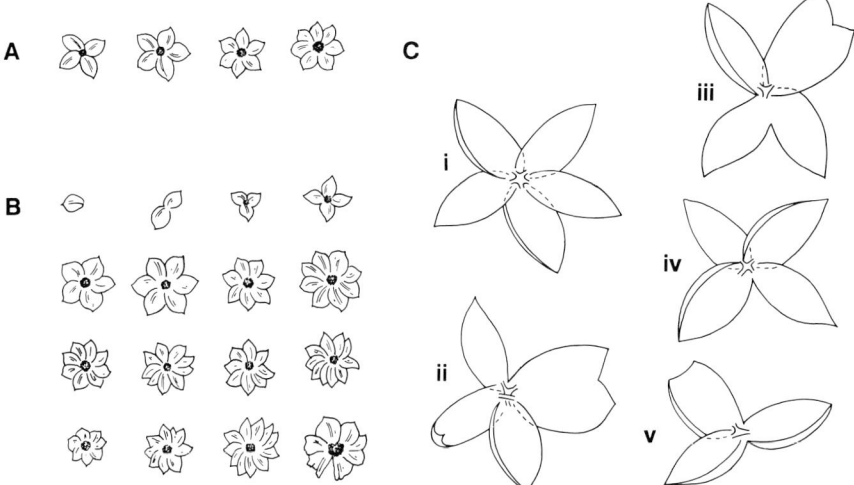

Fig. 4-22A–C. Meristic variation of perianths: evidence of genetic and environmental involvement. (A) Outlines of corollas obtained from natural populations of *Linanthus androsaceus*. Flowers with four to seven lobed corollas are illustrated. The pentamerous example is the most common. (B) Outlines of corollas of *L. androsaceus* obtained through selection from wild plants which produced a high frequency of variability followed by growth under environmentally stressful situations. Corollas with from one to sixteen lobes are illustrated. (A) and (B) from Heuther (1968) with permission of Genetics Society of America. (C) Sketches of perianths from male flowers of *Cannabis sativa:* (i) a normal five-tepal flower from a male plant grown at 22°C. (ii)–(v) examples of perianths obtained from male plants grown at 22°C light (8h) and 10°C dark (16h) (approx. X5). (C) from Heslop-Harrison and Wood (1959) with permission of The Linnean Society of London and authors. (**INTERPRETATION:** In *Linanthus*, the genetic instability of the examples sampled from the wild population was further explored through environmental manipulation. In *Cannabis*, sensitivity of tepal formation to temperature is revealed.)

ing some coherent tepals were observed. The large "within-plant variance" for the experimental sample suggests that, even within this highly inbred line, plant-to-plant variability (genetic components) was also involved.

Horticulture has provided the largest number of examples of variability in organ number, for it is under cultivation where variants are most easily observed. Further, some aberrant forms are frequently desirable commercially. The most

Table 4-3 Mean and Within-Plant Variances for Tepal Number from Male Plants of *Cannabis sativa* Grown Under a 8h Light/16h Dark Photoperiod and a Daytime Temperature of 22°C but at Two Night Temperatures

Night Temp. (°C)	Mean Tepal Number	Perianth Within-Plant Variance	Degrees of Freedom
22	4.98 + 0.00	0.0185	263
10	4.78 + 0.037	0.3289	245

Source: From Heslop-Harrison and Wood (1959) with permission.

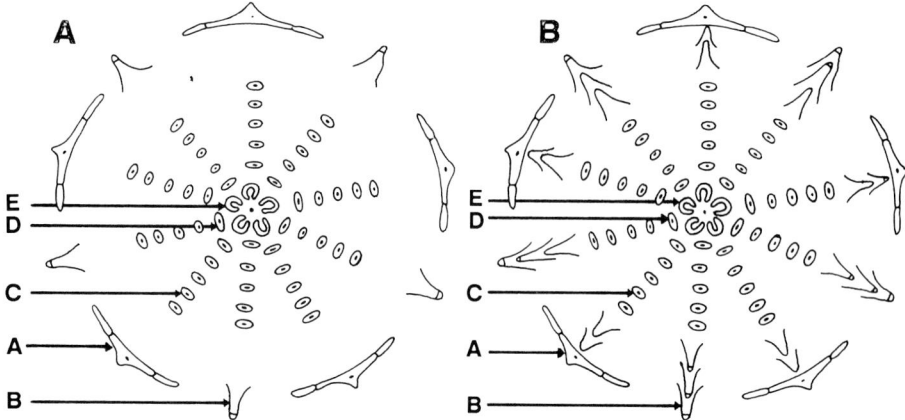

Fig. 4-23A–B. Floral diagrams. (A) A normal five-petal *Aquilegia vulgaris.* (B) A double-flowered horticultural isolate. A = sepal; B = petal; C = stamen; D = staminodium (sterile stamen); E = carpel. From Reynolds and Tampion (1983).

frequent type of number variant in horticulture is described as a "double" flower, in which the number of petals is increased over the normal complement (Meyer 1966). While doubleness might mean only an increase in the number of petals or petal lobes, it can also occur as a result of the **petalody** of organs that would normally develop as stamens or carpels. The phenomenon can arise in a variety of developmental situations. Reynolds and Tampion (1983) recognize five major morphological types of doubleness, with numerous sub-categories.

Petalody resulting solely from an increased number of petals has not always been carefully distinguished from those cases in which it arises through petal development of primordia, where stamens might normally be expected. In fact, one criticism that can be levelled at many studies on the subject is that the materials examined were frequently poorly defined genetically and the data were frequently anecdotal and unsystematically collected.

A study of a "double flowered" cultivar of *Aquilegia* illustrates the complexities involved (Reynolds and Tampion 1983). Even in this case, because of the difficulty of documenting the genetic relationship of the cultivar studied to the so-called 5-petal forms of *A. vulgaris,* many implications must be left hanging. As presented (Fig. 4-23), the mechanism responsible appears to be the addition of four additional whorls of petals. Evidence that this interpretation is oversimplified, however, includes the observation that petal number varied from plant to plant, with position on the plant (Table 4-4) and in response to a number of chemical applications.

Two features about the developmental biology of meristic variation in the perianth remain to be identified: the genetic and the cellular factors. Despite the obvious examples of a genetic component in some cases of meristic variability, the topic is, in general, superficially documented. This is in part due to the fact that few critical selections of material bearing on the topic have been made and maintained for study. Surely, no molecular biology of the topic is possible until carefully selected lines of meristic variability are available.

Table 4-4 Analysis of the Positional Effect on Petal Number in 25 3-Flowered Cymes of *Aquilegia* Counted June 29, 1973

Position	Mean Petal Number ± Standard Deviation	Range
1	16.04 ± 4.036	10–25
2	16.24 ± 4.146	10–25
3	16.64 ± 4.251	10–25

Source: From Reynolds and Tampion (1983).

Despite the lack of critically selected material, where "extra petals" have been distinguished from petalody, it is obvious that a strong genetic component operates in determining the number of organs on a meristem. In some cases, the sequence of genetic expression may be explained in simple Mendelian terms. For example, doubleness in *Nigella damascena* behaves as a simple, recessive trait (Toxopeus 1927; Greyson and Raman 1975). In other cases, inheritance could be based in more complex genetic mechanisms. In the well-studied but incompletely understood "stock" (*Matthiola incana* (Cruciferae)) (Saunders 1913, 1928; Frost 1915), in addition to the demonstration of two alleles, "S" and "s," the segregation and expression are complicated by the observation that one chromosome can have a satellite which contains the "S" locus. The lack of the satellite leads to pollen lethality. Segregation ratios are consequently affected (Reynolds and Tampion 1983). Double flowers of *Matthiola* lack both stamens and a gynoecium.

A second important developmental consideration about meristic variation is its cellular basis. This involves the developmental, nutritional and environmental influences on the size of the flower meristem and the number of organs that it produces. This aspect of variation has been a concern of Stebbins, who has studied it in a number of different settings and discussed it at some length (Stebbins 1974). Since some of the data and examples on this topic are better documented at the production of carpels and the gynoecium, further discussion is reserved for Chapter 6.

Sepal and Petal Abnormalities

For this discussion a distinction is made between "doubleness" arising from an increase in the number of perianth parts and other mechanisms. These variations, traditionally referred to as **abnormalities** or **teratomas,** and their study as **teratology** (Masters 1869; Worsdell 1916), are of interest commercially but should also be viewed as potentially excellent material for developmental studies. Petalody is defined as the situation in which other organs exhibit the shape, anatomy and color attributes of petals. The most frequent situation is the petalody of stamens, and perhaps the least frequently described is the transformation of sepals. For a thorough literature survey and the opportunities available, the reader is referred to Meyer (1966). A more recent discussion of "doubleness" is provided by Reynolds and Tampion (1983).

The most useful examples of perianth variation for the interested developmental biologist will be those which can be explained, maintained and manipulated either genetically or environmentally. Availability of seeds or clones and ease of

78   The Development of Flowers

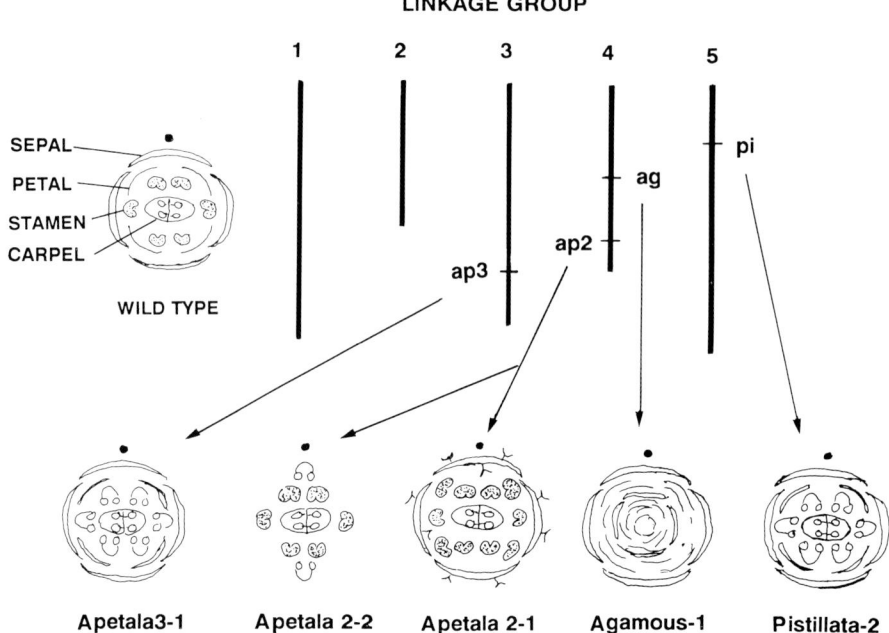

Fig. 4-24. Floral diagrams of the wild-type *Arabidopsis thaliana* and of five mutants in which flower structure is affected. The locus of each of the mutants is shown on the genetic map. Original courtesy of E. M. Meyerowitz.

maintenance are also desirable criteria for research material. Throughout this book a number of proposed "model" species are identified as appropriate for research on flower development. Of these, *Arabidopsis* is currently the material of choice, by some, for in-depth developmental and molecular studies (Fig. 4-24). Our consideration here is restricted to two perianth mutants as described by Meyerowitz (1987) and Bowman, Smyth and Meyerowitz (1989).

Two recessive genes that map to the Chromosome 4 are agamous (ag) and apetala2 (ap2). The flowers of ag plants consist of many sepals and petals and of chimeric organs consisting partly of sepal and partly of petal tissue. There are no stamens or carpels. The mutant flowers have an outer whorl of four sepals, then ten petals (four from petal sites and six from stamen sites) and in place of the gynoecium a variable number of sepals, petals and intermediate organs. The abnormality can be interpreted (Bowman, Smyth and Meyerowitz 1989; Yanofsky et al. 1990) as a new flower developing from the cells that normally form the gynoecium. This process is repeated a number of times so that, at maturity, a single flower may possess more than seventy organs. While this abnormality is known to occur in other Cruciferae, its availability in *Arabidopsis,* which is so admirably suited for experimental and molecular studies (Table 9-1), makes this species a prime candidate for future study.

A second mutant of *Arabidopsis,* apetala2 (ap2), demonstrates another useful developmental attribute: temperature-sensitive expression. As indicated in Table

Table 4-5 Phenotypes of *Arabidopsis thaliana*, Wild-Type (wt) and the Mutant Apetalous 2 (ap2) Grown at 16, 25 and 29°C

Phenotype	Temperature (°C)	1st Whorl	2nd Whorl	3rd Whorl	4th Whorl
wt	16/25/29	sepals	petals	stamens	carpels
ap2	16	leaves	petals and leafy petals	stamens	carpels
ap2	25	stigmoid leaves	staminoid petals	stamens	carpels
ap2	29	carpeloid leaves	absent	stamens	carpels

Source: From Bowman, Smyth and Meyerowitz (1989) with permission of American Society of Plant Physiology.

4-5, depending on the temperature at which the plant is grown, the ap2 gene allows a variety of morphological expressions. Traditional morphologists might argue that such data suggest the homologies or at least the similarities of leaves and sepals. While this argument is provocative, it is also speculative. More cautious developmentalists might limit their explorations to unravelling the molecular ways in which this short section of genetic information can bring about these different expressions. Further, because this gene seems to have no expression in the early stages of plant development and because its products can be traced, assayed and localized, a good deal of very precise information can be obtained. The range of phenotypic expression at a given flower whorl from organs that possess many anatomical and cytological leaf-like features through organs that are distinctly petaloid to nearly normal stamens, depending on the temperature at which the plants are raised, is most impressive (Bowman, Smyth and Meyerowitz 1989). Whether these flower mutants turn out to be "homeotic," in the same sense as genes which allow the development of organs in distinctly abnormal locations and which possess homeobox genetic sequences, as in *Drosophila* and *Xenopus* (McGinnis et al. 1984), remains to be seen. Homeosis is discussed further in Chapter 9.

Because *Arabidopsis* has a short life cycle, can be grown economically under controlled incubator or sterile culture conditions, has the smallest known genome among higher plants and is thus desirable for DNA cloning and mapping, it is a most desirable model research system (Table 9-1). In the eyes of many researchers it has become the *Drosophila* of the plant world, through which the genetic basis of many plant developmental features in addition to flowers are being explored.

Though much less well known genetically, doubleness in *Petunia* warrants consideration because of its variability and its availability from commercial sources. Doubleness in *Petunia* is conferred by a dominant gene Do1 and the recessive do2 (Table 4-6).

Both subtle and obvious variations in expression of the dominant gene affect a number of floral characters and, in some lines, vegetative traits as well (Sink 1973; Natarella and Sink 1971) These include: (i) young Do1/- flower buds are broader than +/+ buds; (ii) sepals of Do1/- flowers are darker green, hairier and broader than of those of +/+ buds; (iii) single flowers produce a sympetalous corolla composed of five petals; (iv) double flowers exhibit dosage effects, so in Do1/- plants (for example) the stamens resemble those of a single flower; (v) the

Table 4-6 Selected List of Nuclear Genes Which Affect Morphological Expression in the Flowers of *Petunia* (2N = 14)

Symbol	Description (chromosome number)	Reference
apt	apetalous. Petals sepal-like (I).	Sink (1973)
ch1	choripetalous-1. Petals not fused (IV).	Wiering et al. (1979)
ch2	choripetalous-2. Petals not fused (III).	Wiering et al. (1979)
ch3	choripetalous-3. Petals not fused (III).	Wiering et al. (1979)
Do1	double-1. Additional petals and stamens. Often female-sterile (VII).	Wiering et al. (1979)
do2	double-2. Additional flower parts, female-sterile (II).	Wiering et al. (1979)
ea	earshaped. Projections on outer surface of corolla (I).	Wiering et al. (1979)
G	grandiflora. Flower stalk and calyx undulated (V).	Plickert (1936)
gp	green petal. Petals same shape and color as sepals (IV).	Wiering et al. (1979)
ms1	male sterile-1. Anthers empty (VII).	Wiering et al. (1979)
ms3	male sterile-3 (VII).	

Source: Compiled by Cornu (1984) and Cornu et al. (1990) with permission.

double flower exhibits a profusion of petaloid structures including stamens, with double flowers most frequently being female sterile; (vi) while no differences were noted between vegetative meristems of double and single plants, differences were noted between the two meristems during perianth initiation (Natarella and Sink 1971).

The relative ease with which *Petunia* can be propagated through cuttings, a feature shared with other members of the Solanaceae, is used in the production of double-flowered hybrid seed. Pollen is produced from the normally female-sterile homozygous double-flowered clones and hand-pollinated on fertile single-flowered emasculated receptors. This ability to propagate clones is extremely useful in developmental studies on flowers of infertile variants. A selected list of nuclear mutants of *Petunia* which affect flower morphology is provided in Table 4-6.

Cell Differentiation in Sepals and Petals

The literature on perianth anatomy remains dispersed and unreviewed, with some features better documented than others: generalizations at this stage are therefore risky. Notwithstanding, perhaps the following comments will prove helpful and can be sustained.

Perianth Vasculature

Perhaps, initially because it was easily accessible through sections and clearings and microscopy, and later because of what anatomists felt those findings suggested, flower vasculature attracted considerable interest. The result was the production of a large body of data and argument detailing the complexity and variability of flower vasculature (Puri 1951; Eames 1931).

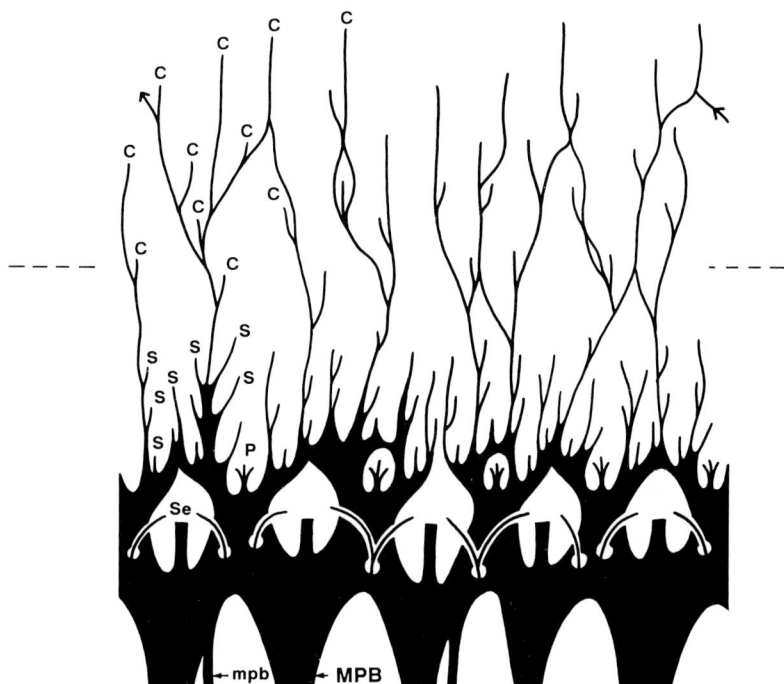

Fig. 4-25. Schematic drawing of the vascular cylinder of the receptacle of *Ranunculus repens* viewed from the side, split lengthwise and spread out in one plane. The traces to the appendages are labelled in the left side of the drawing. The lower part of the drawing was made from whole cleared stained receptacles; the upper part was reconstructed from serial sections. Se = sepals; P = petals; S = stamens; C = carpels; MPB = major pedicel bundle; mpb = minor pedicel bundle. Reprinted from Tepfer (1953) with permission from the University of California Press.

Apart from providing a primitive descriptive base, the results of this effort are, however, rarely of value to developmentalists. Discussions are frequently based on the assumptions and concepts of classical morphology (Mosley 1967); are often based on observations of mature specimens alone (Tepfer 1953; Sporne 1958), and often fail to incorporate the current insights derived from experimental manipulation of vasculature and the physiological factors that influence vascular differentiation (see suggested readings, this chapter).

A description of the vasculature of the flower of *Ranunculus repens* from Tepfer (1953) serves as an introduction. (Recall that the floral organs of *Ranunculus* (Fig. 4-1) are arranged helically on an elongate receptacle). Tepfer's description (Fig. 4-25) presents a complete vascular cylinder at the sepal level which is served by five major pedicel bundles plus some smaller traces. Sepals are served by three traces, each arising from a "gap" in the cylinder. Petals are served by a single short trace that branches into three traces.

Sporne (1958), based on studies of other members of the Ranunculaceae, finds no support for the concept of a vascular cylinder (**siphonostele**) and argues that

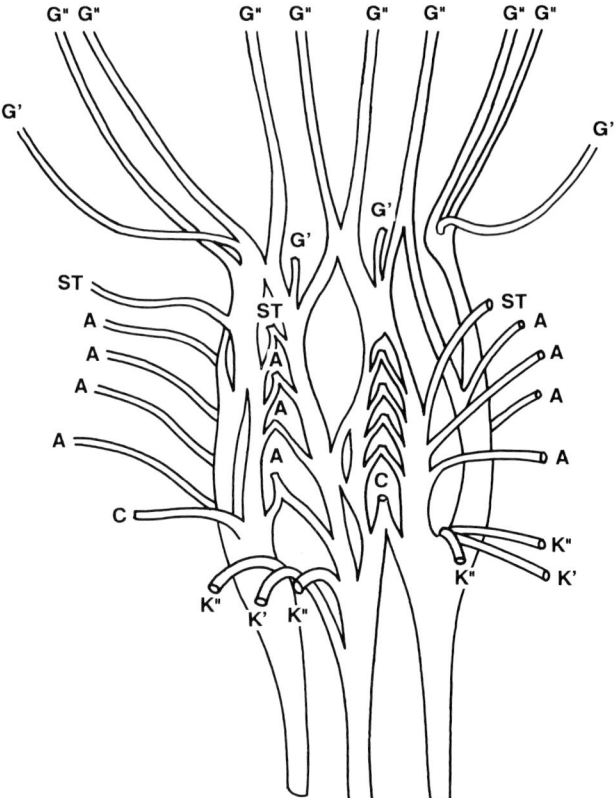

Fig. 4-26. A thick vertical slice through the receptacle of *Aquilegia vulgaris*. The slice was cleared in lactic acid, stained and viewed from the side. K = sepal supply (K' = median, K" = lateral); C = petal supply; A = stamen bundles; G = carpel supply (G' = dorsal, G" = ventral); ST = staminodium bundle. From Sporne (1958) with permission of the Linnean Society of London.

the primary vascular systems of flowers should be viewed as either open or closed but not in terms of steles, gaps and traces. His interpretation of cleared and stained flower buds of *Aquilegia* (Fig. 4-26) indicates that the vasculature at the base of the flower is composed of several separate bundles which branch and recombine in an irregular manner. Three bundles enter each sepal and one enters each petal. The spaces above these traces are unlike leaf gaps in that they extend through the entire stamen region. The stamens and staminodia lie opposite these spaces in vertical rows and receive vascular contributions from both sides.

The vascularization of the lamina of the sepals and petals is generally assumed to resemble the patterns found in leaves (Hickey 1973), though petals are more variable in their venation patterns than are leaves (Arnott and Tucker 1964). Major petal vasculature consisting of solitary veins, open dichotomous systems and closed and open reticulate systems have all been described.

In a quantitative study of the venation of *Ranunculus,* Arnott and Tucker (1963,

1964) found that the predominantly open, dichotomous pattern characteristic of the genus was complicated by anastomoses in 69% of the petals observed from their sample. The non-random distribution of these anastomoses was taken as evidence of morphogenetic control. As yet there is no indication of the validity of this assumption nor any identification of a possible mechanism.

What is lacking, of course, in most discussions of flower vasculature, is an awareness that vasculature formation can be a moderately plastic process with an ability to respond to many different physiological cues. To begin to redress this, studies of vascular differentiation using quantitatively sampled, time-based observations in the manner of Jacobs and Morrow (1957), followed by explorations that approach differentiation from a physiological perspective (Shininger 1979; Sachs 1981, 1986) will be required. Future approaches to vasculature research may incorporate technologies such as: (i) in situ localization of vascular precursors; (ii) study of molecular variation based upon single-gene differences and experiments with novel chimeric forms (Lagrimini, Bradford and Rothstein 1990); (iii) confocal microscopy; (iv) sensitive surgical and graft experiments. These approaches should provide some of the balance that is presently missing in discussions of flower vasculature.

The general morphological similarity of perianth organs to leaves is impressive. Thus, in addition to the general leaf-like morphology of sepals, the presence of epidermal trichomes, stomatal complexes, mesophyll chlorenchyma and vasculature resemble closely those traits of leaves and bracts on the same stem. In developmental terminology, leaves and sepals probably share and express many of the same genetic and developmental sequences or programs during growth and development. The documentation of such an assumption is yet ahead.

Petals, with the exception of their bifacial laminar morphology, share fewer similarities with leaves. Their epidermal layers, while containing stomatal complexes and trichomes, may also have epidermal cells with selectively thickened walls and papillae on their outer surface. The central parenchyma most consistently lacks chlorophyll but is frequently the site of specialized pigment synthesis and accumulation. In some cases epidermal layers also are sites of pigment synthesis. Sclereids and other idioblasts are frequently reported in both sepals and petals.

Pigmentation

It may not be obvious why pigmentation is included in a discussion on flower development. After all, pigmentation is more appropriate to phytochemistry or the investigation of ecological interactions with insects and other pollinators than to discussions of flower ontogeny. Notwithstanding, the synthesis of one or a spectrum of pigments in a cell is as much the consequence of gene expression in an environmental context as is the elaboration of an organ or the differentiation of a tissue. For many plants, the full display of pigment variability is linked to or coincides with the development of the flower or particular flower organs.

In some cases, the display of pigment patterns in the different organs, especially the petals, stimulates exactly the same questions about development as do patterns of organ formation. Frequently, the biochemistry of pigment synthesis is well studied, and in some plants the genetic basis of expression is beginning to be worked out in detail. Further, pigment formation can be a useful marker for following or interpreting developmental experiments. It should be recalled that McClintock's (1950) discovery of transposable elements in maize was based upon the aberrant expression of pigment formation in the maize kernels. Finally, because of the availability of molecular probes for specific genes involved in the pigment pathways, model experiments in molecular genetics are currently fashionable.

Flower color and the intricate color patterns produced in flowers result from synthesis and accumulation of a variety of different light-absorbing molecules and physical factors such as the texture of the cell wall, cuticle and wax. Any individual flower color may be a consequence of the presence of a number of different pigments. Three major molecular pigment classes are recognized: **flavonoids, carotenoids** and **betalains.**

Most of the red-violet-blue coloration of flowers is derived from the accumulation of the most common class of the flavonoids, the **anthocyanins.** These water-soluble molecules are frequently located in the petal epidermal vacuoles. Their diversity is based upon different types of conjugation with sugars and complexing with metals, methylation, hydroxylation, oxidation and reduction and modifications of pH. Less common flavonoids are responsible for some yellow-white pigmentation.

Carotenoids, also common contributors to the yellow-orange-red hues of some flowers, are fat-soluble, osmiophilic materials restricted to plastids and are frequently found in the cytoplasm of the mesophyll cells.

The betalains are a class of water-soluble nitrogen-containing molecules that confer the red-purple hues to a restricted group of families which are mostly included in the Order Caryophyllales. The reddish-purple hues of cacti, *Portulaca* and *Amaranthus* are common examples.

As a model system for studying the molecular biology of plant development, the analysis of flavonoid synthesis is particularly useful. The pigment class consists of thousands of different compounds whose synthetic pathways are well characterized (Heller and Forkmann 1988). The ease with which phenotypes can be detected and measured is a formidable advantage. The attraction for developmentalists lies in the tissue-, organ- and temporal-specific expression, as well as in the fact that genetic control of flavonoid synthesis can be mediated by environmental cues such as light (Koes, Spelt and Mol 1989), temperature (Martin et al. 1987), pathogens (Dangl et al. 1987) and wounding (Smith and Rubery 1981).

The exploration of the molecular biology of anthocyanin pigment synthesis in *Antirrhinum, Petunia* and *Zea* through the genes associated with the key enzyme **chalcone synthase** (CHS) is currently very active. CHS mediates a major early step in the phenylpropanoid pathway (Hahlbrock and Scheel 1989) which is central to the production of phytoalexins, UV protectants, insect repellants, suberin, lignin and many other molecules, including the most obvious but perhaps not the most important: flower pigments.

In *Antirrhinum,* the mutant "nivea," blocks pigment synthesis and the normal allele is known to encode for CHS synthesis (Spribille and Forkmann 1982). The locus is unstable due to the insertion of transposable elements (Sommer et al. 1985; Carpenter et al. 1987). In contrast to *Petunia* (Koes et al. 1987), where a multi-gene family consists of seven complete members (although only one CHS gene is expressed), the CHS gene in *Antirrhinum* appears to be single (Sommer and Saedler 1986). It seems obvious that research on the molecular genetics of this pigment pathway will proceed at a rapid pace, contributing to both basic and applied aspects of the topic. The characterization of the regulatory roles of the other genes in the anthocyanin pathway, the introduction of chimeric sequences into the genes through transposable elements and other techniques should yield a healthy return for plant developmental biologists.

Morphogenetic Factors Involved in Perianth Development

This discussion of perianth development has introduced a number of different cellular phenomena and morphogenetic processes that will arise again in subsequent chapters. They raise a number of questions: (a) What individual pieces of molecular and cellular machinery are responsible for the siting, initiation, organ determination and regulation of growth of the individual organ primordia on the meristem? (b) What dynamic relationships integrate and coordinate the individual developmental components? (c) How can we model or visualize our interpretations? (d) Can experiments be devised to explore these questions?

Site selection on both vegetative and floral meristems has occasionally been modified experimentally. Though the results of individual experiments are subject to various interpretations, they do suggest, within limits, that primordia siting can be interpreted as exhibiting **regulative** rather than **mosaic** developmental behavior. These concepts, derived from animal embryology (Browder 1980), recognize that development can appear to be locally consistent and unaffected by adjacent experimental manipulations (mosaic), or it can recover or the fate of particular parts can be affected by experimental manipulation (regulative). Two types of experiment explore these possibilities with flower meristems.

In the first (Hicks and Sussex 1971), tobacco flower buds of different developmental stages were excised from inflorescences, bisected and cultured on sterile nutrient agar. While organ siting frequently tended to be as expected (Fig. 4-27) on these half meristem explants, some meristems exhibited organs in non-normal positions. In some cases these arose near the cut margin of the cut, and in other cases they were regenerated from the region of the cut itself. A frequent result, often observed in other culture studies, was the production of petals and stamens in unexpected locations. The possibility that, in addition to site modification, alteration in the organ-type determination events is involved will be raised at a later point.

A second example, also from tobacco, is often observed when isolated flower meristems are cultured on sterile nutrient media (McHughen 1982). In this case, meristems with up to two sepals, cultured on a basal liquid medium containing no PGS, produced a single whorl of five sepals, a single whorl of five petals, a single whorl of five stamens and two carpels fused into a pistil. Similar buds cultured on a medium containing kinetin produced several orders of organs which expressed

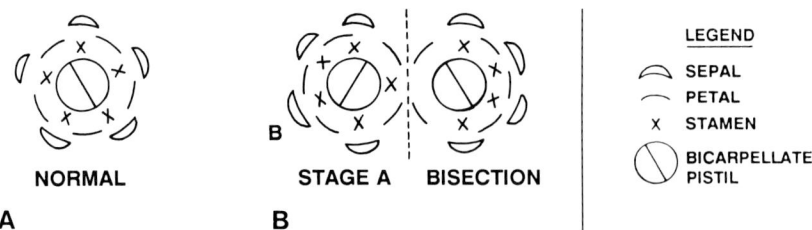

Fig. 4-27A–B. (A) Floral diagram showing the number, kinds and position of organs on a normal tobacco flower bud. (B) Floral diagram of a tobacco meristem, bisected at the pre-sepal stage and the halves attached and cultured. Organs most distant from the cut were unaffected. Near the cut, however, additional sepals petals and stamens are indicated. From Hicks and Sussex (1971) with permission of the University of Chicago Press.

characteristics of both sepals and petals but whose position did not conform to either a whorled or a normal spiral arrangement. This work illustrates a number of developmental phenomena, in addition to the experimental manipulation of siting, which will be raised later. One obvious feature is the plasticity which can be demonstrated under certain experimental procedures, at least on these *Nicotiana* meristems. We conclude that among the factors responsible for the siting of primordia are some involving the cellular environment and that these, in general, would be described as regulative.

Despite the large number of correlative studies involving PGS and the proliferation, determination and differentiation of organs on flower meristems, exacting information of how they interact in the cell-to-cell changes associated with organ siting and initiation is unavailable. Whether these studies are of PGS in liquid or solid media of organ or meristem culture (see reviews by Konar and Kitchlue 1982; Rastogi and Sawhney 1989), exogenous applications or analysis of endogenous levels, the data cannot apply to the fine cell-to-cell concentrations that are in all probability important, though not necessarily primary, features of the siting and initiation of organ primordia. Too few examples have been studied to permit drawing any conclusions. It seems possible, as well, that different plants may use different cellular programs of expression to achieve the same result.

5

The Androecium

Stamens, of all flower organs, have attracted the greatest scientific interest and are the best-studied flower component. Some aspects of stamen development, such as meiosis, microsporogenesis and gametogenesis, have been examined in detail. These are immensely challenging phenomena with general biological appeal and have been explored by a variety of technologies for over a century.

Other aspects of stamen development are only superficially understood. Because, however, of the potential for practical applications that basic studies frequently provoke, even more than with the perianth, future developmental exploration should encompass all stages of stamen development from initiation to maturity. The discovery and analysis of variation, whether in the structure and maturation of the stamen or in the details of microsporogenesis, are important for an overall understanding of this organ system and for a variety of agricultural and horticultural applications.

Arguments are common among flower morphologists as to whether stamens share features with stems or with leaves. Are all angiosperm stamens homologous? To what degree do stamens share morphological and anatomical features with petals or carpels? How are organs with mixed expressions of petal, stamen and carpel morphology to be interpreted? Such typological questions are perhaps foreign to developmentalists, who frequently deal with quantitative variation. Sattler (1967, 1988a) has tackled this topic from the perspective of morphology and argued for quantitative comparisons rather than qualitative, either/or decisions about organography.

These speculations about homologies between organs, based primarily on morphological and anatomical data, are only a prelude to the revelations that undoubtedly lie ahead as biochemical, molecular and genetic details of stamen and microsporangia development are uncovered. For the present, the unifying attribute shared by stamens of nearly all flowers, regardless of variation of other organs (male sterility and teratomas aside), is the development of microsporangia in which the processes of microsporogenesis and pollen production occur. Some of these variations in androecium structure are described in Chapter 2.

Patterns of Stamen Initiation

As with the perianth, stamen primordia are reported, in many cases, to arise individually in a helical and acropetal sequence leading to the characteristic parastichies along the surface of the meristem. One well-studied example of helical initiation is *Ranunculus repens.* (Fig. 4-1). In this study, Meicenheimer (1979) generated, from SEM micrographs, important numerical parameters associated with initiation and confirmed and extended the insights obtained from casual inspection (Tepfer 1953). In *Ranunculus,* and many flowers, stamen primordia arise individually on a floral meristem that in general maintains its volume and shape, at least in the early stages of stamen initiation. From other studies it appears possible that parastichies may result from initiation sequences that are not always based on the "single and helical" pattern. For example, Sattler (1973a) describes a number of features about stamen initiation patterns in *R. acris* which differ from the description for *R. repens.* This topic, like many others raised in this volume, illustrates the provisional and imperfect state of our present knowledge and suggests areas for future detailed study.

The more common arrangement of stamen primordia is a whorl of primordia which arise at the same level coincidentally. For example, *Aquilegia formosa* var. *truncata* (Tepfer 1953) (Fig. 5-1) initiates eight whorls of five stamen primordia each with stunning synchrony before initiating sterile stamens (**staminodia**) and carpels. Different cultivars and isolates produce different numbers of stamen whorls (Reynolds and Tampion 1983) (Fig. 4-23).

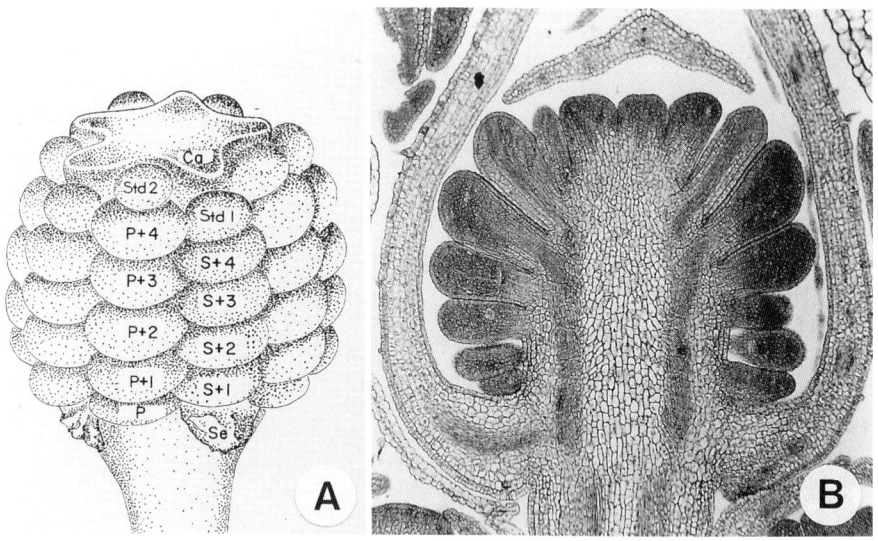

Fig. 5-1A–B. Aquilegia flower apices. (A) Diagram of a flower bud viewed laterally at the time of carpel initiation. Sepals removed for viewing. Primordia are identified by labels in two vertical rows. Se = sepal scar; P = petal; S+1, S+2, S+3, S+4 = stamens in vertical sequence above a sepal; P+1, P+2, P+3, P+4 = stamens in vertical sequence above a petal; Std = staminodium; Ca = carpel, at early grooved stage (approx. X100). From Tepfer et al. (1963) with permission of *American Journal of Botany.* (B) Slightly oblique longitudinal section showing three carpels at the top, and staminodia, stamens and petals on both sides (X110). From Tepfer (1953) with permission of University of California Press.

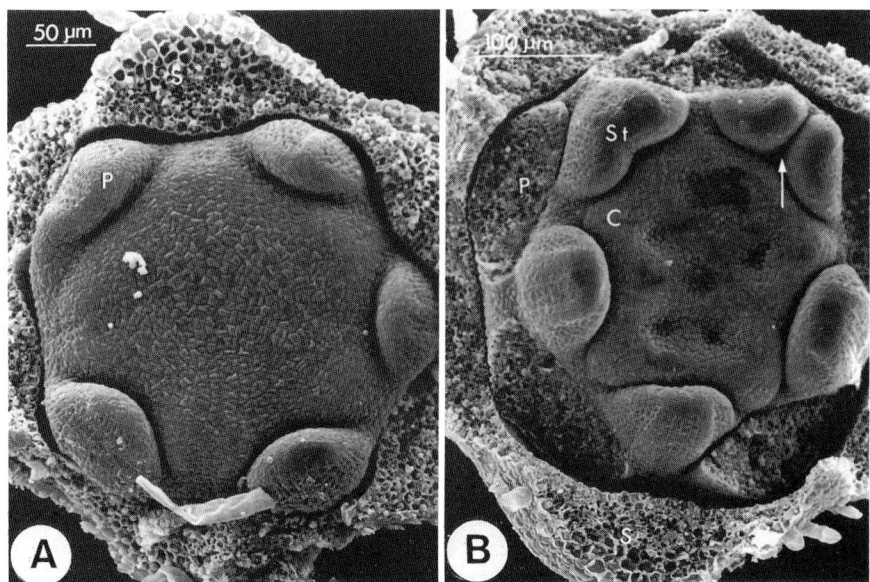

Fig. 5-2A–B. Initiation of petals and stamens on floral meristems of tomato, *Lycopersicon esculentum*. (A) Sepals (S) removed to reveal the simultaneous initiation of petal primordia (P). (B) Sepal and petal primordia removed to reveal the stamen primordia (St). Also visible is the initiation of a carpel primordium (C). Reproduced from Chandra Sekhar and Sawhney (1984) with permission of the National Research Council of Canada from the *Canadian Journal of Botany* 62:2403–2413.

The position of stamens relative to the positions of sepals and petals frequently generates attention in morphological discussions: stamen primordia are said to be **antipetalous** (opposite the petals) or **antisepalous** (opposite the sepals). In *Aquilegia* (Fig. 5-1A), the first stamens (S + 1) arise in antisepalous sites. Stamens also arise in antisepalous positions on the apex of *Lycopersicon* (Fig. 5-2), but with less synchrony. It is obvious that precise information about the level of synchrony and the relative positions of primordia is important for the generation of hypotheses and tests about organ initiation and spacing. Unfortunately, precise objective information is often difficult to obtain. Lyndon (1978) assumes that, while sepals and petals of *Silene* initiate helically, as do leaves, stamens appear to be axillary to the sepals and petals and are therefore stem-like in their position and shape. This interpretation arises frequently in discussions of organ relationships on the flower, though at present it is not possible to generate definitive data for or against it (see Chapter 4). Perhaps the degree of relatedness between organ types is a concept which will be sorted out by future molecular investigations of organ homology.

Brassica (Polowick and Sawhney 1986)(Fig. 4-3C–D) and *Arabidopsis* (Hill and Lord 1989)(Fig. 5-3) illustrate alternative features of stamen initiation. In both, the four long stamens initiate opposite the abaxial and adaxial petals and the two short stamens initiate later in antisepalous positions at a level on the meristem

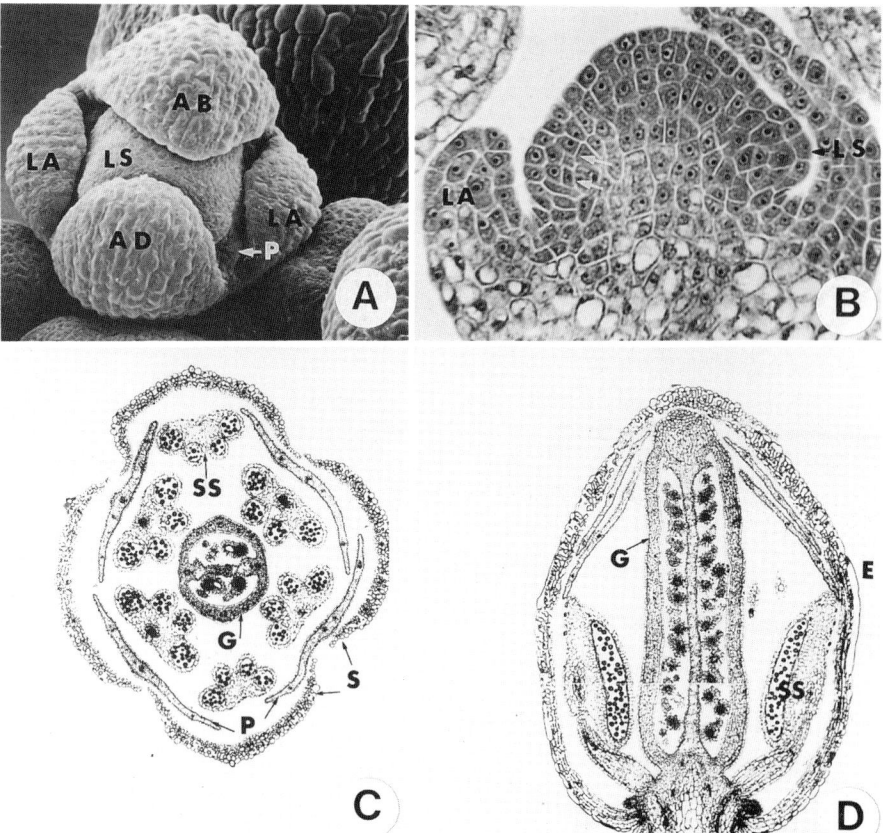

Fig. 5-3A–D. Views of flower organ initiation and nearly mature organs of wild-type *Arabidopsis thaliana*. (A) SEM plan view of a meristem at the time of petal (P) and long-stamen (LS) initiation. Short stamens (not visible) are probably initiating in front of each lateral sepal (LA). AB = abaxial sepal; AD = adaxial sepal (X480). (B) Radial-longitudinal section through meristem at stamen initiation. Short-stamen initiation, opposite lateral sepal (LA), is indicated by periclinal cell divisions in L2 (arrows). The section also passes obliquely through a long stamen (LS) already initiated on the other side of the meristem. FA = floral apex (X620). (C) Transverse section of flower prior to anthesis. Four petals (P) are positioned alternate to four sepals (S). Short stamens (SS) are antesepalous to lateral sepals. The gynoecium (G) is at center (X64). (D) Longitudinal section in the plane of the short-stamens (SS). The gynoecium (G) has a central septum separating the two locules. The sepals have many elongate epidermal cells (E) on the abaxial surface (X72). Reproduced from Hill and Lord (1989) with permission of the National Research Council of Canada from the *Canadian Journal of Botany* 67:2922–2936.

near the petal primordia. In both species, long stamen primordia and petal primordia initiation are nearly coincident.

This initiation pattern, where primordia arise on the meristem below or outside already initiated primordia, is defined as **centrifugal** initiation. It has been demonstrated frequently within the androecium, and more rarely amongst the calyx, corolla and gynoecium. The more common acropetal initiation pattern is described as **centripetal** (Sattler 1967, 1973b; Leins 1975). For additional evidence of stamen variation see Weberling (1989).

These variations in stamen initiation (i.e., non-helical, centrifugal or common primordia) [see also the formation of stamen groups in the Malvales (van Heel 1966) and unidirectional initiation (see Chapter 4)] represent significant deviations from the "single and helical" initiation sequence, frequently exemplified in leaf and sepal initiation. Each variant represents an opportunity to uncover, through careful description, computer simulation or manipulative experimentation, additional components or mechanisms in the developmental biology of primordium initiation.

Cellular Basis of Initiation

As pointed out in Chapter 4, an accurate representation of the cellular dynamics of organ initiation is deceptively difficult to generate. The casual inspection of the position, orientation and apparent frequency of mitotic figures from a few, or even many, microscopic sections is no substitute for proper cell cycle kinetic studies (Corson 1969; Francis and Lyndon 1978). The details of a dynamic process cannot be inferred satisfactorily: they must be analyzed directly. Since no thoroughly sampled examples exist, only general preliminary assumptions about initiation can be prepared.

On this basis it had been generally agreed that the cellular events of stamen initiation are similar to those involved in the initiation of perianth members, involving contributions from the hypodermal layer (T2) with accompanying anticlinal activity in the protoderm (T1) and with little contribution from the outer corpus. An example of this pattern of initiation is illustrated in *Arabidopsis* (Hill and Lord 1989) (Fig. 5-3B).

It was subsequently shown in some monocotyledons (Barnard 1957, 1958) (Fig. 5-4), that, like flowers and vegetative buds, but unlike leaves and glumes, stamens initiated from periclinal divisions in the outer (sub-hypodermal) layers of the corpus tissue. Further, in *Nicotiana,* Hicks (1973) demonstrated that stamen initiation also involved contributions from the outer corpus (Fig. 5-5) as well as from the T2. It appears, therefore, that no single, simple association between the pattern of initiation and mature morphology of the stamens is presently possible. Clearly more extensive and more critical quantitative studies are warranted.

Observations on periclinal chimeras of *Datura* also suggested that stamens were different in their histogenesis from leaves and perianth members. Whereas sepal and petal initiation depended primarily on the activity of the L-II, stamen primordia arose primarily through activity in the L-III (Satina and Blakeslee 1941).

From a developmental perspective these differences, regardless of how carefully documented, may be of secondary significance. Stamens are apparently produced regardless of the sequence of cell divisions in the different meristem layers from these different situations. Conversely, the first visible evidence of the absence of the PISTILLATA (PI) gene in *Arabidopsis* (Hill and Lord 1989) is abnormal division patterns in the T2 and T3 of stamens which lead to variable, sterile structures fused with the gynoecium.

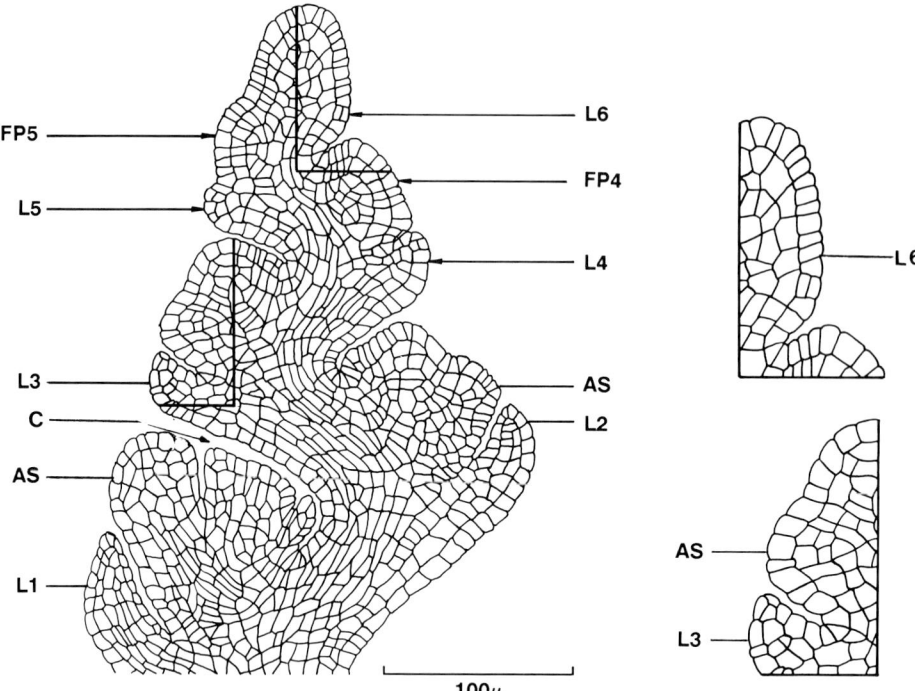

Fig. 5-4. Evidence supporting a different view of stamen initiation. A median longitudinal section of a spikelet of *Lolium multiflorum* showing the origin of the lemma at lemma 6 and its progressive development in lemma 5 (L5) to lemma 1 (L1); the origin of the flower primordium at FP5 and its early development in FP4; the origin of the palea and anterior stamen (AS) in the flower primordia in the axils of the third and second lemmas; and the origin of the carpel (C) in the most advanced flower. (**COMMENT:** One interpretation is as follows: While lemma initiation involves periclinal divisions of T1, stamen initiation involves periclinal divisions in the sub-hypodermal layer (outer corpus) and T2. Stamen initiation never involves periclinal divisions of the T1.) From Barnard (1957) with permission *Australian Journal of Botany.*

The notion of stem-like as opposed to leaf-like affinities for stamens remains debatable. The apparent axillary position of some stamens, the site of first periclinal divisions and the generally radial symmetry are consistent with stem-like affinities. Determinate growth and absence of an apical meristem, coupled with the exceptions to the first three traits, are more leaf-like. This debate seems to be one which developmentalists, especially those with molecular technologies, might explore with profit. Choosing sides in the argument should probably be delayed until more data are available.

As with the initiation of leaves, sepals and petals, fundamental questions can be asked: When, to what degree and in what form is the organ-specific information expressed during development? Explicit answers to these questions will require the use of molecular technologies applied to well-characterized research materials. Preliminary observations such as allometric comparisons between growing flower organs (Greyson and Sawhney 1972), which suggest early expression, re-

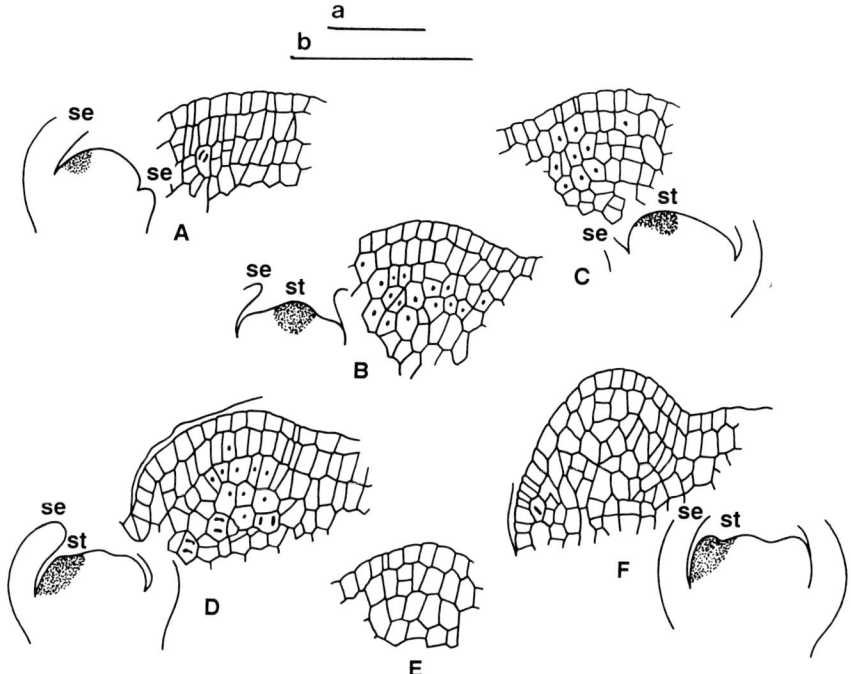

Fig. 5-5A–F. Stamen initiation and emergence in *Nicotiana tabacum*. (**General Orientation:** Each figure letter is for a pair of drawings: the smaller of each pair is an outline of a longitudinally sectioned floral bud, with stippling indicating the area from which the larger drawing was taken. Scale "a" above designates 200μm for all low magnification drawings. Scale "b" designates 100μm for the larger magnification views. Symbols for the particular primordia: se = sepal; st = stamen. All buds sectioned longitudinally). (A) Before emergence. Periclinal division of corpus cells at stamen site. (B) An emergent stamen. Periclinal division of corpus cells but not of T2 cells. (C) Tangential longitudinal section of emergent stamen. (D) Emergent stamen showing probable recent anticlinal divisions of T2 cells and corpus and also division of deeper corpus cells. (E) Adjacent section to (D) showing periclinal T2 division. (F) Later stamen showing predominance of anticlinal divisions in T2–derived cells. Note files of cells in corpus on adaxial flank of primordium. Reproduced from Hicks (1973) with permission of the National Research Council of Canada from the *Canadian Journal of Botany* 51:1611–1617.

quire confirmation by these more explicit tests. The developmental question of "determination" was raised in Chapter 3 (see also Chapter 9).

Stamen Growth and Differentiation

Considering the interest in microsporogenesis and pollen development and the importance of anther position relative to stigmas in pollination studies, it is surprising that there is so little precise information on the cellular basis of the early growth of the stamen and its differentiation into the mature anther and filament. Information is typically limited to qualitative description based upon observations of a limited number of microslide preparations. We assume, however, that stamen growth after initiation involves a generally diffuse meristematic activity and accompanying cell enlargement which lead, in many cases, to an elongate, more or less radially symmetric organ. This interpretation, while differing slightly from Boke's view (Boke 1949) that the stamen grows in length by means of apical and sub-apical initials until it is 200μm high, must be provisional. As Poethig (1984) explains, conclusions about the cell dynamics of organs can only be derived from direct analyses of the processes involved.

This primarily radial and elongate growth pattern of most stamens is in contrast to the planar symmetry generated in sepals and petals. Through differential development, two different areas of growth, the **anther** and **filament** become established. In *Vinca* (Boke 1949), however, anther development is discernable before a filament can be identified. The *Vinca* filament subsequently develops through intercalary activity at the base of the anther. It seems obvious that this aspect of stamen growth and maturation varies considerably both within and between families.

Filament Growth and Differentiation

Filament elongation in the disc flowers of *Gaillardia* (Koning 1983) exhibits a number of interesting features (Fig. 5-6). First, the number of epidermal cells along the filament (forty-three) was surprisingly constant. On a 2mm filament the approximate length of the eleven short cells was 30μm, and of the thirty-two basal long cells was approximately 50μm. Over 24h, prior to anthesis, during which filaments elongated threefold, the long cells elongated to an average of 120μm; the short cells did not elongate.

In *Cleome* (Koevenig 1973), however, filament elongation to a mature length of 40–60mm, appears to be based upon different patterns of cell activity. While cell division ceased when the filament was about 5–6mm, the last divisions were localized towards the base. This localization appears to be associated with the formation of an abscission layer. At maturity, the shortest cells were at the base of the filament and the longest cells were near the mid-region.

In *Nigella* (Greyson and Tepfer 1966), cell division with accompanying cell expansion in the epidermis was distributed throughout the filament until it was approximately 3mm long. At this stage the epidermis is composed of eighty to ninety epidermal cells, each 30–40μm long, along its length. Filament elongation to a final length of 12mm was accomplished through cell elongation. The longest cells (180–200μm) were at the base of the filament, while the shortest (70–100μm) were near the filament tip. Though their analysis was similar, but not exactly equivalent, Hess and Morre (1978) conclude that the base-to-tip pattern of elongation in *Lilium* is similar to that in *Nigella*.

Comparison of the cellular basis of filament growth in *Gaillardia, Cleome* and *Nigella* reveal striking differences. Despite the limited data from which the comparisons are made, one can assume that the physiological components regulating the cellular activities are also different. More direct and complete methods of observing organ growth will inevitably yield new information on the timing and patterns of growth and their regulation in these and other stamen filaments.

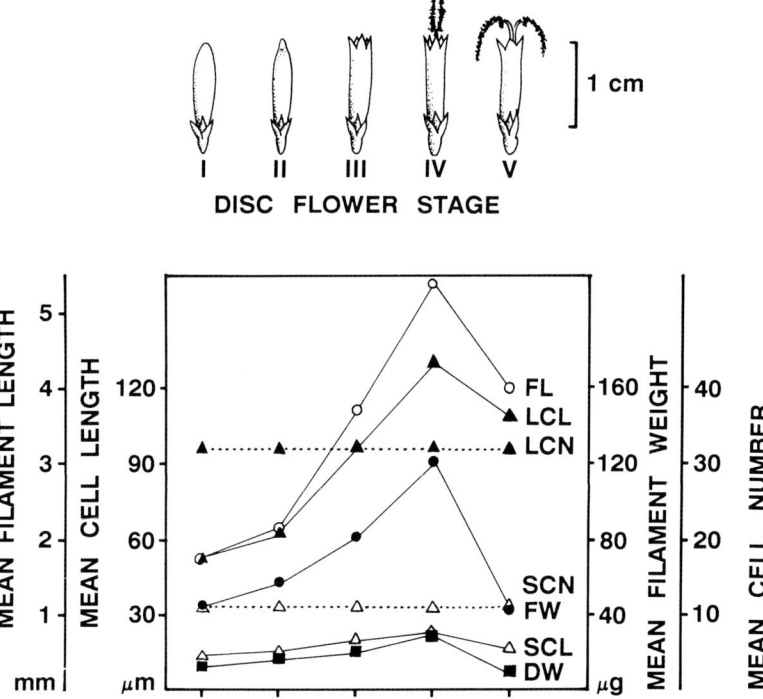

Fig. 5-6. Parameters of *in vivo* filament elongation in *Gaillardia* through five developmental stages (I = 15 days; II = 3 days; III = day of anthesis; IV = afternoon of anthesis; V = day after anthesis) of opening disc flowers. The parameters represented are: filament length (FL), epidermal cell lengths (long cells [LCL], short cells [SCL]), epidermal cell number (long cells [LCN], short cells [SCN]), filament weight (fresh [FW], dry [DW]). From Koning (1983) with permission of *American Journal of Botany*.

This final phase of stamen filament growth is frequently rapid, involving cell elongation alone. In the grasses—for example, maize (Schaeverbeke 1965) and wheat (Ledbetter and Porter 1970)—this final ten- to twentyfold elongation occurs over a few hours and is probably cued to light, temperature and humidity signals. The data available regarding the physiological concomitants of this cell elongation to maturity suggest that a number of components are involved and that considerable variability exists between different species in the mechanisms actually used.

The first insights concerning the physiological coordination of stamen filament growth came from observations on elongation dynamics (see above), emasculation experiments (Kiss and Koning 1990) and from observations on unisexual flowers (Plack 1957) [see also reviews by Lang (1961) and Raab and Koning (1988)]. The corolla, as described in Chapter 4, frequently responds in parallel with the stamens but often is retarded compared to stamens until shortly before anthesis.

Elongation of stamen filaments, in the few cases studied, can involve auxins, gibberellins, ethylene and cytokinins as regulating factors. This potential for multi-factor physiology is further complicated by the realization that, in addition to the endogenous concentrations of relevant PGS, their action can be affected by the relative rates of synthesis and destruction of active forms, combined with the formation and hydrolysis of inactive conjugated forms.

The availability of PGS-acceptors and the permeability of membranes to these metabolites are also recognized as important components to PGS physiology. Without taking sides in the arguments between physiologists (Trewavas and Cleland 1983) about the relative importance of each of these potential modes of action of PGS, it is enough to reflect that the potential for multiple sites of control make the design and interpretation of experimental analyses difficult (Guern 1987). In the absence of exhaustive studies, only provisional interpretations of what actually takes place are possible. The discussion here is limited to the best studied case, *Ipomoea nil*.

A correlation between the level of gibberellin-like substances from the anther and cell elongation of the filament cells during the last 36h before anthesis in *Ipomoea* was first described by Murakami (1973) and followed up by Koning and associates (Fig. 5-7). More difficult to verify is the involvement of ethylene, but a series of careful *in vivo* and *in vitro* studies incorporating various combinations of inhibitors and metabolites suggests strongly that normal filament growth is mediated by an interaction of ethylene with gibberellins (Koning and Raab 1987; Kiss and Koning 1990). In view of contemporary ideas of the ways in which ethylene and other PGS interact, a reworking of earlier, one- or two-molecule correlations (Schaeverbeke 1965; Greyson and Tepfer 1967; Koevenig 1973) with filament elongation seems warranted. A sample of the potential complexities of what might actually be operating is suggested by the hypothetical model (Fig. 5-8) prepared as an interpretation of the involvement of PGS during corolla and filament elongation in *I. nil* (Raab and Koning 1988).

Filament Anatomy

Reflecting the specialized role of stamens for the nutrition, positioning and ultimate release of pollen, anatomy is less variable and perhaps simpler than it is in other flower organs (Schmid 1976). The epidermis may bear stomata (occasionally permanently open), trichomes and nectaries and is covered by a cuticle. In *Lilium* (Heslop-Harrison, Heslop-Harrison and Reger 1987) the cuticle is continuous throughout the whole period of extension and remains impermeable. The ground tissue is of parenchyma, typically with large inter-cellular spaces. In *Triticum*, collenchyma also surrounds the inner parenchyma (Ledbetter and Porter 1970). The central vascular strand is frequently rich in phloem cells, surrounding a limited primary xylem of annular and helically thickened tracheary elements. Bundle sheaths, sometimes interpreted as an endodermis, frequently surround the strand. Crystals, secretory structures, collenchyma and schlerenchyma are rare in filaments.

Stamens generally receive a single vascular trace, which remains unbranched throughout the length of the filament. Some branching of the trace may occur in the anther connective (Johri 1951). In some flowers in which anthers are united

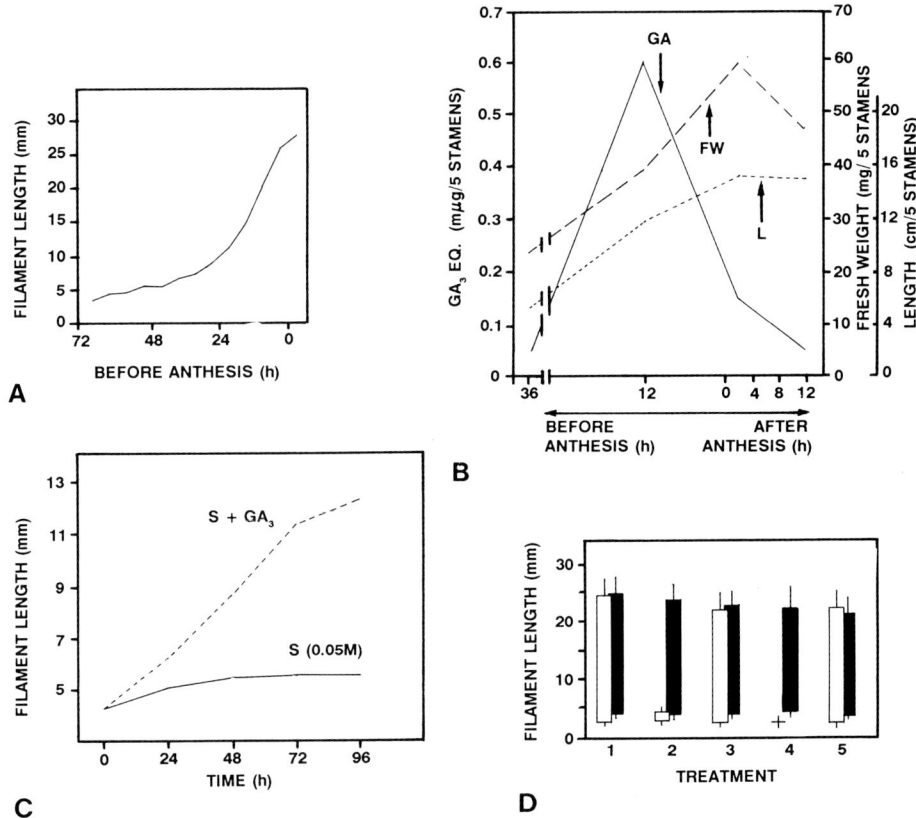

Fig. 5-7A–D. Observations on PGS status and filament growth in *Ipomoea nil.* (A) Observations on length of filaments from intact flowers during the 72h prior to anthesis. Each point is a mean value calculated from two separate samples. (B) Changes in the endogenous GA-like substances from stamens at different stages of flowering. (C) The time course measurements of filaments cultured *in vitro* + and − GA$_3$. Vertical lines indicate +/− SD. (D) Growth responses of filaments to various organ excisions at 69h (white bars) and 45h (black bars) before anthesis in the presence of and the absence of 10^{-4} AVG (aminoethoxyvinylglycine). The lower and upper ends of the bar for each treatment represent the initial (at the time of treatment) and the final (at the time of anthesis) mean filament lengths, respectively, for the two experiments. Vertical lines indicate +/− SD. Treatments (1) intact; (2) − calyx, − corolla; (3) − calyx, − corolla, + avg; (4) − calyx, −corolla, − anther; (5) −calyx, −corolla, − anther; + avg. **COMMENT:** These selected observations illustrate the accumulation of GA-like substances just before anthesis, the sensitivity of the filament to GA$_3$ and the reversal of filament inhibition by a known ethylene antagonist. The time course of filament elongation is provided in (A).) Parts A and D redrawn from Kiss and Koning (1990) with permission of *American Journal of Botany;* part B from Murakami (1973) with permission of *Plant Cell Physiology;* part C from Koning and Raab (1987) with permission of *American Journal of Botany.*

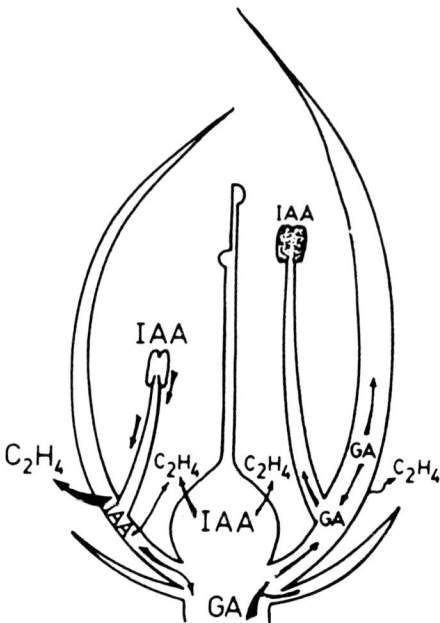

Fig. 5-8. A hypothetical model of regulation of corolla expansion and filament elongation by growth substances in *Ipomoea nil*. The diagram depicts two stages of development: a younger corolla on the left and an older corolla on the right. The control of growth appears to consist of a balance between levels of the growth inhibitor (ethylene) and the growth promoter (gibberellin). Ethylene produced in the younger corolla prevents premature expansion. As the corolla matures, ethylene production declines, apparently allowing the increased gibberellin level to induce rapid corolla expansion. Figure and legend from Raab and Koning (1988) with permission. Copyright © 1988, American Institute of Biological Sciences.

into much-branched fascicles, the fascicle vasculature is also branched but is connected to the receptacle vasculature by a single trace (Sporne 1958).

In *Gasteria* (Keijzer, Hoek and Willemse 1987), the cellular changes associated with filament elongation include a base-to-tip progression of elongation, the thickening of the epidermal tangential wall and its cuticle, increasing amounts of starch and increasing turgidity. Beginning with the pollen-mitosis stage, a progressive cytoplasmic degeneration, from tip to base, is observed in the central parenchyma and the epidermis. At anther dehiscence, the filament tip shrivels and the starch has disappeared from the entire filament. In *Lilium longiflorum*, K^+ ion concentration accounts for approximately 1/4 of the measured osmolality of the sap expressed from turgid stamen filaments at the period of rapid extension (Heslop-Harrison, Heslop-Harrison and Reger 1987). Though not determined, the remainder of required solute is assumed to be soluble sugars.

There are, unfortunately, too few data from which to draw general conclusions about how these events might relate to the physiological control of filament extension and anther dehiscence. They do, nevertheless, provide a glimpse of the interplay of physiological factors that are involved in these important growth responses included in the process of **anthesis.** The identification of all the connections of these processes to gene expression and to environmental cues has yet to be deciphered.

Anther Growth

Our knowledge about the quantitative aspects of anther growth is limited. In *Lilium longiflorum,* Erickson (1948) established: (a) bud growth under well-controlled environmental conditions is probably exponential; (b) based upon allometric, log-length comparisons between bud and anther, two phases of anther elongation may be identified, namely, an early phase from anther length 4–25mm of rapid elongation, followed by a smaller and slower growth phase from bud length 40–140mm; (c) comparisons of fresh and dry weight of anther to bud length gave similar results; (d) microsporogenesis (meiosis) and subsequent microspore mitosis are nearly simultaneous within the anther, which suggests that they are subject to common controls. The precision and consistency of this synchrony evidently varies with different cultivars (Walters 1976, 1980). The anthers of *Lilium* are relatively large and obvious candidates for this type of study. It is therefore encouraging to see comparable data obtained from the much smaller anther of rice (Raghavan 1988) (Fig. 5-9).

A study by Gould and Lord (1988) adds another complexity to our understanding of the growth dynamics of the *Lilium* anther. By marking individual immature anthers, and following the length changes of the individual segments, they identified peaks and troughs of growth which passed over the anther from tip-to-base. These data suggested that waves of cellular activity, corroborated by the detection of waves of mitotic activity (Fig. 5-10), pass along the anther, from tip to base, as it grows. While the stimulus for the coordination of these waves is unknown, this observation is substantial and original. Its extension to the dynamics of anther growth in other species, and in fact to other organs, in combination with computer analysis and display (Delozier et al. 1987), should provide some fresh views of organ growth and development.

Anther Differentiation

A number of morphological and cellular changes accompany the overall growth of the anther. The most obvious morphological change on a typical anther primordium is the development of, usually, four **microsporangia** along the length of each anther. Cross-sectional views of these microsporangia during development in *Oryza* are illustrated in Figs. 5-11 and 5-12 and at maturity in *Arabidopsis* in Fig. 5-3. Bisporangiate anthers and other variations are relatively rare (Davis 1966).

Since there appears to be considerable variation, especially between families, in the precise details of microsporangia formation (Davis 1966) we have chosen to follow the details of a single case. Discussions of microsporangium development in other species are available from primary sources and from reviews (Maheshwari 1950; Bhandari 1984).

In rice, *Oryza sativa* (Raghavan 1988) (Fig. 5-11), differential mitosis and

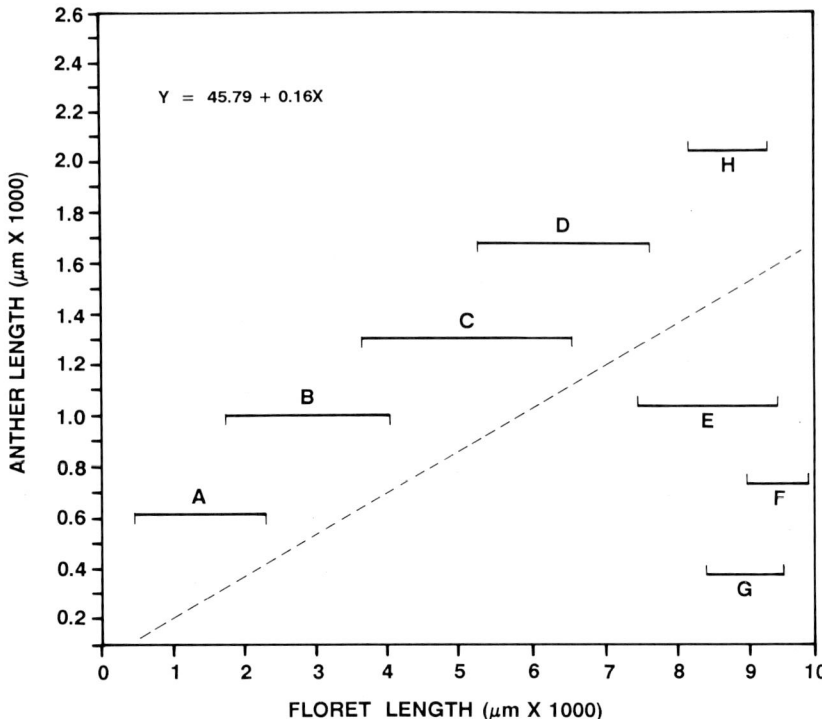

Fig. 5-9. Linear plot of length of rice anthers (to 2.0mm) against length of florets. The calculated regression line and its equation are given. Ranges of floret lengths in the sampling when specific cytological events of microsporogenesis occur in the anther are indicated at A to G. (A) Archesporial initials—primary sporogenous cells. (B) Secondary sporogenous cells. (C) Meiosis. (D) Non-vacuolate microspores. (E) Vacuolate microspores. (F) Bicellular vacuolate pollen. (G) Bicellular, starch-filled pollen. (H) Starch hydrolysis in pollen. From Raghavan (1988) with permission of *American Journal of Botany*.

accompanying cell enlargement in four corners of the anther primordium (approx. 100μm long × 80μm wide) generate four longitudinal lobes (**microsporangia**). Divisions in the epidermis are restricted to the anticlinal planes. Hypodermal differentiation leads to a single row of distinctive cells, the **archesporial initial cells**, along the length of each microsporangium. Each initial divides periclinally to give rise to a small cell toward the periphery (**primary parietal cell (PPC)** of the anther lobe and a large cell (**primary sporogenous cell (PSC)** towards the interior. The **primary parietal layer (PPL)**, the layer surrounding the PSC, is derived in rice, according to Raghavan (1988), from the PPC and the other adjacent cells. The cells of the PPL then divide periclinally to form two layers of cells: the outer layer is the forerunner of the **endothecium**, while the cells of the inner layer divide once more, periclinally, to form an outer **middle layer** and an inner **tapetum.**

The sequence described above is consistent with the usual description for other monocotyledons (Davis 1966) (Fig. 5-13) with the exception that not all the cells

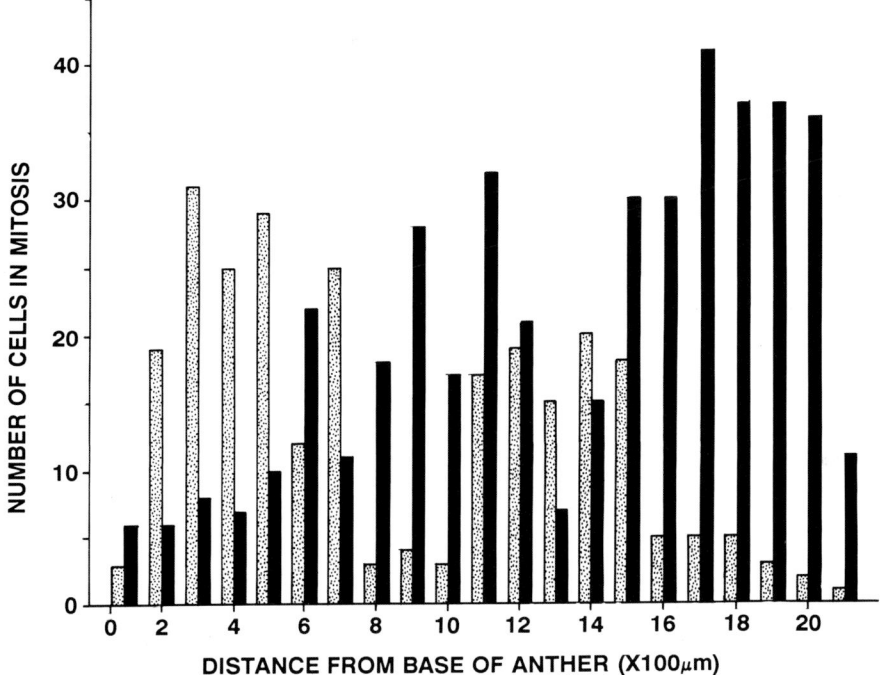

Fig. 5-10. Observations on the mitotic activity in 100μm segments of two anthers from a single bud of *Lilium longiflorum*. The pattern of mitotic activity along anther A (stippled bars) possessed peaks of mitosis at 300–500 μm and at 1200–1500 μm. Anther B (solid bars) exhibited peaks at 900–1100 μm and at 1700–1900 μm. Bud length was 2.1mm. (***COMMENT:*** Since these anthers were from the same flower and at the same stage of development, these data are interpreted as indicating that the mitotic activity proceeds in waves along the anther.) From Gould and Lord (1988) with permission of Springer-Verlag Heidelberg.

of the wall were derivatives of the archesporial cell. From a developmental point of view, Raghavan's description of the origin of PPL in rice is most intriguing. It suggests that a cell's "location" can be as important to what it becomes as its "lineage." This dual origin of some of the layers in the anther wall is also discussed by Bhandari (1984) with reference to other species.

The scheme, described as the Dicotyledonous Type (Fig. 5-13), differs from that described for monocotyledons in the manner of formation of the middle layer and the tapetum. Other patterns are known and no doubt other subtleties and details are yet to be identified as more precise studies are performed.

The **epidermis, endothecium, middle layer**(s) and **tapetum** constitute the wall of the anther. Anticlinal divisions with accompanying enlargement generate the mature anther walls of the four microsporangia (Fig. 5-14) which remain attached to the vascularized connective. In maize, the long axis of the epidermal cells is parallel to the long axis of the anther. The long axis of the other wall layers parallels the circumference (Fig. 5-14). During anther dehiscence, the ten-

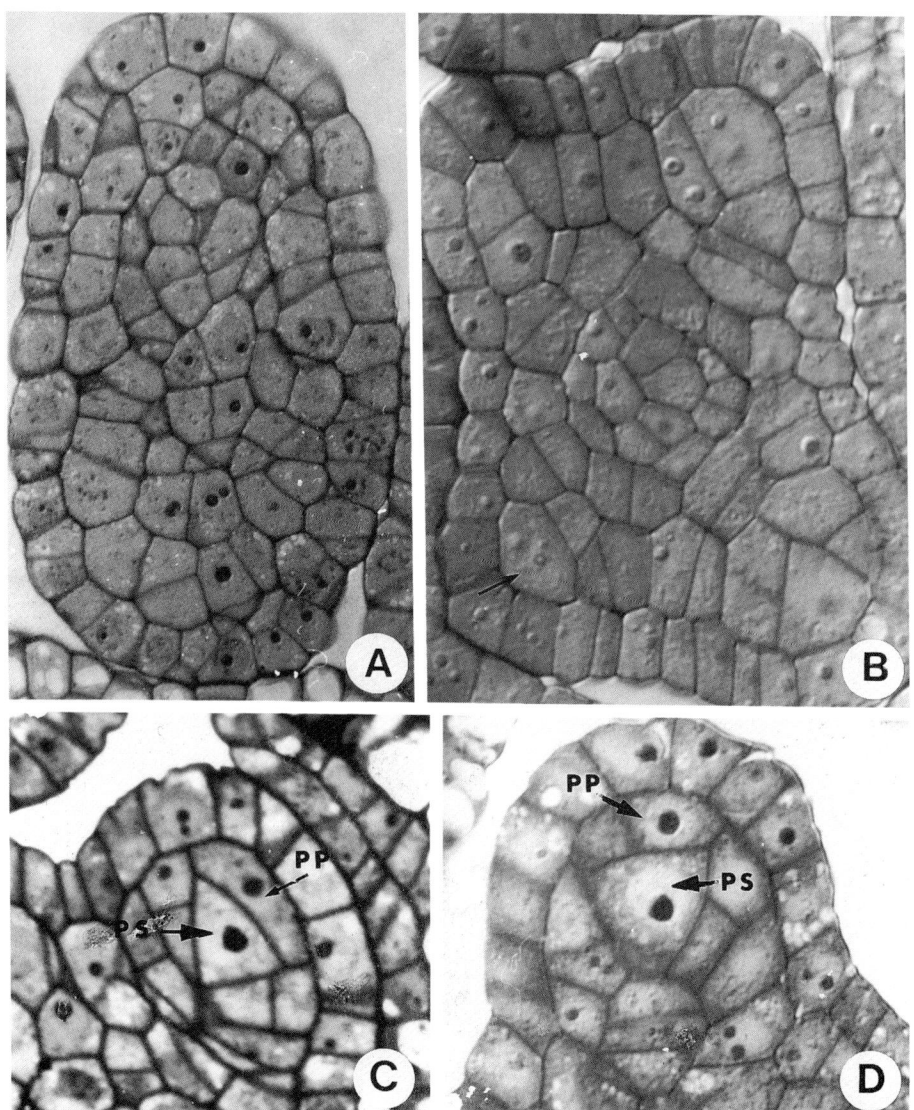

Fig. 5-11A–D. Selected microscopic cross-sectional views of early anther development in rice. (A) An anther primordium showing a protodermal layer and a homogenous mass of internal cells (X1200). (B) An anther at the stage where the archesporial initials (arrow) begin to differentiate. The anther has become four-lobed (X1350). (C) An anther lobe showing the formation of the primary parietal cell (PP) and the primary sporogenous cell (PS). (D) An anther at a slightly later stage to illustrate that the primary parietal cell has divided anticlinally (X500). From Raghavan (1988) with permission of *American Journal of Botany*.

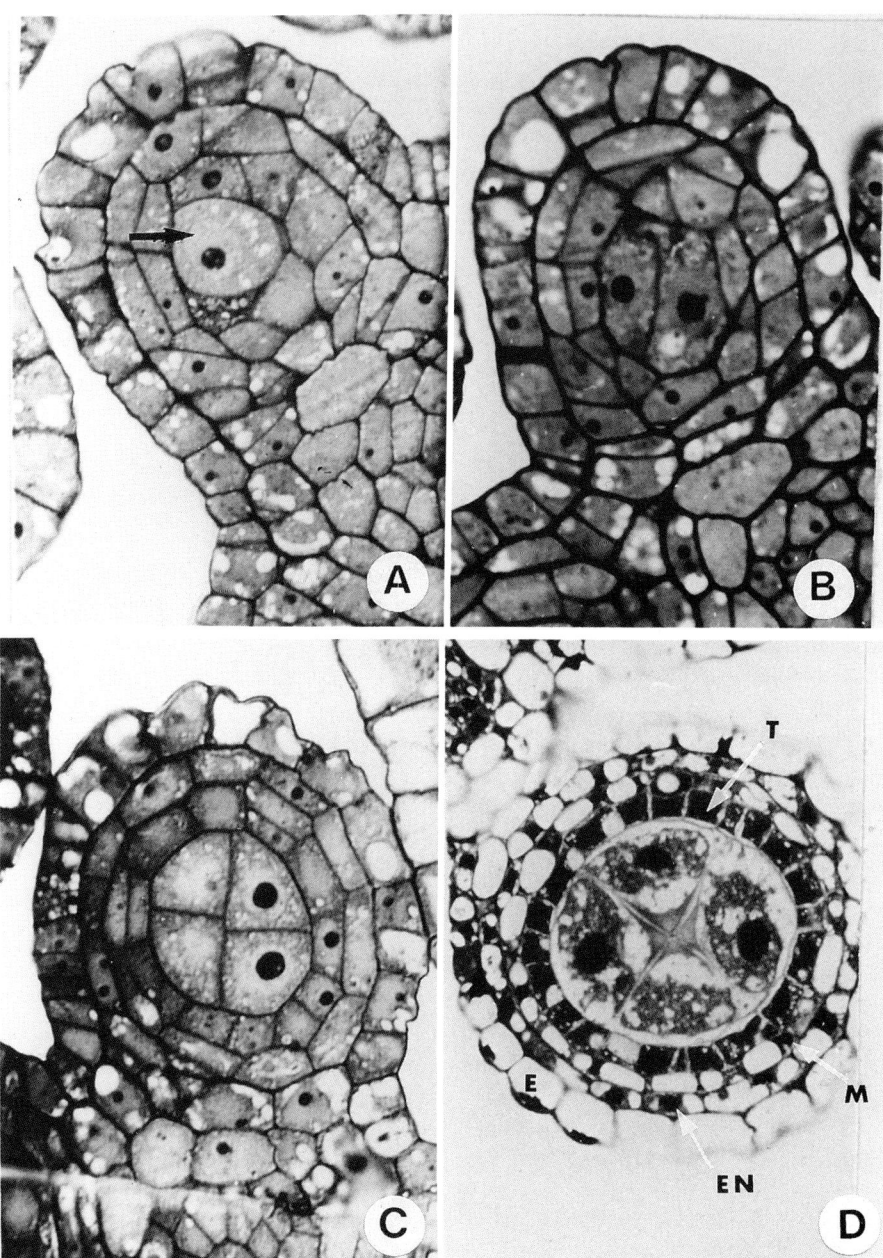

Fig. 5-12A–D. Microscopic cross-sectional views of wall formation and development of the microsporocytes in rice. (A) An anther lobe showing the primary sporogenous cell (arrow) surrounded by a three-layered wall (X 1200). (B) An anther lobe in which the first division of the primary sporogenous cell has occurred (X1200). (C) An anther in which the middle layer and the tapetum have formed from the innermost wall layer (X1250). (D) Anther lobe showing the anther wall, constituted of an epidermis (E), endothecium (EN), middle layer (M) and the tapetum (T). The microsporocytes (arrow) are about to go through meiosis (X750). From Raghavan (1988) with permission of *American Journal of Botany*.

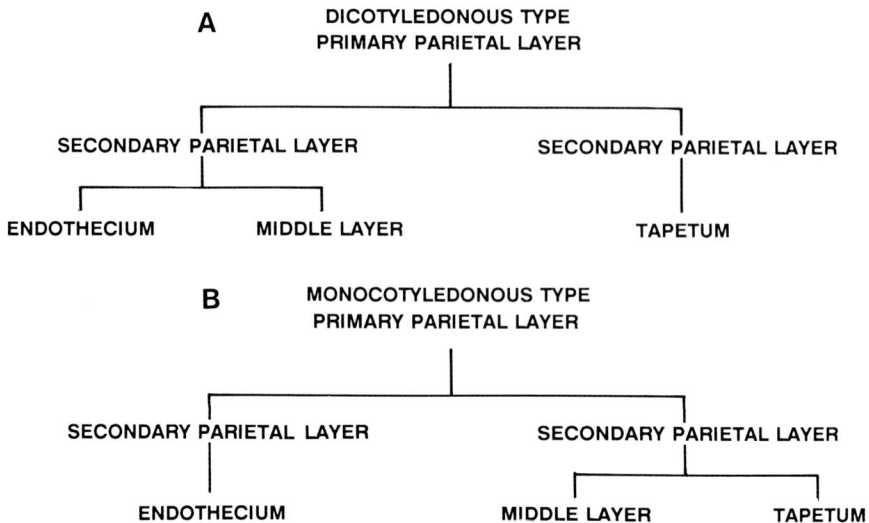

Fig. 5-13. Summary charts of assumed origins of the wall layers in anthers of a dicotyledonous type (A) of anther and a monocotyledonous type (B). From Davis (1966). Reprinted with permission of John Wiley & Sons, Inc. Copyright © 1966.

sions on these layers due to desiccation leads to separation of the cells between the microsporangia (inter-microsporangial stripe 1) (Fig. 5-16). This results in formation of two large pollen sacs at the time of pollen shedding. Other variations of anther structure and pollen shedding are described in Chapter 2.

The Microsporangium

The anther epidermis, occasionally dried and absent at maturity, is a single layer. On some anthers—for example, *Sorghum* (Warmke and Lee 1977), maize (Cheng, Greyson and Walden 1979), *Brassica* (Polowick and Sawhney 1986) and *Arabidopsis* (Kunst et al. 1989)—the epidermis develops an ornate, ridged cuticle. Many of the cells of the endothecium, also a single cell layer, develop cellulosic wall thickenings. In some species, the majority of endothecial cells develop wall thickenings. In those with poricidal dehiscence (e.g., maize), thickenings are restricted mainly to those cells in the vicinity of the pore (Cheng, Greyson and Walden 1979). The number of middle layers of parenchyma vary among different species and as well, in some species, within the individual anthers. Many of these traits plus their variations are associated with the maturation of the pollen and anther dehiscence, which is generally considered to be a carefully regulated desiccatory process (Schmid 1976).

The tapetum, because of its unique character and its location surrounding the sporogenous tissue, is, deservedly, the best studied anther wall tissue. Despite this attention (Chapman 1987), tapeta from only a handful of species have been studied in detail, and only the most obvious developmental features of this single-cell layer are discussed here.

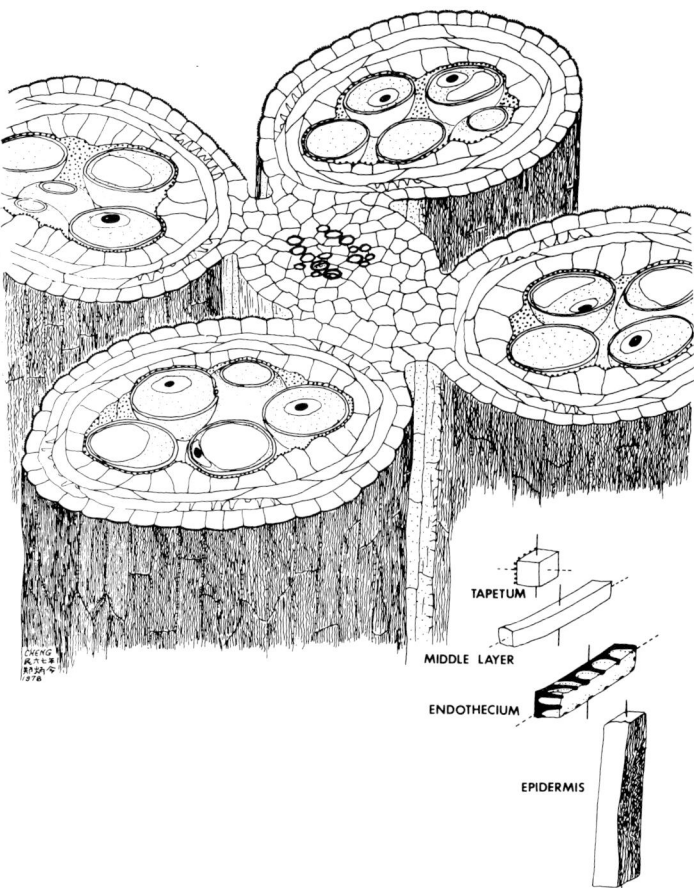

Fig. 5-14. Diagram illustrating the arrangement of microsporangia around the connective of an anther at the young pollen stage of development. The orientation of the cells in the anther wall is also portrayed. Intermicrosporangial stripes are indicated. The site of separation between the epidermis and the parenchyma of the connective is indicated by arrows. In the isolated diagrams of wall cells, the solid vertical lines indicate the long axis of the anther and the dotted lines indicate the long axis of a particular cell. Reproduced from Cheng, Greyson and Walden (1979) with permission of the National Research Council of Canada from the *Canadian Journal of Botany* 57:578–596.

Two types of tapeta, **amoeboid** and **glandular** (secretory), are recognized. In amoeboid tapeta the individual wall-less protoplasts enlarge and move among the cells of the locule to produce a periplasmodium which surrounds the developing pollen grains. In glandular tapeta, walls can remain intact but in some cases are digested. The glandular tapetum is eventually resorbed, and its inner wall becomes appressed to the middle layer. It is generally assumed that the tapetum: (i) provides nutrition for the sporogenous tissue; (ii) secretes a callase during the breakup of the tetrads; (iii) is responsible, in some cases, for cytoplasmic and genic male sterility; (iv) produces Ubisch bodies; (v) secretes Pollenkitt, tryphine [collectively 'pollencoat' (Knox 1984b)], enzymes, recognition proteins and other substances on the pollen grain. As more detailed studies using a range of microscopic technologies and detailed analyses are performed, many of these generalizations will undoubtedly require modification.

The tapetum undergoes a number of complex but difficult to analyze cytological changes throughout its brief existence. Initially, plasmodesmata connect the tapetal cells with the sporogenous cells and are assumed to be a route for the transport of molecules from the anther to the sporogenous cells. During meiosis these connections are lost, thereby raising questions about the type and amount of nutrition available to the cells during the meiotic process. During early prophase, tapetal cells of both types undergo mitosis without cytokinesis. This commonly results in the formation of binucleate cells in secretory tapeta. Endoreplication is also common. The adjacent walls between the tapetal cells are frequently missing, and in the case of maize large holes arise in the membranes, thereby making the tissue into a syncitium. Ubisch bodies (Echlin 1971) (orbicules) consisting, in part of sporopollenin, are a common feature of the inner surface of secretory tapetum. The presence of abundant rough endoplasmic reticulum, dictyosomes and a variety of vescicles suggests a highly synthetic cytoplasm.

The recent demonstration by *in situ* localization techniques of a tapetum-specific chimeric ribonuclease (Goldberg 1988) which induces male sterility (Mariani et al. 1990) is an exemplary model of one analytical route which should in future yield fresh insights about the role of the tapetum in anther function. Further variations of anther structure and dehiscence mechanisms are discussed by Weberling (1989).

Microsporogenesis and Gametogenesis

The collection of cellular events that occur within the anther **locule** (the cavity internal to the tapetum) from the initiation of the **microsporocyte** (microspore mother cell) to the production of microspores is called **microsporogenesis.** In angiosperms, it combines at least three individual biological events: (a) meiosis; (b) the switch from the sporophytic to the gametophytic phase of the life cycle; (c) the first stages of gametogenesis. In lower plants such as ferns these three events are temporally and spatially separated (Dickinson 1987), but in angiosperms they occur within the same cells, simultaneously, over short time periods. Understanding the total process, which is filled with intricate detail and considerable potential for variability, represents a stimulating challenge for developmental cytologists (Mascarenhas 1975).

In studying microsporogenesis, researchers divide the total process into a series of time-based stages. For example, in canary grass *(Phalaris)* (Vithanage and Knox 1980), anther length and the stage of pollen development were followed over an 18-day period (Fig. 5-15). Meiosis within the microsporangia occurred between days 1 and 5, with the first microspore mitosis occurring at the midvacuolate stage (about day 8). For *Lolium* and *Phalaris* (Vithanage and Knox 1980), eight stages of pollen development were recognized over 13 days. In *Zea*

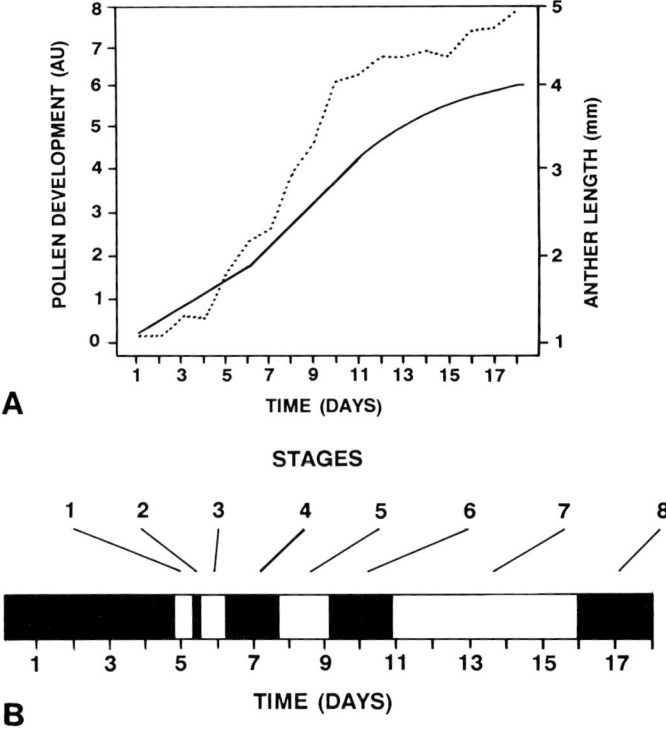

Fig. 5-15A–B. Observations on the relationship between the growth of the anther and the stage of development of the pollen in *Phalaris tuberosa* L. (A) Anther length (right ordinate axis, —) and pollen developmental stage (left ordinate axis, ······) are plotted against time. The data are derived from three plants. (B) Relationship between eight stages of development and the number of days is portrayed. This index was developed for *Lolium* and *Phalaris*. The stages identified (arbitrary units (AU)) include: (1) Tetrad; (2) Spore release; (3) Pre-vacuolate; (4) Early vacuolate; (5) Mid-vacuolate; (6) Late vacuolate; (7) Early maturation; (8) Late maturation. From Vithanage and Knox (1980) with permission of *Annals of Botany*.

mays, a nine-stage outline beginning with anther primordia occupied 25 days to pollen maturity (Warmke and Lee 1977; Cheng, Greyson and Walden 1979). Such time-based descriptions, and others (Sakate 1976; Horner 1977) are essential for subsequent rigorous cytological, cyto-histochemical and molecular studies. Many of the studies on meiosis in *Lilium* (Stern and Hotta 1973) were based on the original correlative study of Erickson (1948).

It is now obvious that, coincident with the nuclear and chromosomal rearrangements during microsporogenesis, other significant cytoplasmic alterations also occur. First, the meiocyte ribosome population and extractable RNA disappear during leptotene and zygotene. This change includes both ribosomal and mRNA (Porter et al. 1984). Second, in most species so far examined, both mitochondria and plastids undergo rapid dedifferentiation to become small inclusions. Organelle degeneration is also believed to occur during prophase. Third, there is a redifferentiation of mitochondria and plastids during late tetrad and microspore stages (Dickinson 1987).

These observations, subject to experimental confirmation, could answer some perennial questions about the causal basis of the alternative potentialities of spore and zygote. In the past, the organisms of choice for probing these questions were ferns and other lower tracheophytes (Morel 1963; Ward 1963; Bell 1975), but they can now be studied in angiosperm anthers (Dickinson and Heslop-Harrison 1977) and manipulated during anther culture experiments (Heberle-Bors 1985). More discussion of anther culture is provided later in this chapter.

Despite considerable effort with both plant and animal systems, no decisive answer is yet available as to what controls the switch from mitosis to meiosis (Riley and Flavell 1977). Perhaps the anther might yet provide some basic insights into this perennial problem.

The Microspore

The events that transpire between the product of meiosis, the **microspore,** and the mature **male gametophyte,** the pollen grain, and its germination and growth have attracted biologists from many persuasions. This interest arises from the realization that pollen: (i) is a relatively simple organism of one, two or three cells (depending upon the stage and the plant); (ii) is available in large populations; (iii) can represent a population of variable forms; (iv) can be maintained and cultured in vitro; (v) relies for some aspects of differentiation upon interaction and communication between cells and tissues; (vi) offers opportunities to explore the time and site of gene action; (vii) is open to exploration through a variety of technologies. Further, it is commonly assumed the accumulation of knowledge about pollen will have practical spin-offs in agricultural practice. Because many fine reviews of the details of microspores exist (Knox 1984a; 1984b; Mascarenhas 1975, 1989), only a brief discussion of selected developmental topics is required here.

Following telophase of meiosis II (Fig. 5-16A–F), the individual microspores are released from the callose wall surrounding the tetrad. Maturation over a few days involves changes in shape, size, the cytoplasm and the wall, and it ends with the mitotic division of the microspore nucleus and the beginning of the male gametophyte (pollen grain). In many anthers, microspore and pollen maturation is synchronous and produces morphologically similar pollen at dehiscence. It is now recognized, however, that, in addition to the genetic variation generated by meiosis, some species exhibit pollen dimorphism in which a variety of anomalous pollen is formed (Sunderland and Huang 1987). One possible consequence of this dimorphism is discussed below in the section on anther culture.

Two apparently antagonistic mechanisms operate during production and maturation of microspores, which enhance communication between microspores and later contribute to their isolation. In the early stages of microsporangium development, all microspores are interconnected by plasmodesmata, but these degenerate by early prophase 1 (Shivanna and Johri 1985) (Fig. 5-17). During meiosis, beginning at leptotene the accumulation of callose, accompanied by primary wall

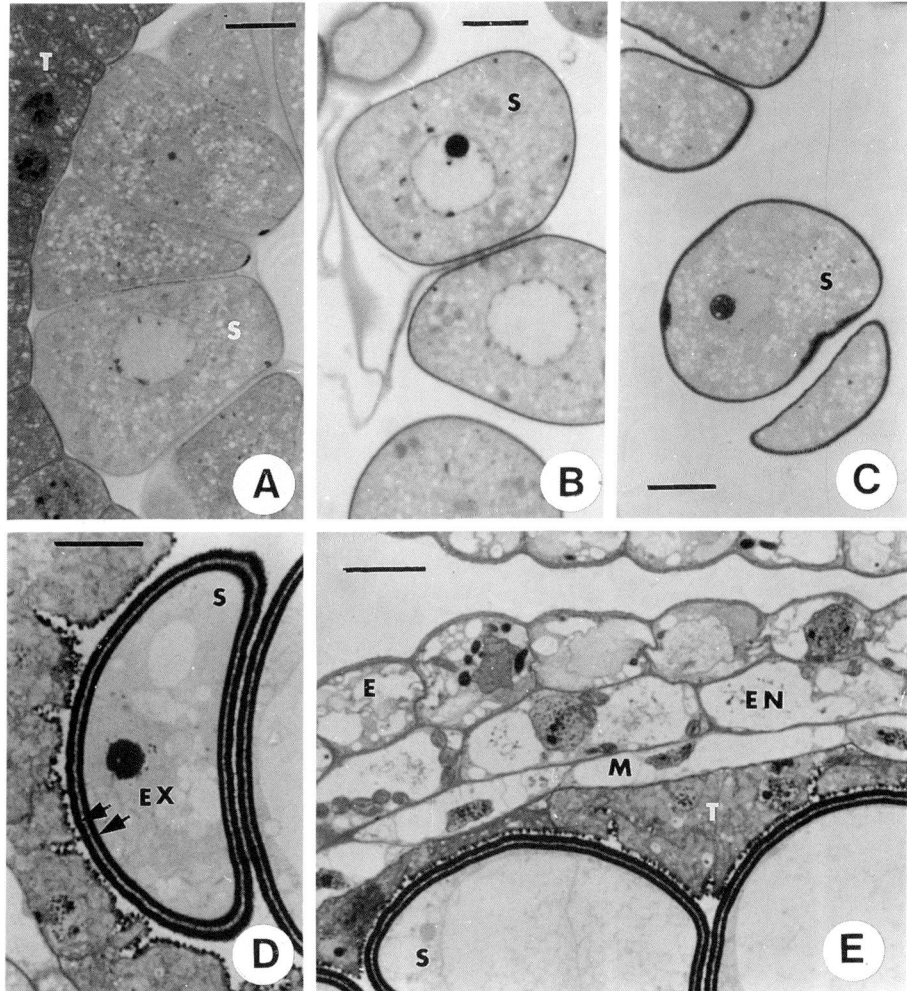

Fig. 5-16A–E. Light microscope views of stages in the development of microspores of maize. (A) Section of anther with spores (S) early in development. T = tapetum. (B) Microspores with the initial development of exine. (C) Later stage of microspore. (D) Microspore at a later stage is enlarged with further development of the exine (EX). (E) section of the anther wall and the microspores. E = epidermis; EN = endodermis; M = middle layer; S = spore; T = tapetum. Bars = 10µm. Reproduced from Cheng, Greyson and Walden (1979) with permission of the National Research Council of Canada from the *Canadian Journal of Botany* 57:578–596.

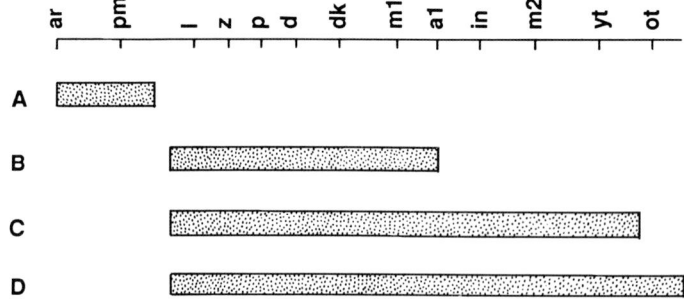

Fig. 5-17. A generalized representation of the period of appearance and disappearance of plasmodesmata and cytoplasmic channels in meiocytes and tapetum. Events: A = plasmodesmata between meiocytes, tapetum and wall cells; B = cytoplasmic channels between meiocytes; C = channels between tapetal cells; D = callose accumulation around the meiocytes. Meiotic stages: ar = archesporium; pm = pre-meiotic; l = leptotene; z = zygotene; p = pachytene; d = diplotene; dk = diakinesis; m1 = metaphase I; a1 = anaphase I; in = interphase; m2 = metaphase II; yt = young tetrad; ot = old tetrad; Modified from Shivanna and Johri (1985).

degradation between the microsporocytes, begins to isolate the microspores. Initially, large cytoplasmic channels persist between microspores. By the end of M1, these have disappeared and callose surrounds each microspore. These changes are assumed to explain, at least in part, why the events of meiosis in many plants are synchronous, while later post-meiotic events tend to become out of phase.

The wall of the pollen grain has attracted the curiosity of a variety of different biologists. The accumulation and persistence of wall remnants in sedimentary deposits is the basis of the science of palynology and is used in geological, anthropological and evolutionary explorations. The intricate detail and variability of wall structure of extant forms represents an abundant source of information for taxonomic distinctions. The wall is also a challenge to developmentalists, presenting a number of intriguing problems in cell biology.

Mysteries surrounding wall development of angiosperm pollen, resulting from the variety of forms and confusion about the array of detail presented by mature walls, have been alleviated in recent years through concerted histochemical and ultrastructural studies in a number of laboratories. More complete treatments of these data are available elsewhere (Knox 1984a; Mascarenhas 1975; Dickinson 1987; Heslop-Harrison 1971, 1980). It is sufficient for this presentation to cite a few of the major conclusions of these studies: (i) A generalized illustration of the wall layers of a mature pollen grain is presented in Fig. 5-18. Wall variation at maturity arises from the extent to which these different components develop in individual species. (ii) Transcription of wall pattern genes takes place in the pollen mother cells. (iii) Translation of these messages to gene products, organelles and membranes takes place during meiosis and tetrad formation. (iv) During the tetrad stage, before the dissolution of callose, cellulosic elements are secreted from the protoplast which give rise to the **primexine** (exine precursor). This layer, together with its lamellae, becomes the matrix of receptors on which protosporopollenin/

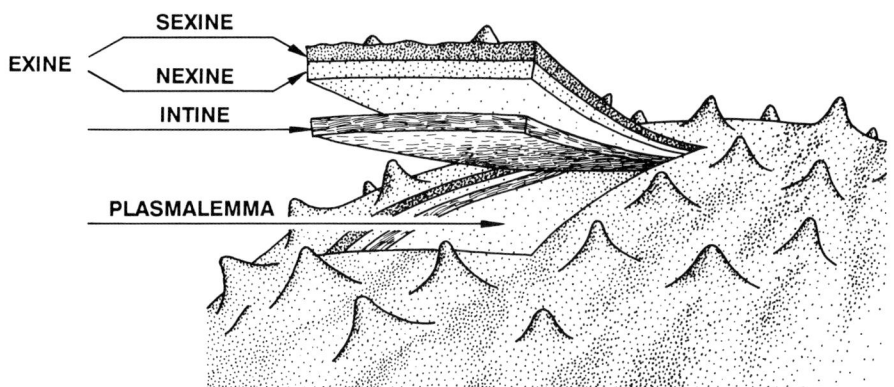

Fig. 5-18. Generalized structure of pollen wall. Sexine has species-specific sculpture, and inside the intine the cytoplasm and nuclei are enveloped by the plasmalemma. From Iwanami, Sasakuma and Yamada (1988) with permission.

sporopollenin polymerizes to form the **exine.** (v) Initially, at least, sporopollenin is derived from the microspore. It is also synthesized by the tapetum and accumulates on the spore and pollen wall. Other tapetum-derived substances include sporopollenin (acetolysis-resistant polymer), pollenkitt (carotenoids and other lipids), tryphine (complex mixture of hydrophobic substances) and proteins (including enzymes and allergens) to the external and internal surfaces of the wall. (vi) Intine synthesis causes the final wall to be synthesized and includes the elaboration of the components surrounding the **apertures.** The intine, composed of different polysaccharides, also contains proteins.

In addition to establishing the timing and sequence of genetically based signals that regulate wall synthesis, efforts have been expended to determine what genes are specifically expressed during the gametophytic phase (Mascarenhas 1989, 1990). At the level of ultrastructural cytology, work is beginning (Willemse and Reznickova 1980; Reznickova and Willemse 1980) on a quantitative analysis of changes in organelles during development. The opportunities for more objective morphometric analyses, in the manner of Havelange and Bernier (1974) and Mauseth (1980), remain for future studies.

The Male Gametophyte

The wealth of general information and specific detail about the angiosperm male gametophyte is so large that it is only possible to record here brief discussions of selected developmental topics and to record some references. The amount of primary data and the number of reviews are truly impressive. These include Mascarenhas (1975, 1989, 1990), Knox (1984*a*, 1984*b*), Shivanna and Johri (1985), Heslop-Harrison (1987), Dickinson (1987) and Blackmore and Knox (1991). It is now apparent that, beyond the purely descriptive features of pollination and fertilization, many developmental aspects involving complex phenomena between the developing male gametophyte and the gynoecium, including some quite sophisticated recognition reactions, are now available for study.

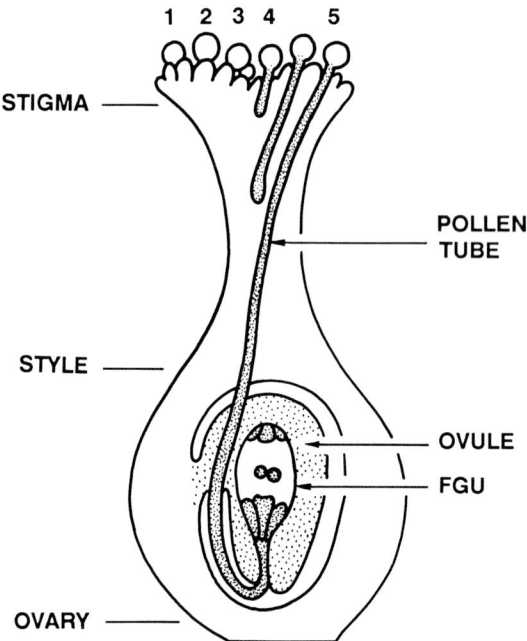

Fig. 5-19. Schematic summary of the stages in pollen germination and growth. (1) Pollen is received and adheres to the stigma. (2) Hydration. (3) Germination. (4) Pollen tube penetrates stigma. (5) Pollen tube grows through style discharges sperm into the embryo sac. FGU = female germ unit. Adapted from Knox and Clarke (1976), Knox (1984*b*) and Gaude and Dumas (1987).

Following meiosis, the microspore achieves its mature condition both through its own metabolism and through the accumulation of contributions from the sporophytic tissue, primarily the tapetum. In 70% of the angiosperms studied (Brewbaker 1967), it undergoes one mitotic division producing the **generative** and **vegetative** cells. In these species, pollen is shed as a two-celled gametophyte. In the remainder, the generative cell undergoes a mitotic division to produce two sperm cells. In these, pollen is shed in the tricellular state. Ultrastructural studies have now concluded that, in a least some species, the generative cell is indeed a cell with cell wall, its own cytoplasm and nucleus existing within the vegetative cytoplasm.

Knox (1984*b*) has summarized the main events and possible interactions between the male gametophyte and the gynoecium into five steps (Fig. 5-19). Germination (#1) and tube growth (#3 and #4) are discussed here. Fertilization (#5) is considered in Chapter 6.

Pollen Germination and Tube Growth

The initial hydration of pollen on a receptive stigma is followed within minutes or hours, depending on the species, by germination. This involves the growth of the pollen tube through one of the apertures in the grain wall. Estimates of growth rates of pollen tubes vary widely: some *in vivo* estimates suggest rates between 1 and 2μm/s (Heslop-Harrison 1987). The main events in the life of the male gametophyte were conveniently summarized by Dumas et al. (1984) (Fig. 5-19). Each step is worthy of careful examination by itself, and many individual references are available.

These growth activities take place at and near the tip of the tube involving an active metabolic tube cytoplasm, uptake of metabolites from the surrounding matrix and cell wall extension and synthesis at the tip. In solid styles (possessing transmitting tissue (Chapter 6)), the tube grows along and between the surface of cells of the transmitting tissue. A number of different variations have been noted, including the presence of pectin-containing intercellular secretions from the cells of the transmitting tissue. In hollow styles, tubes grow along the surface of the stylar canal, which also can be coated with a secretion.

Despite the consistent orientation of pollen tube growth through the style towards the ovule, no satisfactory chemotropic mechanism has been uncovered. While sensitivity to Ca^{++} has been detected in some pollen, a complementary and necessary Ca^{++} gradient along the style has not been demonstrated. Some argue that none need be postulated, at least until the tube nears the micropyle.

Though incompletely studied, the male germ unit, consisting of the vegetative cell and two sperm nuclei, has been shown in a few species to possess considerable organization and differentiation. The two sperm cells may differ in their attachment to the vegetative cell and in the number of mitochondria. In *Zea*, nondisjunction of B- chromosomes frequently generates two unequal nuclei (2B and 0B), of which the 2B cell fertilizes the egg approximately twice as often as it (the 2B cell) justifies the polar nuclei [see Roman (1947, 1948) as cited by Carlson (1988)].

The convergence of suitable technologies, interested and imaginative researchers and practical applications means that much new information about the male germ unit and its role in fertilization will be available in the near future. A sample of the state of research is available in Cresti, Gori and Pacini (1988).

Self-incompatability

Beyond the failure of pollen germination and gamete fertilization between different species because of some morphological, cytological or chemical maladaption between pollen and gynoecium (incongruity (Hogenboom 1975)), the developmental phenomenon of self-incompatibility is much studied but incompletely understood. Self-incompatability within species and populations is estimated to be present in more than 1/2 of all angiosperm species (Haring et al. 1990). Incompatibility can be distinguished in **heteromorphic** (morphologically distinct flowers: see heterostyly in Chapter 6) and **homomorphic** flowers. In this latter case, two mechanisms—**sporophytic** self-incompatibility (SSI) and **gametophytic** self-incompatibility (GSI)—are recognized. The following two examples illustrate the phenomenon.

In *Brassica campestris,* an SSI example, the following data have been assembled: (1) SSI is controlled by a single-gene locus, the S-locus (Ockendon 1982); (2) the locus is highly polymorphic, with over fifty alleles identified; (3) the locus codes in the stigma for polymorphic S-locus-specific glycoproteins (SLSGs) (Nasrallah and Nasrallah 1984; Hinata 1982); (4) dominance and reciprocal differences affect expression; (5) self-incompatibility is displayed in the abortion of selfed pollen at the stigma surface (Kandasamy et al. 1989); (6) SSI operates whenever the S-allele is present in the stigma; (7) the recognition reaction is between the SLSGs of the stigma and the sporophyte-derived proteins of the pollen wall.

More common are the GSI mechanisms (Haring et al. 1990) which are characteristic of bicellular pollen, as illustrated by *Nicotiana alata*: (1) GSI is controlled by a single-gene S-locus which is polyallelic; (2) pollen behavior is controlled by its own haploid genome; (3) there is no dominance of S-alleles in the style; (4) pollen tube development is arrested in the style, involving contact between the pollen tube and the mucilage secreted by the cells of the transmitting tissue; (5) the products of the S-locus expressed in the gynoecium are ribonucleases (Gray et al. 1991) and may function as S-allele-specific cytotoxins (McClure et al. 1990).

While the molecular details of both SSI and GSI are being unravelled, the basic mechanism of self-recognition remains obscure. Some assume that recognition involves an "inhibition" of pollen tube growth by the gynoecial tissue with like genotypes. Another view assumes that compatibility is conferred by interaction of products from the different genotypes. In developing experiments and analyses, it is convenient that self-incompatibility expresses developmentally. In some situations it is not expressed in immature gynoecia (Cornish, Pettit and Clarke 1988). There is also some question at present of the genetic and molecular homology between the two forms. Incompatibility, therefore, because of the basic intellectual questions which it raises and the potential application of knowledge to agriculture, represents a major research front for developmental studies. Many of the technologies necessary for its exploitation are now at hand.

In Vitro *Culture of Pollen*

The opportunity to isolate and culture, *in vitro*, a population of haploid organisms and to study them, both individually and as a population, is noteworthy. Though the results vary with pollen source, techniques and knowledge are available so that reasonable success should be anticipated if this technique is chosen to explore developmental questions (Vasil 1987; Shivanna and Johri 1985). The cautionary note of Heslop-Harrison (1987), to the effect that the *in vitro* situation may be irrelevant to some questions about *in vivo* development and physiology, should be weighed carefully.

The nutritional conditions for successful germination and short-term growth of many pollen are relatively simple. Common requirements for either liquid or solid (agar) culture include sucrose, boron, frequently calcium (Brewbaker and Kwack 1963) and other elements. In particular cases, other organic and inorganic additives are beneficial. For individual species and cultivars, the optimal conditions may, nonetheless, be exacting.

Beyond questions of optimal nutritional and physical conditions, *in vitro* techniques can be used to explore developmental questions. For example, the demonstration (Gray et al. 1991) that an enzyme, in this case the self-incompatibility RNAase, was taken up by the developing pollen tube (as demonstrated by radioactivity and *in situ* immuno-labelling) and remained functional, is an elegant model system which no doubt will be exploited with other proteins, enzymes and molecules.

Variations and Abnormalities in Stamen Development

Variations from so-called "normal" stamens are common and have been studied and catalogued for many years (Masters 1869; Worsdell 1916; Meyer 1966). These include stamen sites within the flower that are occupied by other organs such as phyllody (leaf-like), petalody (petal-like) and carpellody (carpel-like). Petalody is the most common; the resultant double flowers are frequently prized horticulturally. In some cases, variation in stamen morphology is induced by environmental or disease stimuli in sensitive strains. In other cases, obvious genetic components can be tied to the aberrations.

Stamens that possess portions or mosaic patches of cells that express the phenotypes of other organs are also common. Heslop-Harrison's (1952) pre-molecular but essentially developmental review is a useful introduction to the topic of flower teratology. Male sterility (Kaul 1988) is also an important variable in stamen development.

Variations in Stamen Morphology

Developmental Variation

The identification of "developmental variation" as a special category requires some explanation. As a topic in developmental botany it was first used by Goebel (1900), who distinguished homoblasty (small, gradual differences) from heteroblasty (major, phasic differences). Subsequent authors, when faced with a wider display of variability, have tended to obscure this distinction such that heteroblasty has come to be synonymous with "developmental variation." The main focus of previous studies of heteroblasty was with leaf shape (Ashby 1948*a;* Allsopp 1967; Kaplan 1973), but the concept can apply in a general way to any feature of a plant which presents different expressions of the same attribute at different stages in its existence (Greyson, Walden and Smith 1982). Poethig (1990) presents a more comprehensive analysis of phase-related variation.

Developmental variations in stamen morphology can include: (1) variations of stamen structure within a single flower; (2) variations of stamen structure among flowers that are formed at different times and at different locations on the same plant; (3) variations that are formed in response to changes in environmental cues.

As a consequence of a generally plastic organogenesis in plants, the number of potential examples of this type of variation is vast. Examples of such developmental variation previously mentioned include the long and short stamens of *Brassica* and *Arabidopsis* flowers, the fertile and sterile (staminodia) stamens of *Aquilegia,* and the different stamens of the CH and Cl flowers of *Lamium*. Many more examples can be uncovered in taxonomic treatments of different species. Additional opportunities for fundamental study of this topic are numerous (Meyer 1966); but because of the sometimes obscure causal basis of the variation, experimental work with this category is often difficult.

Developmental variation, then, includes those variants, both large and small, that are generated from a single genome, both nuclear and cytoplasmic, under different conditions, at different times in the life of the plant and at different locations on the organism. Further refinement of this definition to recognize different causal mechanisms and to take into account up-to-date understandings of cell and molecular biology must await more data and analysis. While there is an assumed genetic involvement in the elaboration of developmental variation, it seems worthwhile to distinguish it from those cases where variation is demonstrably based in genetic variability.

Meristic Variation

Developmental studies, and especially this manuscript, frequently concentrate on the simpler examples and more easily understood events in development. Those that can be directly related to specific genes are highly prized. More complex phenomena should not be ignored, however, for they point to levels of regulation and control which have yet to be addressed.

An example of developmental complexity involving different numbers of stamens is provided by the study of stamen variability in the European annual *Scleranthus annuus* (Svensson 1988, 1990), which, unlike most members of the Caryophyllaceae, which bear two whorls of five stamens, produces variable numbers with variable fertility. In the most common variant, two of the inner five stamens were fertile; the others and those of the outer whorl were usually reduced to varying degrees. The interesting and enigmatic demonstration of this research was that the variability in the number of functional stamens and their site within the flower correlated significantly with the distance the pollen travelled. Selfed progeny showed the lowest total male fertility. The flowers of progeny from crossing distances of 6m and 12m had significantly more stamens and more staminodes developing into fertile stamens (Svensson 1988). These observations were obtained during ecological attempts to evaluate an "optimal outcrossing distance." Developmental analysis of such a phenomenon would require the isolation of a number of different lines and carefully controlled experimental conditions.

Teratomas and Genetically Based Stamen Morphology Variability

Genetically based aberrations in stamen development are probably common, though few have been studied carefully enough to be useful for serious developmental study. In fact, abnormal stamens must be the most common flower abnormality recorded. Meyer's (1966) review lists 173 genera in which petalody of stamens is reported. Carpelloid stamens were listed as being reported in 62 genera. These reports include environmentally or disease-induced situations as well as genetically based variation. Thus, the list of suitable candidates for gene-based developmental studies is considerably shorter. A personal short-list of candidates includes *Antirrhinum, Arabidopsis, Lycopersicon, Nicotiana, Petunia, Pisum, Nigella,* and *Zea*. For each, a number of mutations are available in appropriate backgrounds, and with some the first steps in a molecular interpretation of development have begun. Two species, *Arabidopsis* and *Lycopersicon,* are cited here to illustrate the range of expressions observed and to outline some opportunities for research.

To date, many "floral mutants" of *Arabibopsis thaliana* [not cited by Meyer (1966)] have been identified, though few have been characterized in detail (Koornneef et al. 1983; Meyerowitz 1987). Haughn and Somerville (1988) list thirteen isolates which they feel are characterized sufficiently well for further developmental and molecular studies. Of these, seven exhibit some alteration in stamen structure or position. Additional identifications and updated lists (Koornneef 1990) are frequent. The general justifications for considering *Arabidopsis* for developmental studies were discussed in Chapter 4 (see also Table 9-1), and some corolla mutants were identified in Fig. 4-22. Table 5-1 describes nine mutant genes that affect stamen development.

Bowman, Smyth and Meyerowitz (1989) selected four mutants for in-depth developmental studies. Of these, ap2 and ap3 are especially revealing (Fig. 5-20). In the following descriptions, the *Arabidopsis* flower is viewed as consisting of four concentric whorls, one each of sepals, petals, stamens and gynoecium. For this discussion, the authors' interpretations have been retained (see also earlier discussion on organography of *Brassica*).

apetala2-1 (ap2-1): Chromosome map position 4-63.5 (Koornneef 1990). The ap-2 locus exhibits a number of alleles (Kunst et al. 1989). This mutant is temperature-sensitive, with the morphology of the organs in the first two whorls varying depending on temperature. No other pleiotropic effects were noted. At 25°C, the organs of the outer whorl resemble leaves, bearing trichomes (Figs. 5-21A and B), stipules and senescing-like leaves. Their sepal-like characters include long ($>100\mu$m) epidermal cells on their abaxial surface and often bear stigmatic papillae. The second whorl demonstrates the greatest variability. The range of expression from phyllody and petalody to nearly normal stamens, depending on the temperature, is illustrated in Fig. 5-21A–G.

apetala3-1 (ap3-1): Chromosome map position 3-73.9 (Koornneef 1990). Like ap2, this is also a temperature-sensitive locus affecting the organs in the second and third whorl (Fig. 5-22C). At 25°C or 29°C, ap3-1 homozygotes develop flowers in which the organs of the second whorl are sepals, indistinguishable from wild-type sepals except for their slightly smaller size. These sepal-like organs are situated in the sites that in a wild-type flower would be occupied by petals. A range of organs developing in the third whorl is illustrated in Fig. 5-22A–E.

At 16°C the organs of the second whorl of ap3-1 flowers exhibit a number of petal-like traits including smooth margins and white pigmentation. They are, however, smaller than normal petals and have sepal-like epidermes.

Table 5-1 Selected List of Nuclear Genes Which Affect Morphological Variation in the Stamens and Floral Organs of *Arabidopsis thaliana* (2N = 10)

Gene Symbol	Gene Name	Nature of Lesion	Map Position	Phenotype
ag1	agamous[a]	NR	4-33.3	Nested perianths, each with four sepals and approx. 10 petals.
ap1	apetala-1	NR	1-102.4	Apetalous; bracts replace sepals; flowers initiate in sepal axils.
ap2	apetala-2[a]	NR	4-58.7	Bracts replace sepals; petals replaced by anther-like and petal-like organs.
pi1	pistillata[a]	NR	5-24.1	Sepals replace petals; stamens absent or with carpel-like features; carpels often mishapen.
clv1	clavata-1	NR	1-113.5	Club-shaped gynoecium with four fused carpels.
clv2	clavata-2	NR	1-88.4	Similar to clv2.
lfy1	leafy inflorescence	NR	ND	Floral organs leaf-like; carpels under-developed; internodes elongated; lateral flowers produced in all axils.
sup-1 (also alleles -2, -3 and -4)	superman-1 (synonymous with flo series)[a]	NR	ND	Defects primarily in the inner whorls: both numbers and types of organs affected: excess of staminoid organs with partial/complete loss of gynoecium[b]
pin	pinformed	NR	ND	Most floral stalks lack flowers; flowers have thin elongate pistils, multiple petals, no stamens, reduced sepals.

Abbreviations: NR = Nuclear recessive. ND = Not determined.
[a] Illustrated by Meyerowitz, Smyth and Bowman (1989).
[b] Bowman et al. (1992).
Source: Most items from Haughn and Somerville (1988) reprinted with permission of John Wiley & Sons, Inc.

The third whorl of organs of ap3-1 flowers are more variable. At 16°C, they develop as functional stamens (Fig. 5-22A) and at 29°C as carpel-like organs (Fig. 5-22E). At intermediate temperatures, organs possessing features of both stamens and gynoecia are formed. Tissue mixtures, common in mutant strains such as these, are described as **chimeric** or in some cases **mosaic.**

Chimeric organs, often found in the genetic strains but rarely in typical wild-type selections, represent a considerable challenge to developmental biologists. Material such as this should eventually reveal how certain elements of genetic

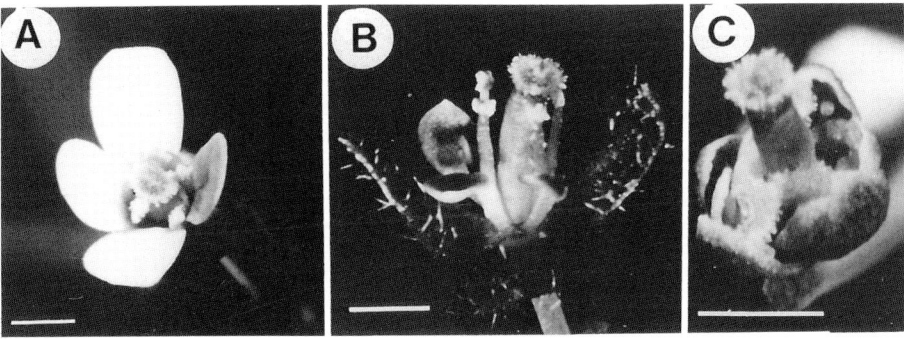

Fig. 5-20A–C. Photos of *Arabidopsis* flowers. (A) Wild-type flower. (B) Flower of the mutant apetala2. (C) apetala3. The plants were grown at 25°C. Bar = 1mm. From Bowman, Smyth and Meyerowitz (1989) with permission.

information are normally restricted to expression within a clone of cells and to a single organ type. A further discussion of this topic, as well as consideration of the term **homeotic** to characterize some of these mutant forms, is provided in Chapter 9.

The **stamenless-2** (sl-2) gene of tomato represents another developmental mutant that may help unmask some of the detail connecting gene with phenotype (Table 5–2). It behaves in breeding experiments as a simple Mendelian recessive trait. It has been mapped to Chromosome 4; map position 36 (Rick 1977). It is

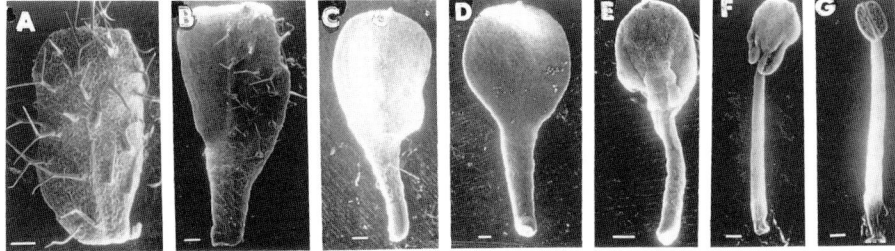

Fig. 5-21A–G. SEM micrographs of organs produced in the second whorl of apetala2-1 flowers. (***COMMENT:*** Petals and leaf-like organs are common at 16°C; stamen-like petals are typical at 25°C; petal-like stamens or no organ occurs at 29°C.) (A) Organs with no trace of white pigmentation were classified as cauline, leaf-like. (B) Those with a small amount of white pigmentation were classed as petaloid leaves. (C) Mostly white or all-white organs possessing trichomes were classified as phylloid petals. (D) Morphologically wild-type petals were typical at 16°C. (E) White petal-shaped organs possessing rudimentary locules were termed staminoid petals. (F) Organs classified as petaloid stamens were shaped like stamens but had some white pigmentation usually near the top of the organ. Misshapen stamens and filaments lacking anthers were observed at a low frequency. (G) Morphologically wild-type stamens occur at a low frequency at 29°C. Outer surfaces are shown in A, B and C; inner surfaces in D, E, F, and G. Bar = 100μm. From Bowman, Smyth and Meyerowitz (1989) with permission.

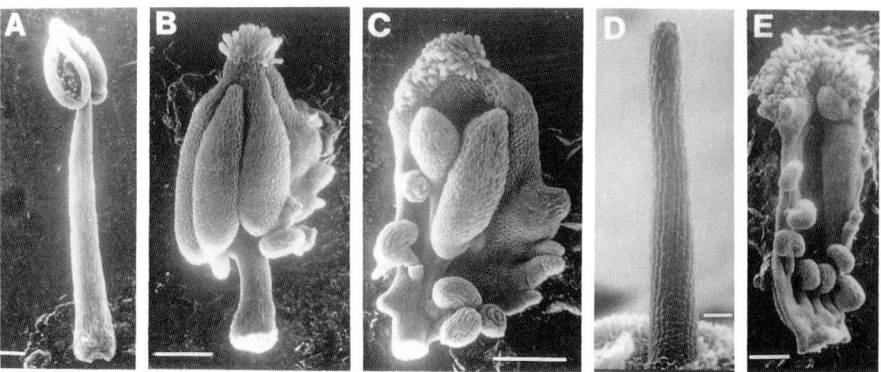

Fig. 5-22A–E. SEM micrographs of organs produced in the third whorl of apetala3 flowers. (A) Morphologically wild-type stamens observed at 16°C. (B) Organs with anther region on filament. Anthers have microsporangia capped with stigmatic tissue. Ovules are present along the anther; the most-developed examples are at the anther base; those along the anther are less well developed. These organs were termed carpelloid stamens. (C) Organs shaped like carpels, capped with stigmatic papillae, possessing ovules. Some microsporangial tissue is present. These were termed staminoid carpels. (D) Filaments with no anther but sometimes topped with a stigma. These may form at higher temperatures. (E) Carpel-like organs typically formed at 29°C. Bar = 100μm. From Bowman, Smyth and Meyerowitz (1989) with permission.

allelic to sl-1, sl-3, sl-4, sl-5 and sl^{cor} (corollaless). Stamens of sl-2/sl-2 flowers are laterally free, twisted, shorter than normal and pale green; they possess abnormal non-fertile pollen and bear naked ovules on their adaxial surface (Sawhney and Greyson 1973a) (Fig. 5-23B). Compare with Figure 5-23A.

Flowers of sl-2/sl-2 plants that initiate and develop following applications of GA_3, applied as a spray or dropwise to the stem at the base of the inflorescence, develop stamens with near-normal morphology (Fig 5-23C). Pollen from these reverted flowers, when used to pollinate a sl-2/sl-2 stigma, produces fertile, homozygous sl-2 seed (Sawhney and Greyson 1973b). Bioassays of the extractable GA-like substances suggest that vegetative and floral tissues of sl-2/sl-2 plants possess lower levels than are found in wild-type tissues (Sawhney 1974).

Further evidence of a gibberellin involvement of this genetically mediated phenotype comes from *in vitro* flower bud culture (Rastogi and Sawhney 1986, 1988). Stamenless-2 flower buds require GA_3 in the culture medium for the initiation and growth of floral organs. Wild-type buds require only a cytokinin in the medium for somewhat normal development of flower organs. That gibberellins alone, or at least the ones so far applied, are not sufficient of themselves to overcome the genetic deficiency is suggested from the fact that the sl-2 flower meristems cultured *in vitro* bear only sl-2 type-stamens.

Two other pieces of data (Greyson unpublished) support the conclusion that gibberellins alone are not sufficient. First, carefully prepared grafts of small sl-2 scions to vigorous wild-type stocks did not produce normal flowers on the scion. Further, dihybrid crosses involving stamenless (sl-2) and yellow-green (yg-6), a

Table 5-2 Selected List of Nuclear Genes Which Affect Morphological Variation in the Stamens and Floral Organs of *Lycopersicon esculentum* (2N = 24)

Gene Symbol	Gene Name	Nature of Lesion	Map Position	Phenotype
ap	apetalous	NR	11-	Most or part of corolla lacking.
cl-2	cleistogamous-2	NR	6-113	Flowers open only slightly.
def	deformis	NR	6-57	Flowers deformed; fertility reduced; leaves progressively reduced.
gas	gamosepalous	NR	1-	Sepals partly connate; plants and leaves highly modified.
f	fasciated	NR	11-95	Fruits many-loculed.
Lpg	Lapageria	D	1-16	Leaves small, dark green, glossy; flowers campanulate.
ms-	male steriles (12 listed)	NR		Anthers shrunken; little or no pollen.
og	old gold	NR	6-106	Corolla tawney orange.
sf	solanifolia	NR	3-111	Primary leaves entire; segments of later leaves entire, concave; flower parts filiform.
sl	stamenless-1	NR	4-89	Stamens usually absent; GA_3 revertable.
sl2	stamenless-2	NR	allele of sl	External ovules on stamens; anthers sterile; GA_3 revertable; temperature-sensitive.[a]
sl5	stamenless-5	NR	allele of sl	
slcor	corollaless	NR	allele of sl	
vg	vegetative	NR	4-89	Flowers highly deformed, usually functionless.
vms	variable male sterile	NR	8-24	Anthers abort at high temperatures.
wf	white flower	NR	3-41	Corolla white to buff colored.

Abbreviations: D = dominant. NR = nuclear recessive.
[a] Sawhney and Greyson (1973a,b) and Sawhney (1983).
Source: Compiled from *Tomato Genetics Cooperative (TGC) Reports* 37 (1987) and Tanksley and Mutschler (1990) with permission.

high-gibberellin strain produced the double recessive combination in the F_2 generation at near expected rates.

Additional evidence of the complexity of the physiological features associated with expression of this gene is the evidence that polyamine levels may also be implicated in the expression of this mutant (Rastogi and Sawhney 1990*a*, 1990*b*)

Stamenless-2 is also a temperature-sensitive trait. Normal phenotypes with fertile stamens develop when plants are grown on a low-temperature regime (18°C day, 15°C night), while flowers raised on a high-temperature regime (28°C day, 23°C night) produced carpelloid structures which possessed little evidence of stamen expression (Sawhney 1983). This observation is immensely important when attempting to plan and interpret experiments such as the graft and double-mutant

Fig. 5-23A–C. Tomato flowers. (A) Wild-type flower with normal stamens (X3). (B) Flower of stamenless-2 (sl-2) plant illustrating the sterile stamens bearing the external ovules (EO) (X3). (C) Flower from a sl-2 plant treated with GA_3. The flower initiated and grew following the application (X4). Parts A and B from Sawhney and Greyson (1973a) with permission of *American Journal of Botany.* Part C reproduced from Sawhney and Greyson (1973b) with permission of the National Research Council of Canada from the *Canadian Journal of Botany* 51:2473–2479.

combinations (described above). It is possible that the negative results in these experiments could have resulted from non-optimal temperature regimes.

The use of genetically based morphological and biochemical abnormalities in research strategies that attempt to unravel the developmental sequences involved in the elaboration of a biological organ is a common feature of developmental studies. The successful use of model animals such as *Drosophila* and *Coenorhabditis* supports this claim (Wilkins 1986). The use of genetically based variability in the analysis of plant development is less common and more recent (Marx 1983). Despite earlier calls for their use (Sinnott 1960; Stebbins 1974; Postlethwait and Nelson 1964), the serious use of flower mutants in developmental studies has just begun (Haughn and Somerville 1988).

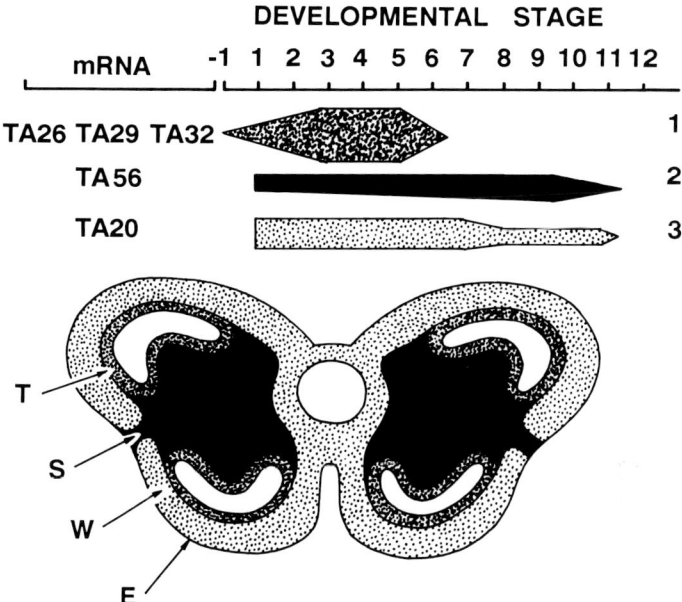

Fig. 5-24. The temporal and spatial regulation of five stamen-specific mRNAs during anther development. Summaries of RNA blot and *in situ* hybridization experiments. The tapering of each colored bar shows periods when the mRNA accumulates or decays. Bar thickness approximates the prevalence of each set averaged over the entire anther mRNA population. The shading of each bar corresponds to the anther region in which each mRNA is localized. Three shades were used for TA56 because this mRNA is localized in several anther regions at different developmental periods. T = tapetum; S = stomium; W = wall; E = epidermis. Adapted from Koltunow et al. (1990) with permission.

Stamen-specific Genes

This topic is a glimpse of the future rather than a report of past efforts. Nevertheless, the beginnings of what will no doubt become an important body of data about organ-specific genetic expression is beginning to accumulate.

In general terms, the measurement of nuclear RNA (Kamalay and Goldberg 1984) and messenger RNA (Kamalay and Goldberg 1980) of the vegetative and floral organs of tobacco indicate that both contain organ-specific fractions. For example, about 1/4 of the nuclear RNA sequences from anthers are organ-specific, and approximately 1/3 of the anther-derived polysomal mRNA sets are unique to the anther.

More specifically, the demonstration of expression of specific genes in stamens and pollen through *in situ* localization of mRNA is being demonstrated. Koltunow et al. (1990) identified eight anther-specific mRNAs of which three probes produced identical, tapetal-specific localization patterns (Fig. 5-24). The localized expression of mRNA for the flower genes AP3 (Bowman et al. 1992) and LFY (Weigel et al. 1992) are additional examples of this research approach which will be more common in future. Elsewhere, Mascarenhas (1989) discusses a variety of pieces of evidence from a number of studies to the effect that 10–20% of the genes that are expressed during the development and maturation of the male gametophyte are not expressed elsewhere during the life cycle. The identification and study of the expression of these represent another important future research front.

Variations in Microspore Development

Given the complexity of the sequence of timed and genetically programmed developmental events in pollen production, coupled with the fragile and exposed position of the microsporangium and the range of environmental conditions in which it develops, it is not surprising that the process frequently fails with the result that most pollen samples contain some proportion of abnormal pollen grains. Kaul (1988) outlines and describes in considerable detail the information available on the environmental, developmental and genetic factors responsible for sterility.

Beyond these gross morphological variants are those which represent a restricted failure of anther or microspore development and frequently result in **male** sterility (mst). In many species, evidence of **genic male sterility** (g-mst) is well established and can be used in developmental studies. **Cytoplasmic male sterility** (c-mst), also well documented in many species, is exploited in the preparation of hybrid seed for agriculture but can also serve as useful model material for basic developmental research.

Cytoplasmic Male Sterility

Cytoplasmic male sterility arises through the action of a sterility-inducing (S) cytoplasm. Kaul (1988) distinguishes two categories: (1) in cases of cytoplasmic male sterility (c-mst), nuclear fertility restorers (fr) are unknown; (2) more common are those cases of gene-cytoplasmic male sterility (gc-mst), which includes those that are of research and agricultural interest, for which nuclear fertility restorer genes are known.

The main stimulus for research on gc-mst will continue to arise from the agricultural and horticultural interest in mechanisms for hybrid seed production. Basic developmental studies will be restricted to a small number of particularly revealing models and will increasingly concentrate on the molecular biology of nucleus-cytoplasm interaction. The exploration of the interaction between the nucleus and the T and S cytoplasm of *Zea mays* (Laughnan and Gabay-Laughnan (1983)) serves as a useful model.

At the morphological and anatomical levels, as the need for descriptive studies (reviewed by Laser and Lersten (1972)) diminishes, research will be supplanted by careful analysis of specially prepared nucleus-cytoplasm combinations [see the example (Kofer, Glimelius and Bonnett 1990) discussed in Chapter 9]. A review of some of the materials available for research is provided by Kaul (1988).

Genic Male Sterility (g-mst)

In its classic form, g-mst is based on the expression of nuclear, often recessive, genes. Their expression is frequently restricted spatially to the tapetum and sporogenous tissue and temporally to different stages during microsporogenesis. Kaul (1988) identified more than 180 species and varieties of angiosperms in which spontaneous genic male-sterile mutants have been identified. We assume that the developmental basis of expression is either a deficiency or an over-production of an essential gene-regulated metabolite. This precision of expression of timing and location, plus the fact that some of these mutants provide insights into the genetic control of meiosis, make these single-gene mutants formidable tools for developmental studies. More than 50% of these genes have been shown to be recessive and fewer than 10% are dominant. In addition to spontaneously arising forms, many induced mutants have been recovered and studied.

Besides possessing desirable features for basic developmental studies, genic male steriles of agricultural crops are in considerable demand as potential components of hybrid seed production protocols. The desired characteristics of a male sterile for use in hybrid seed production include: (1) high level of sterility under normal conditions of expression; (2) ability to express sterility in all cultivars of the crop; (3) capacity to revert to normal expression by chemical or environmental cues at a high level so that homozygous seed of the g-mst trait can be increased.

While the total assemblage of desired characters is rarely available, efforts to find ways to use g-mst in those crops where alternative methods of hybrid seed production are unavailable or inefficient [for example, in the perfect flowered grasses (wheat, rice, barley, oats)] are in demand (Hockett, Beanziger and Steffens 1978) though as yet are in limited use. The rationale in this research is to uncover a chemical or environmental condition with which the trait can be reverted. Ideally the objective is to provide a condition which will override a genetic bloc. The reversion of sl-2 in tomato (Sawhney and Greyson 1973b) by GA_3 has already been discussed. The msg6 mutant of barley, which fails to develop an aperture, might be reverted by the application of the missing substrate or the blocked enzyme, poral esterase (Ahokas 1976). Another provocative report claims that a photoperiodically sensitive g-mst of rice which exhibits complete male sterility under long photoperiods but is fertile under short photoperiods has been identified (Deming, Zebing and Jingmeng 1988). This topic is discussed further in Chapter 8.

Many g-mst traits exhibit little pleiotropic expression in other organs. This feature, plus their recessive condition, complicates some research strategies and precludes others due to the inability to identify the sterile plant until sterility is expressed.

In maize, two alternatives are available to circumvent this situation. Patterson's (1975) duplicate-deficient chromosome system generates seed that yields 99% g-mst of either ms-2 or ms-10 material. A second strategy would be to identify an early expressing phenotype which is closely linked to the g-mst. Male sterile-silky (sil) is one such example in maize. This gene is located 1.3 map units from Y (Yellow seed) on chromosome 6. This convenient linkage allows the identification of seed that proves to be >98% sil (D. W. Dales, personal communication). It is to be expected that a variety of new molecular techniques will be available to manage these mutants more efficiently and to study their developmental messages more sharply.

For some basic research purposes, clones of g-mst plants of species such as rice, tomato and soybean can be maintained by traditional methods of vegetative propagation. For those species which cannot be readily propagated, clones of desired g-mst individuals might be established and maintained through variations of axillary bud culture (Raman, Walden and Greyson 1980). The protocol for this procedure would include: (1) establish single-seed lines of plantlet-generating explants from segregating populations of g-mst seed; (2) allow some of the plantlets of each clone to mature; (3) identify the lines of plantlet that form g-mst plants, and retain and increase these clones.

Experimental and Manipulative Procedures

Anther Culture for Observations on Meiosis

The relative ease with which anthers and sporogenous cells can be placed in sterile culture made them a logical object for studies on meiosis (Narayanaswamy and George 1982). Considerable effort was therefore expended between 1940 and 1970 to this end. While many biochemical details about portions of meiosis were uncovered, the complete process from pre-prophase sporocytes to microspores was never accomplished with explanted anthers or isolated sporocytes (Vasil 1967). Newer more direct and sensitive technologies, in large measure, circumvent the need for culture in the solution of many questions about meiosis. Meiosis in cultured flowers, as opposed to anthers, has been achieved a number of times. Thus, meiosis occurred and pollen was produced from *in vitro* cultured flower primordia of *Begonia, Cucumis, Zea* and others [see Pareddy, Greyson and Walden (1987) for citations].

Anther Culture for Embryogenesis

The fortuitous discovery (Guha and Maheshwari 1964) that cultured anthers of *Datura* produced microspore-derived haploid embryos has subsequently produced a formidable research area and simultaneously generated significantly new insights and questions regarding the developmental biology of microsporogenesis. An intense practical interest arises from the potential opportunity to shorten the time to generate isogenic, diploid inbred strains of agronomically important crops for the production of hybrid seed. Fundamental developmental interest arises from the apparent opportunity to explore, under experimental conditions, the factors that regulate the switch between sporophytic and gametophytic development. In addition to producing haploid embryos, cultured anthers can generate embryos of different ploidy levels and also disorganized calli.

While both practical and basic research objectives have been generously rewarded, expectations of future findings are more realistic due to an awareness of the complexity of the phenomenon and the variety of its products. For example, of the 200^+ species that have successfully produced microspore-derived embryos, the majority were from two families: Solonaceae and grasses (Dunwell 1985b; Sangwan and Sangwan-Norreel 1987). In most cases, the number of microspores/

anther that respond to optimum conditions is low; and on a per-microspore basis, extremely low.

One interpretation is that these poor results reflect the failure to develop optimal nutritional, PGS and physical conditions. Attempts, therefore, to remedy these deficiencies are widely reported and reviewed (Bajaj 1983). Another interpretation suggests that the switch from gametophytic to sporophytic expression is induced through stress. Cold, heat (Keller and Armstrong 1979; Arnison et al. 1990), water stress, reduced atmospheric pressure and anaerobiosis (Harada, Kyo and Imamura 1988) and other forms of stress have been shown to enhance results in selected cases.

An alternative interpretation (Heberle-Bors 1985), based on the observations of pollen variability (Horner and Street 1978; Sunderland and Huang 1987), argues that the determination to follow the embryogenic (sporophytic) rather than the gametophytic pathway occurs in a relatively small number of cells during meiosis. These cells become, it is argued, the abnormal pollen of normal anther and the embryos when cultured. The justification for this interpretation is as follows.

Rarely is a pollen sample uniform. The individual grains within a population vary in size, staining reactions, nuclear status, germination ability, etc. Recently (Sunderland 1974; Wenzel and Thomas 1974; Sunderland and Huang 1987) two otherwise unrelated observations have been claimed to be related. TEM studies of pollen from some species of plants have revealed the existence of two populations of pollen: normal and anomalous. Normal pollen is recognized by the densely staining, starch-filled vegetative cell and the obvious spherical or elongate generative cell. These grains are functional male gametophytes. Anomalous forms, by contrast, lack starch and contain a variable number of nuclei of different sizes. Anther culture studies revealed, as well, that the recovery of anther haploids was unrelated to the environmental conditions but was based upon the existence of two populations of pollen: embryogenic and non-embryogenic. It is claimed that anomalous forms represent the source of the embryoids of anther culture (Heberle-Bors 1985).

It is unnecessary to resolve the differences between these three alternative interpretations in order to recognize that microspore embryogenesis, at least in some selected species, represents superb experimental material for basic exploration of the factors that are involved in the alternative potentialities of pollen and the dichotomy between interpretations of totipotent regulative development and a more rigid deterministic interpretation.

In Vitro *Culture of Flower Buds*

While it has not achieved the most optimistic expectations of its practitioners, *in vitro* culture has clarified a number of questions surrounding the developing floral meristem. Beginning with the exploratory work of White (1933) and La Rue (1942), and the studies of Galun, Jung and Lang (1962), Tepfer et al. (1963) and Blake (1966) and followed by more sharply focussed reports [reviewed by Konar and Kitchlue (1982) and Rastogi and Sawhney (1989)], the isolation and culture of floral meristems can be credited with documenting a considerable nutritional and developmental autonomy for many flower meristems. Accepting the probability that many of the media conditions used were sub-optimum, this autonomy is often limited by an obligatory requirement for some particular blend of PGS.

Surprisingly, there are no universal formulations. Thus, auxins, cytokinins, gibberellins and other PGS have been shown in different combinations and at different concentrations to benefit growth and development *in vitro*. For example, despite many attempts to solve the riddle of the control of sexuality, no single combination of PGS is known which unequivocally stimulates androecium or gynoecium development. The requirements frequently turn out to be species- or cultivar-specific. This observation probably reflects the variety of parallel but different physiological ways these organs achieve maturity and as well something of the rather generalized, non-specific fashion in which PGS operate during growth and development.

The expected insights about flower development from *in vitro* culture can be approached most sharply when experimental comparisons are made between meristems with close but recognizably different developmental fates. Thus, the culture of meristems of the genetically related flower forms of *Nigella damascena*, double (d/d) and single (+/-), uncovered the fact that the two forms metabolize gibberellins in quite different ways (Greyson and Raman 1975; Raman and Greyson 1978). Further, the culture of flower buds of the stamenless-2 (sl-2/sl-2) mutant of tomato, discussed earlier, demonstrated the importance of added gibberellins in the physiology of the flower, but revealed that added GA_3 did not generate normal stamen phenocopies (Rastogi and Sawhney 1988)

Unfortunately, *in vitro* studies have not provided much insight into the basic morphogenetic mechanisms which regulate the numbers, kinds and arrangements of flowers on the meristem. While tantalizing suggestions occasionally arise, no consistent trends are obvious. This failure perhaps indicates that these features of organogenesis on the flower meristem involve a high degree of cell autonomy and that their molecular mechanisms are, in general, protected from manipulation by these *in vitro* procedures.

Regenerative and Proliferative Experiments

Many plant tissues, when excised from the parent plant and cultured *in vitro*, proliferate shoots, roots, embryos and flowers depending on the condition of the plant source and the conditions of culture. The proliferation of flower meristems directly from the epidermal peels from the stems of florally induced *Nicotiana* and other species is a striking and important example of this capacity [see review by Tanimoto and Harada (1985)]. Data from these experiments and from non-induced plants are important in the development of an understanding of the induction of flowering (McDaniel et al. 1987).

Flower tissues also exhibit proliferative capacities, though the number of reports is much smaller than for vegetative tissues (George and Sherrington 1984; Ammirato 1983). The most cited organ source is the pedicel or flower stalk, but other organs can also be regenerative. For example, aseptically cultured perianth segments of immature *Hyacinthus* flowers proliferated tepals, stamens and ovules, depending on the developmental stage of the source inflorescence and depending on the concentration of cytokinins and auxins in the culture medium (Lu et al. 1988). Stamens, primarily anthers, can also be proliferative. The stamen filaments of *Actinidia* (Kiwi fruit), for example, produced root primordia and, less frequently, callus and buds. Interesting differences in the responses of stamens (fertile or infertile) from either male or female flowers were reported (Brossard-Chriqui and Tripathi 1984). While the details of these and other reports are beyond the scope of this text, it is enough to record that the general organogenic and embryogenic potential of plant tissues can generally be assumed for floral tissue. The opportunities for the further exploration of materials such as these, especially in the context of genetically based variability, seem substantial.

6

The Gynoecium

The **gynoecium,** the collective term used for the ovules and the structures that bear or enclose them, is probably the flower's most controversial but least understood developmental feature. This is because the category includes such variety and complexity and because many of the specialized parts develop internally and are hidden. As well, gynoecial literature and terminology are frequently charged with classical interpretations and controversy.

Interest in the gynoecium is well placed. While angiosperms appear to possess a clearly identifiable dominance in the post-Cretaceous flora, there is no single trait that unequivocally distinguishes them from other land flora. If uniquely angiosperm developmental sequences underlying novel processes and structures do exist, they are probably expressed during the initiation and development of the gynoecium. The carpel, the ovule integuments, double fertilization and the endosperm could be candidates for such recognition. A description of some of the variation in gynoecial morphology, along with appropriate terminology, was presented in Chapter 2.

Gynoecium Initiation

Except for teratogenic situations, gynoecia arise on floral meristems distal to, and usually after the initiation of, the androecium. At initiation, individual carpels or the primordia of compound gynoecia are obvious. Surface views of single carpel initiation often show a crescent-shaped ridge on the surface of the floral meristem at the site of the differential growth, which results in carpel initiation. In *Delphinium elatum* (Ranunculaceae) (van Heel 1981) (Fig. 6-1), three primordia elongate through intercalary growth, and the carpels fuse along the margin of the adaxial groove. Later development, in this case, leads to a basal ovarian portion in each of the three carpels. The extension of the distal portion forms a style and stigmatic region. Some interprimordial growth also occurs to fuse the carpels at their base. At maturity, the *Delphinium* fruit dries and dehisces along the sutures formed at the carpel margins.

In *Pisum sativum* (Fabaceae), the single-carpelled gynoecium arises from the conversion of the floral apex (Tucker 1989). A cleft, not initially present, forms on the adaxial side when the carpel is about 70μm high (Fig. 6-2). Various inter-

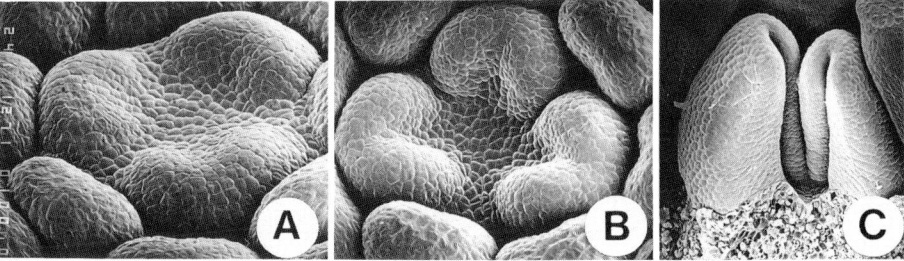

Fig. 6-1A–C. Carpel initiation in *Delphinium elatum*. (A) Initiation of three carpel primordia on the floral meristem distal to the stamen primordia (X15). (B) The early grooved stage of carpel initiation (X12). (C) Elongating carpel stage. One carpel has been removed to reveal the length of the inner grooves which extend to the base of the carpels (X8). From van Heel (1981) with permission.

calary growth activities of the stylar and stigmatic regions complete the mature morphology. Similar SEM views of gynoecium initiation in other legumes are provided by van Heel (1981) and Moncur (1981) and other papers from Tucker's laboratory [including Tucker (1987, 1988*a*)].

Surface views of the early stages of initiation of compound ovaries also demonstrate considerable variability. In some flowers, individual carpel primordia are obvious [e.g., *Butomus* (Fig. 6-3A–B)] (see also *Aquilegia,* Fig. 5-1A). The extent of subsequent interprimordial growth to connect the individual primordia and of intercalary growth throughout the whole organ determines the degree to which a single compound gynoecium is formed. In others, a ring of gynoecium initiation with only subtle suggestions of individual carpels initiates (Fig. 6-3C–D). Assumed vestiges of the individual carpel origins sometimes are found in the structure of the wall, the vasculature, the styles and stigmas and the cavities (locules). These vestiges have been popular pieces of data for interpreting and extending classical carpel theory.

In some species, especially members of the woody Ranales or flowers whose carpels develop into single-seeded achenes, carpels are interpreted as initiating as radially symmetric (or nearly so) **ascidiate** or **peltate** forms (Fig. 6-4). While the differences between the two interpretations—bifacial versus radial primordia—may appear to be slight, and the judgement uncomfortably subjective, considerable importance has been attached to them. In this context van Heel (1981) notes that the base of the primordium of *Lupinus* possesses radial rather than zygomorphic symmetry.

The crescent-shaped image of the carpel initial has, for many morphologists, suggested leaf-like homologies. The radially symmetric gynoecium primordium with an adaxial depression suggests, to others, stem-like or cauline affinities (see Franck (1976) and van Heel (1981, 1983) for reviews of these arguments).

The varieties of initiation patterns of individual carpels on the surface of the meristem are similar to those described for other organs. Helical patterns are derived from a sequence of individual initiations, as in *Ranunculus* (Fig. 6-4). Coincident initiation of whorls of individual carpels is illustrated in *Delphinium* (Fig.

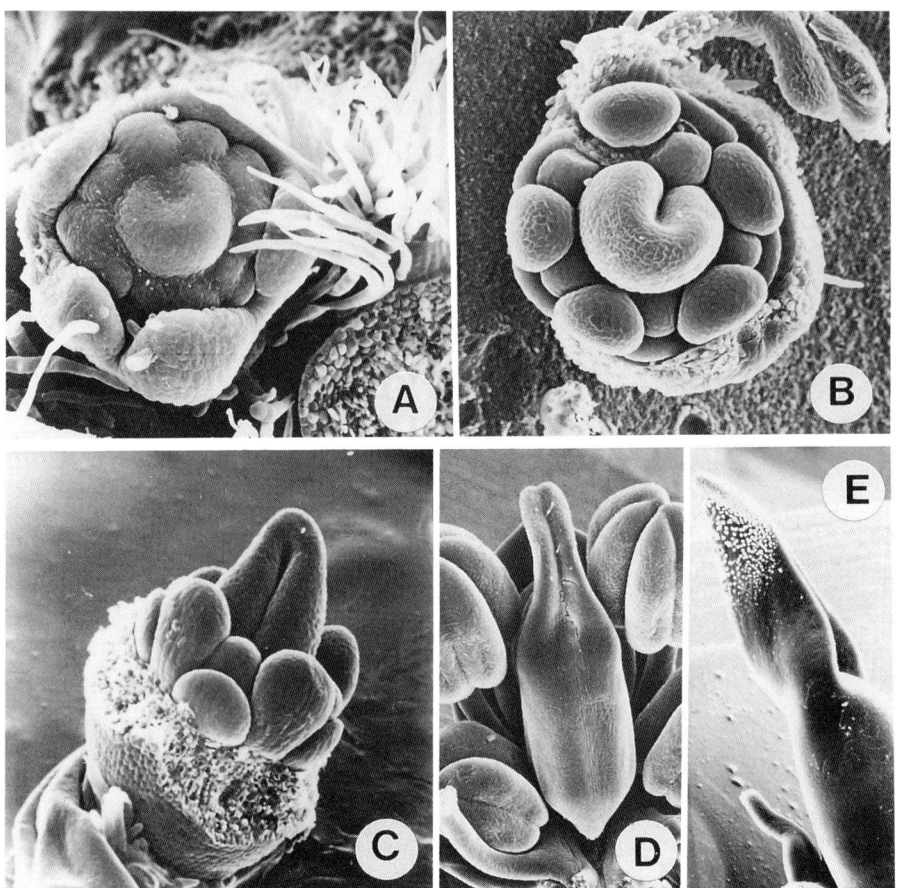

Fig. 6-2A–E. SEM views of carpel initiation in *Pisum sativum*. (A) Polar view of a flower bud at carpel initiation (X130). (B) Later stage of carpel with distinct margins (X140). (C) Lateral view of an older flower to illustrate the elongation of the carpel (X10). (D) Later stage in carpel development. Style with stylar groove, the fused suture on the abaxial side of the carpel, is illustrated (X35). (E) Later stage of carpel development viewed from the side. Differentiation of stigma, style and ovarian portions is illustrated (X25). From Tucker (1989) with permission of *American Journal of Botany*.

6-1). As with the initiation of other organs, only primitive notions of the factors that govern the positioning of carpel initiation are available.

To illustrate, by way of SEM views, the development of a compound gynoecium, we continue the description of the tomato flower (Fig. 5-2) prepared by Chandra Sekhar and Sawhney (1984), in which ovary initiation is indicated (specifically Fig. 5-2B). It is not clear from this figure whether initiation sites are isolated or are continuous around the shoot apex. Histological analysis of a number of samples would be required to determine the precise timing of the sequence. Regardless, a cone of tissue with an open center (Fig. 6-5A) is produced. The

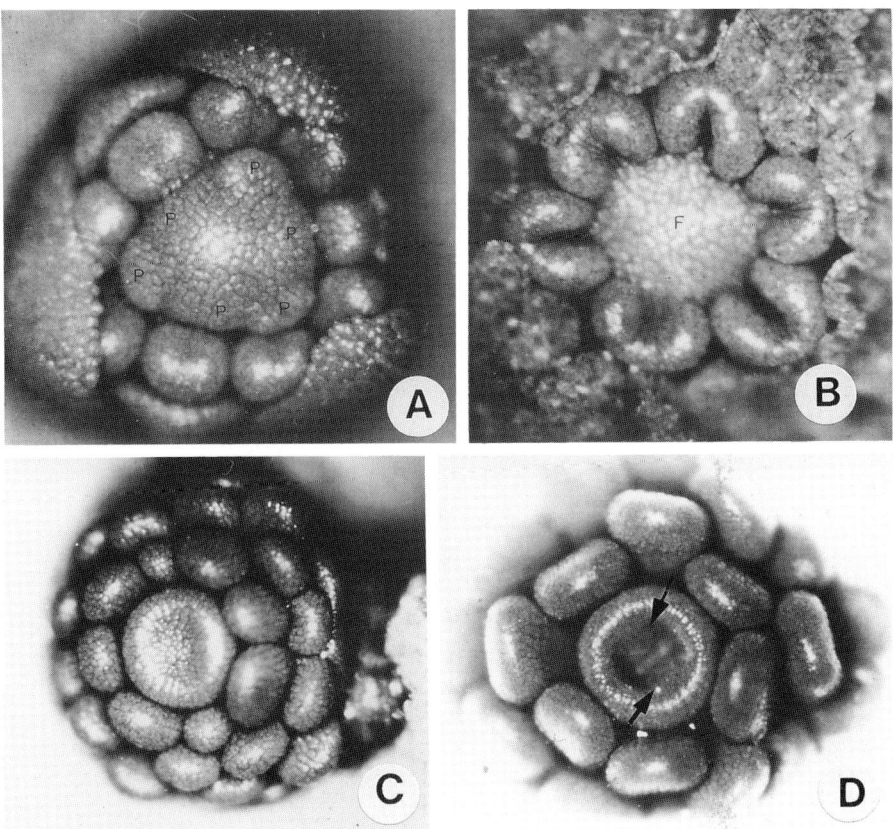

Fig. 6-3A–D. Macrophotographic views of initiation of compound gyneocia. (A) and (B) *Butomus umbellatus* (Butomaceae). Plan view of floral buds showing the initiation of three (A) and six (B) gynoecium initiation sites. F = floral apex (X120). Reproduced from Singh and Sattler (1974) with permission the National Research Council of Canada from the *Canadian Journal of Botany* 52:223–230. (C) and (D) *Chelidonium majus* (Papaveraceae). Plan views of meristems at the initiation (C) and early growth (D) of the compound gynoecium. In D, initiation of placentae is indicated by arrows (X120). From Sattler (1973) with permission.

growth of this cone, probably throughout as well as at its margin, gradually covers the apex and begins the formation of the style and stigma (Fig. 6-5B–D). No studies of the details of the growth dynamics are available, so we must speculate about the cellular events underway. The over-growth of the shoot apical meristem by the ovary wall involves intercalary activity. Less obvious is the extension of septa and the resultant formation of locules within the ovary wall. The generation of the placental surfaces in this multi-locular cultivar is complex. In some cultivars, placentae may appear to arise from the residual floral apex. The origin of placentae and ovules from the apical meristem rather than from the margins of the carpels is the observation which is used by some (Moeliono 1970; Sattler 1974;

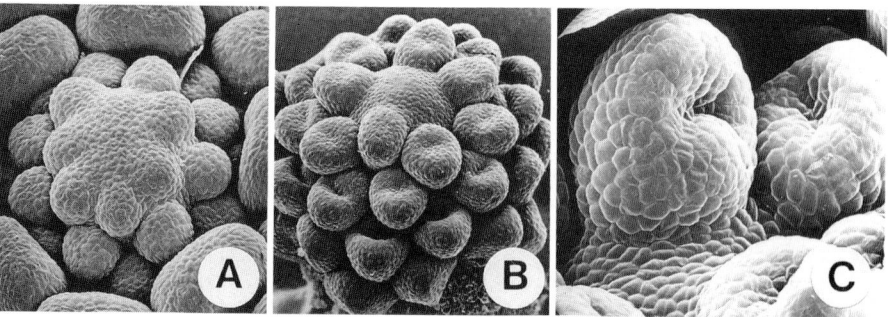

Fig. 6-4A–C. Carpel initiation in *Ranunculus acris*. (A) Helical initiation of carpel primordia on the floral apex distal to the stamen primordia (X100). (B) Development of a slight cavity on the adaxial surface of the carpel primordia (X80). (C) Enlarged view of the carpel to illustrate the beginning of the cavity as well as the ascidiate base (125). From van Heel (1981) with permission.

Sattler and Lacroix 1988) to argue for a cauline (stem-like) interpretation of ovules in some flowers.

Once again, some of the morphological distinctions claimed for gynoecia represent embellishments on the main developmental program. Regardless of morphological complexity, however, all functional carpels and ovaries generate ovules and, ultimately, receptive embryo sacs. Just how the genome is arranged and expressed in order to accomplish this complex feat remains in the domain of developmental genetics and whole-plant physiology. Some aspects of this fascinating topic are looked at in more detail in Chapter 9.

Cellular Basis of Initiation

As discussed in Chapters 4 and 5, three approaches have been used to uncover the cellular events involved in primordium initiation: (a) direct observation of carefully prepared microscopic sections through the primordia; (b) studies of the periclinal chimeras; (c) quantitative analysis of cell cycle kinetics and orientation of division planes.

Studies of cellular dynamics of carpel or ovary initiation have been attempted, though the number of cases of systematic (as opposed to casual) study is low. In *Nicotiana*, Hicks (1973) documented reorientated division patterns, associated with initiation, in the hypodermal layer (T2) and also in the outer corpus (Fig. 6-6). Barnard (1955) illustrated the periclinal contributions of both T1 and T2 to the initiating ovary of *Triticum* (wheat) (Fig. 6-7). This identification of differences in the cellular patterns of initiation between dicotyledons and monocotyledons was cited earlier, though in both cases the patterns in the formations of gynoecia resemble leaf initiation in their respective groups.

Another view of the relative contributions of the different layers of the floral meristem to gynoecium initiation comes from the analysis of periclinal chimeras

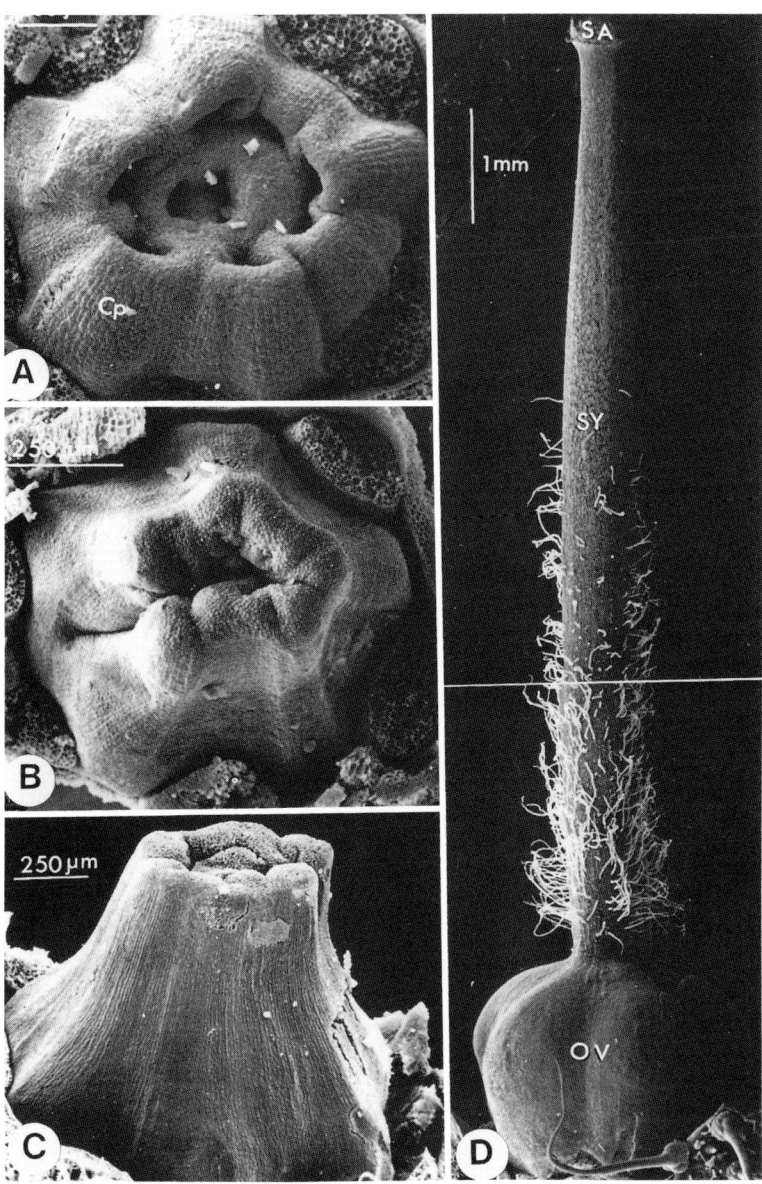

Fig. 6-5A–D. SEM images of stages in the initiation and early growth of the gynoecium of tomato (*Lycopersicon esculentum* cv. Pearson). For earlier stages, see Fig. 5-2. (A) Plan view of developing gynoecium in which individual carpels (Cp) are obvious. (B) A longitudinal section of a young gynoecium (same stage as in part A) indicating the development of septa and probably some placenta tissue, the locule and gynoecial wall (Cp). (C) A lateral view of a developing gynoecium indicating the continuous gynoecial wall and an early stage in style and stigma development. (D) A mature gynoecium with differentiated ovary (OV), style (SY) and stigma (SA). Hairs have developed on the lower half of the style. Reproduced from Chandra Sekhar and Sawhney (1984) with permission of the National Research Council of Canada from the *Canadian Journal of Botany* 62:2403–2413.

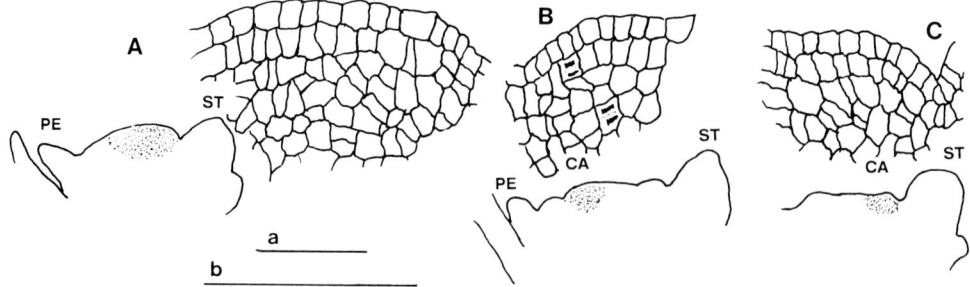

Fig. 6-6A–C. Gynoecium initiation and emergence in *Nicotiana tabacum*. Each figure includes a low magnification sketch of the outline of the flower apex from which the detail was taken. (A) The status of the floral apex before gynoecium initiation: two layers (T1 and T2) are indicated. Periclinal divisions in T2 are indicated at the right flank. (B) and (C) Adjacent sections of one bud. Gynoecium emergence: left flank in B and right flank in C. Periclinal T2 divisions are evident in B only. Periclinal divisions in corpus are evident in both A and B. PE = petal primordium; ST = stamen primordium; CA = carpel primordium. Scale "a" designates 200μm for all smaller drawings. Scale "b" designates 100μm for the larger drawings. Reproduced from Hicks (1973) with permission of the National Research Council of Canada from the *Canadian Journal of Botany* 51:1611–1617.

(Satina and Blakeslee 1943). In this study of *Datura*, the first indications of carpel initiation including the wall, the septa and placentae are in the innermost layer of the meristem, the L-III. Later, cells from L-II contribute to the building of the carpel wall. L-I contributes to the epidermis (Fig. 6-8A–B). These authors suggest that their observations support a stem-like rather than a leaf-like affinity of carpels.

Regardless of what these differences might mean to our interpretations of the morphological relationships between the different flower organs, it seems obvious

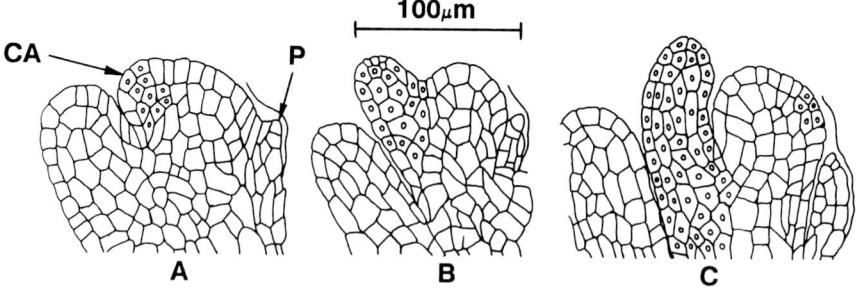

Fig. 6-7A–C. Tracings of radial longitudinal sections through flower meristems of *Triticum* to indicate stages in gynoecium development. (A) Evidence at "CA" of periclinal division of the dermatogen (T1) and hypodermal cells (T2) during gynoecium initiation; P = palea. (B) Immature gynoecium. (C) Margin of the gynoecium has encircled the growing point, as indicated by periclinal divisions in the dermatogen on the adaxial side of the meristem. From Barnard (1955) with permission of the *Australian Journal of Botany*.

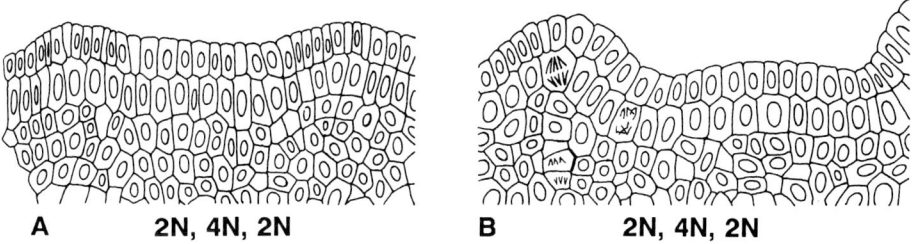

Fig. 6-8A–B. Early stages of gynoecium initiation in a periclinal chimera of *Datura* (2n, 4n, 2n). (A) At initiation. Periclinal activity in L-III. (B) Young carpel primordium showing first periclinal divisions in L-II and periclinal activity throughout the L-III of the meristem. From Satina and Blakeslee (1943) with permission of the *American Journal of Botany*.

that there is considerable variability in the cellular sequences and contributions, despite the similarity of the final product. One practical consequence of this observation is that developmental researchers, whether attempting to study clonal analysis or attempting to introduce novel genetic plasmids into meristems, need to know with some precision just what the situation is for the organ and organism of their study. Other features of this phenomenon are explored by Poethig (1987).

Proliferation and Reversion

The above observations document, as well, the determinate nature of most flower meristems. Gynoecium formation usually encompasses the remainder of the flower meristem. This feature has been commonly documented from sectioned apices—for example, in the almond, *Amygdalus* (Brooks 1940).

This determinate feature of most flowers contrasts with those teratological specimens in which, following petal or stamen production, the meristem produces additional flowers-within-the-flower [for example, the *agamous* mutant of *Arabidopsis* (Yanofsky et al. 1990) and the Mgr21 line of *Nicotiana* (Malmberg et al. 1985)]. Other examples include the proliferation of vegetative shoots from flowers [e.g. *Rosa* (Worsdell 1916)]. The terms **proliferation** and **reversion** are used, sometimes interchangeably, by different authors (Heslop-Harrison 1952; Battey and Lyndon 1990) to describe these variants.

Some examples of reversion which resemble the "agamous" flower of *Arabidopsis* mentioned above, are gene-based. Others, with interesting developmental implications, are environmentally based—in particular daylength-mediated reversions. For example, *Anagallis arvensis,* a long-day (LD) plant (Brulfert, Fontaine and Imhoff 1985) with a critical daylength of 12h (Ballard 1969), when given 1–6 LD inductive cycles followed by a return to SD, forms shoots with varying degrees of flower traits and organs interposed between vegetative portions. Both quantitative and qualitative reversion responses to different daylength combina-

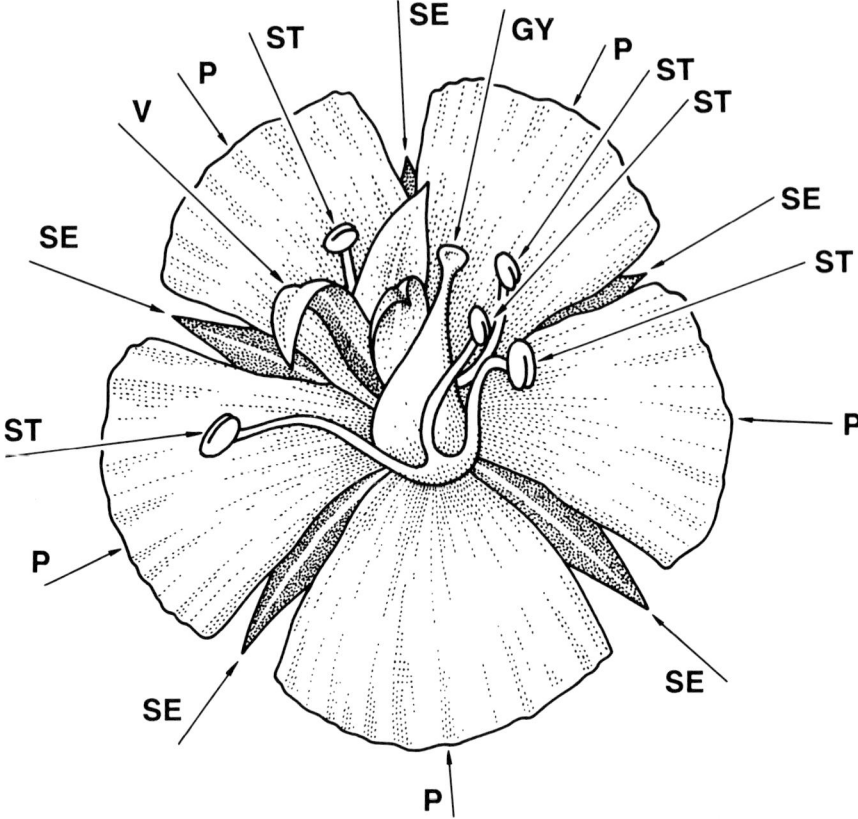

Fig. 6-9. An example of flower reversion in *Anagallis* produced by inductive LD periods followed by return to SD in which the flower possessed 5 sepals (SE), 5 petals (P), 5 stamens (ST), a carpel-like structure (GY) and a vegetative apex with leaves (V). From Brulfert (1965) with permission.

tions are described (Brulfert 1965; Brulfert, Fontaine and Imhoff 1985). An example of reversion in *Anagallis* is illustrated in Fig. 6-9.

In *Impatiens balsamina,* a short-day (SD) plant, a range of different responses were obtained from plants that were transferred back to LD after 3–10 SD. These ranged from plants that showed no sign of flowering and subsequent reversion to those with a reversion zone in which a variety of flower organ traits were expressed (Battey and Lyndon 1984). The data from these two studies of *Impatiens* and *Anagallis* bear directly on two important developmental aspects of flower expression: (a) the nature of the inductive stimulus; and (b) the question of flower organ determination. As such, these and similar materials deserve further exploration, especially at the molecular levels underlying these important processes. The collection, maintenance and distribution of lines with defined genetic variability in these traits should be a major feature of future investigations of these materials. Additional discussion of this topic is presented in Chapter 9.

Gynoecium Growth

A number of mature gynoecium types are recognized (see Chapter 2), though, unfortunately, the growth patterns that produced them have rarely been studied directly and their classification has little developmental content. It is also obvious that there are many variations. In addition, differences in initiation, including circular initiation patterns [c.g., Primulaceae (Sattler 1973a) or *Myrica* (McDonald 1980)] and different patterns of growth, are observed.

Intuitively and from observations on the growth of other organs, we can assume that, following initiation, gynoecial growth involves a variety of growth patterns. In some, a period of interprimordial growth connects the independently initiated primordia. Variations of intercalary cell division and cell enlargement then are responsible for the subsequent elaboration of the organ and its parts. In many cases, "fusion" of tissues is a normal feature of development (see below).

In *Lilium,* gynoecium length over 40–50 days (Crone and Lord 1991) correlates allometrically with bud (i.e., tepal) length (Fig. 6-10A). This growth can be divided into three phases: Phase I (gynoecium <10mm), Phase II (10–100mm) and Phase III (100–135mm). From an analysis of marking studies, an indication of the local relative growth rates (LRGR) during this period was calculated (Fig. 6-10B). These data indicate that different portions of the gynoecium have different maximum growth rates at different times. Phase I gynoecia exhibit non-steady growth in which LRGR profiles rise and fall with no predictable pattern. Phase II growth was mainly confined to the style. Phase III growth, while mainly in the upper style, also took place in the basal region. The study of the lily gynoecium also corroborated the observations made on the anther (Gould and Lord 1989) of a non-steady distribution of cell division along the organ during Phase I. We presume that data such as these will ultimately form the basis for experiments which can uncover the cues which regulate these cellular activities.

The Inferior Ovary

Both the ontogeny and phylogeny of inferior ovaries has stimulated considerable controversy (Douglas 1944, 1957; Kaplan 1967; Leins 1972). In recent years, with the availability of SEM and a more critical evaluation of the cellular basis of organ growth and fusion, a more objective interpretation is possible.

Hufford (1988), in a work with broad evolutionary objectives but data applicable to ontogeny, provides a useful view of the development of an inferior ovary (Figs. 6-11 and 6-12). In this example, gynoecium initiation begins coincident with the initiation of the second tier of stamens. It begins as a ring of inward growth at the level of the sepal attachment on the receptacle. The ring grows inward to cover the ovarian cavity, leaving a central opening. Five placentae initiate as decurrent, antipetalous ingrowths and become semicircular in transection. Ovule initiation begins in the upper portion of the cavity and subsequently extends over the placental surfaces. Following the establishment of the placentae, the edge

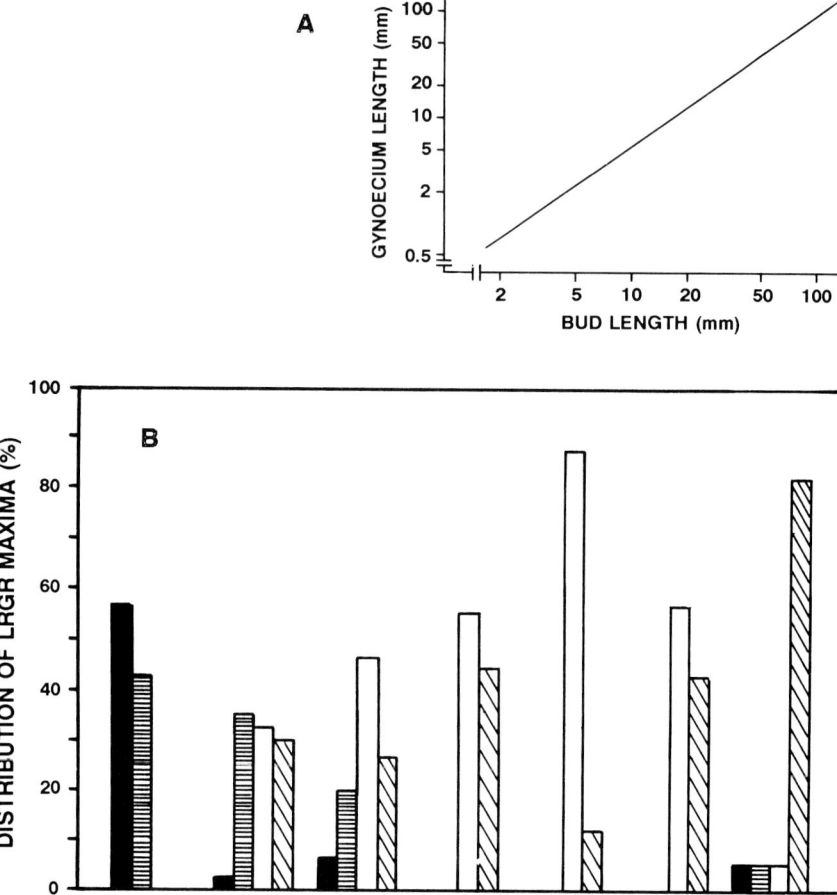

Fig. 6-10A–B. (A). Allometric plot of *Lilium* gynoecium length versus tepal (bud) length. N = 119. $r^2 = 0.977$. (B) Distribution of Local Relative Growth Rate (LRGR) maxima in gynoecia of increasing size. Histogram bars: black = lower ovary (lo); cross-barred = upper ovary (uo); white = lower style; hatching = upper style (us). From Crone and Lord (1991) with permission of Academic Press.

of the ring forms the primordium from which the style develops. Additional observations on gynoecium initiation, in the Rosaceae, are given in Steeves, Steeves and Olson (1991).

Another view of inferior ovary formation, with generally similar interpretations, comes from the work from Leins's laboratory (Leins and Erbar 1985; Erbar 1986). In these studies, in addition to the preparation of outlines of the flower

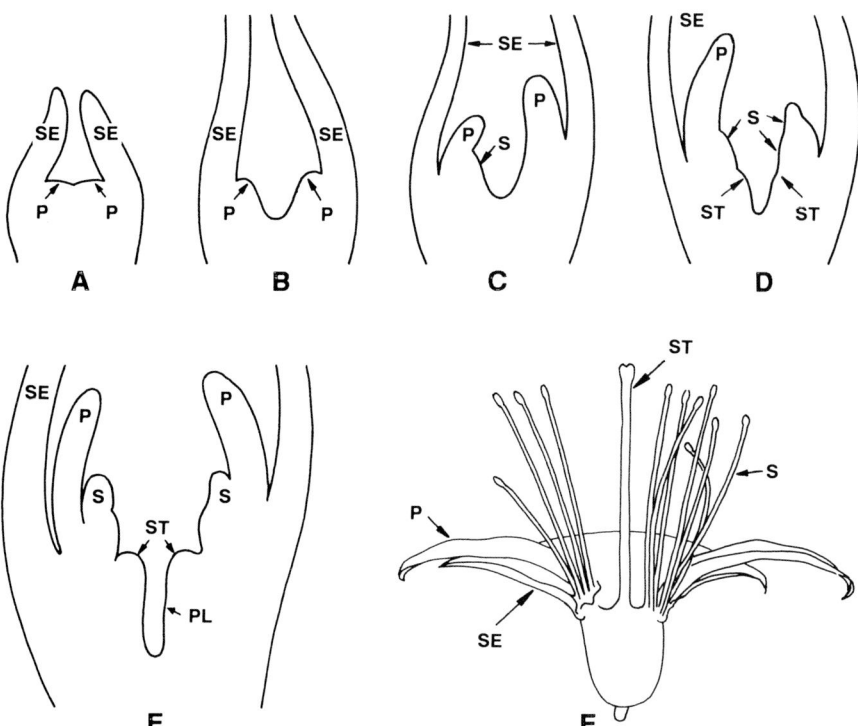

Fig. 6-11A–F. Generalized summary of flower bud development in species of *Eucnide*. Outlines of median longitudinal sections of floral buds at different stages of development. (A) During early corolla organogenesis. (B) Floral tube forming by intercalary growth, which forms an interior concavity prior to androecium initiation. (C) Early androecium initiation. (D) The extension of the petal/stamen confluent region and the beginning of gynoecial development delimit an inner scarp-like surface where stamens arise. (E) Calyx, corolla, androecium and gynoecium inception completed. Maturation and differentiation underway. (F) Outline of a mature flower. Magnifications: A–E, approx. 30X; F, approx. 1/2. Abbreviations: P = petal; S = stamen; SE = sepal; ST = style; PL = placenta. From Hufford (1988) with permission.

during organ formation, a semi-quantitative interpretation of the degree of intercalary activity is suggested by the indication of the cell files (Fig. 6-13).

Studies of vasculature patterns in mature ovaries have traditionally been used to support the contention that two major types of inferior ovaries exist (Fig. 6-14). Despite compelling presentations based primarily on mature vasculature, such as Puri (1952), Douglas (1957), Eames (1961) and Kaplan (1967), it is not clear that a developmental basis for such a distinction is tenable. In the light of more recent observations on ovary development and modern understandings about vascularization and the physiological factors that affect it, there is a need for a re-evaluation of the formation of vasculature of inferior ovaries. In this regard, it would help if researchers could identify and explore materials in which a range of expression from inferior to superior ovaries is presented.

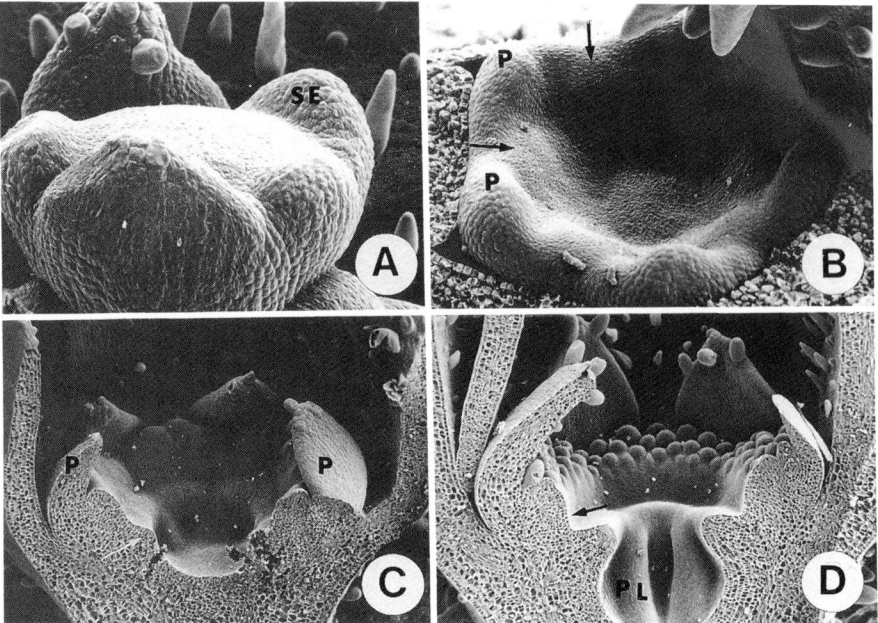

Fig. 6-12A–D. SEM views of early gynoecium development of *Eucnide urens* (Loasaceae). (A) Floral meristem following sepal initiation possesses a convex dome (X140). (B) Initial androecial mounds in antesepalous positions (X140). (C) Longitudinal section of a young flower possessing first three sets of stamen primordia (S) in each antesepalous group. Arrow indicates the upper rim of the gynoecium (X65). (D) Longitudinal section of a young flower showing the inner scarp of the confluent petal/stamen region (arrow) where stamen initiation occurs. Stylar upgrowth is underway, and placentae (PL) have formed (X50). SE = sepal; P = petal. From Hufford (1988) with permission.

The Role of Fusion

It is reassuring to relate that ''carpel fusion,'' traditionally surrounded by controversy and confusion (Cusick 1966), has achieved a degree of intellectual rigor and acceptance. First, it is now increasingly acknowledged that the term ''fusion,'' as applied to phylogeny (as in ''congenital fusion''), is ill-advised (Sattler 1974; Leins and Erbar 1985). Thus, ''fusion'' should be restricted to ontogenetic situations where it can be observed and studied directly.

A second piece of evidence is the demonstration of the cellular events involved in observable carpel fusion in *Catharanthus*. This elegant model system, originally described by Boke (1949), uncovers a number of significant components of this developmental phenomenon. These include: (i) fusion is unrelated to pressure or cell confinement (Walker 1978); (ii) epidermal cells at the site of fusion undergo organelle activity, wall reorganization and periclinal cell division (Walker 1975); (iii) fusion is triggered by water-diffusible factors (Seigel and Verbeke 1989). Differences in the abilities of the organ surfaces to generate and respond to this initiator were detected. This phenomenon is documented in Fig. 6-15.

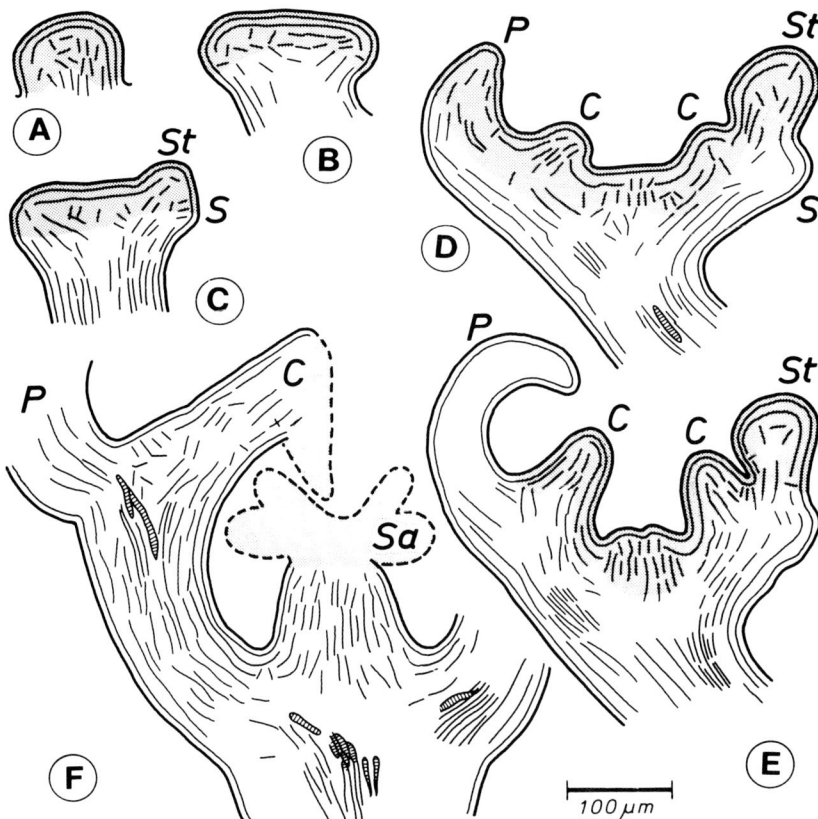

Fig. 6-13A–F. Sketches of longitudinal sections from flower buds of *Levisticum officinale* (Apiaceae) to illustrate the formation of the inferior ovary. Outlines include location of rows of cells (drawn as lines) which represent the main directions of division orientations during intercalary growth. (A) and (B) Early flower meristems. (C) Sepal (S) and stamen (St) primordia being formed at the right of the section. (D) and (E) Petal (P) and upper marginal ring of the gynoecium (C) are indicated. (F) The extension of the gynoecial ring will ultimately form the style a stigma. The placentae form by the upward growth of the base of the ovarian cavity. Ovules initiate on the placentae. From Leins and Erbar (1985) with permission.

The significance of fusion as a component of flower development has been identified in many species (Nishino 1978, 1982, van Heel 1978). The discoveries in *Catharanthus* should stimulate an enhanced research effort to identify the morphogenetic triggers and the physiological, developmental and genetic sequences that connect the triggers with the final phenotype. While the number of flowers in which fusion, of the sort described for *Catharanthus,* occurs should be large, there is no reason to assume that all fusions invoke the same developmental sequences. One should expect a variety of developmental mechanisms. Further, fusions involving other organs and mutant-based variations (Chandra Sekhar and Sawhney

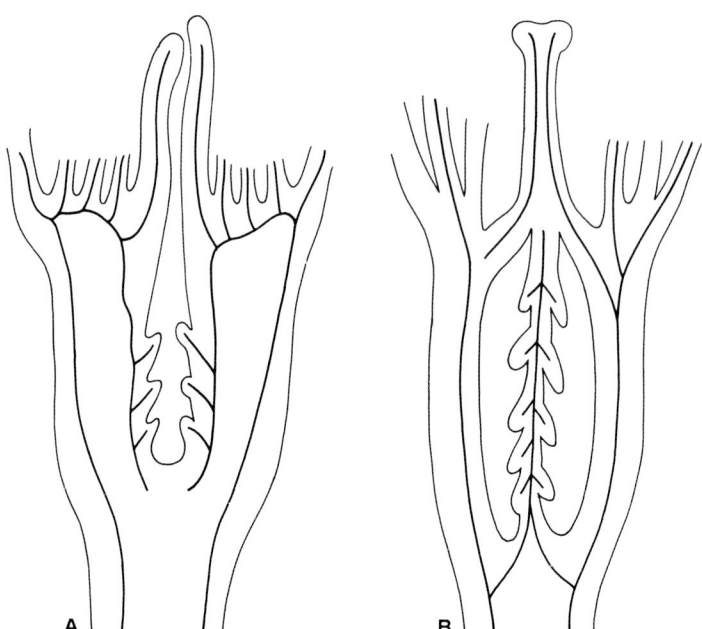

Fig. 6-14A–B. Diagramatic median sections of representative flowers to illustrate the vascular systems found in inferior ovaries. (A) A flower in which the inferior ovary is interpreted as having a "Receptacular" origin. (B) A flower with an inferior ovary that is interpreted as having an "appendicular" origin. From Kaplan (1967) with permission of *American Journal of Botany*.

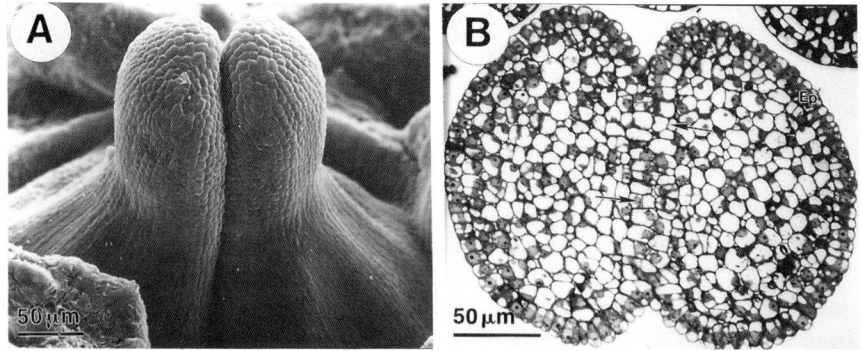

Fig. 6-15A–B. Documentation of the fusion of carpels of *Catharanthus roseus*. (A) SEM view of the gynoecium tips. Two prefusion carpels arose as separate primordia and then grew towards each other. (B) Light microscopic view of a cross section of carpels indicating epidermal redifferentiation in the fused region. From Seigel and Verbeke (1989) with permission of *Science*. Copyright (1989) by the AAAS.

1987) and experimental manipulation (Chandra Sekhar and Sawhney 1990) should be anticipated.

Cell Differentiation in Gynoecia

Gynoecia are obviously complex organs made up of a wide spectrum of cell types and tissues common in other organs characteristic of the species. Three of these tissues—vasculature, transmitting tissue and nectary secreting tissue—should attract the attention of developmentalists because they can be readily identified and can be used as developmental markers.

Gynoecium Vasculature

The literature on the vasculature of the gynoecium greatly exceeds that dealing with the vasculatures of perianths and androecia [see bibliography of Rao (1951, 1961)]. The high variability plus the central position of the carpel in the "classical theory" has provided the stimulus for these studies (Puri 1951). Further, vasculature in morphological structures has frequently been interpreted as a conservative trait which can provide evidence of atavism and vestigial traits when these are not obvious at other levels or in other tissues (Puri 1951).

There are, however, difficulties (Sporne 1958). In the main, interpretations are based upon serial sections rather than whole-mount clearings. Distinctions between primary and secondary vasculature are not always maintained. Finally, observations are frequently limited to mature rather than developing vasculature.

Individual free carpels are vascularized by one, three, five or more traces. In a commonly presented view, three veins (one dorsal and two ventral) arise from a single carpellary trace (Fig. 4-25). Two possibilities are illustrated in Fig. 6-16. In the conduplicately folded model (Fig. 6-16B), the ventral strands are aligned with the primary xylem of both oriented internally: phloem is external on both ventral strands. In the involutely folded model (Fig. 6-16A), the two ventral bundles are juxtaposed with phloem external to both xylem traces. Many smaller strands, derived from these three, vascularize the tissues of the carpel and the ovules.

The vasculature of compound gynoecia is generally interpreted as providing evidence for the gynoecium phylogeny through the cohesion and subsequent modification of individual carpels. The number, position and orientation of vascular strands have been used with considerable ingenuity in attempts to understand the history of the flower and the gynoecium (Puri 1951; Eames 1961; Fahn 1982). According to Sporne (1958), there is no reason to view the vascularization of the flower and the gynoecium in terms of a stele with leaf traces, but rather it should be interpreted as being composed of individual open or closed traces. To our knowledge, neither descriptive nor experimental studies of gynoecial vasculature development are available.

The vasculature of the crucifer gynoecium is of interest because *Arabidopsis* and *Brassica* are popular research model systems; and independently, it generated

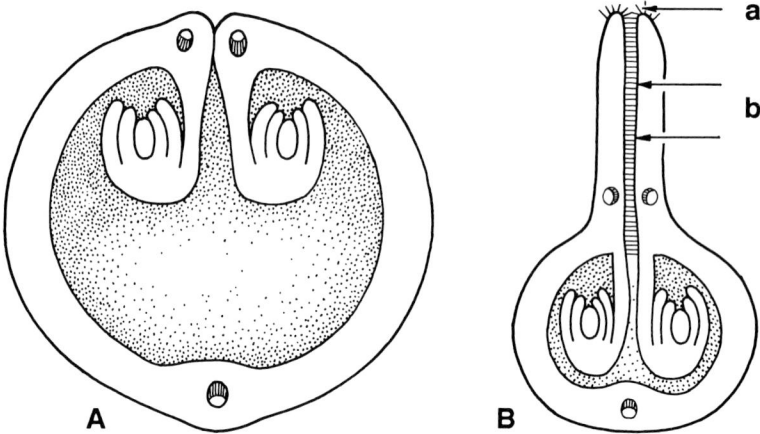

Fig. 6-16A B. Two generalizations regarding closure and vascularization of carpels. (A) An idealized section through the ovuliferous portion of a closed carpel, illustrating involute closure with the orientation of primary xylem (black) and phloem (dotted) in the ventral bundles. (B) An idealized section through the ovuliferous region of a carpel interpreted as being folded in the conduplicate manner. a = stigmatic region; b = short cross bars representing transmitting tissue derived from stigmatic tissue. Redrawn from Mosley (1967).

considerable discussion and an abundant literature. The question at issue was the number of carpels that are assumed to be present in the crucifer gynoecium: two, four or more? Further, how does the presence and arrangement of vasculature bear on this question? A short introduction to the kinds of complexity of the situation and the nature of the discussions is provided in Meyerowitz, Smyth and Bowman (1989), who cite a number of early papers.

Without reworking the arguments (Saunders 1926; Eames and Wilson 1928; Puri 1941; Merxmuller and Leins 1967), we present only a brief description of a typical crucifer gynoecium and its vasculature (Fig. 6-17). As depicted, ten veins diverge from a cylinder of vasculature at the base of the ovary. The orientation of the phloem and xylem of the ventral fertile bundles is inverted from the expected orientation. Thus, the number of strands and their orientation are difficult to harmonize with traditional expectations.

Once again, an opportunity exists to use these observations as the basis for some authentic developmental explorations. Of the mutant forms of *Arabidopsis* that involve flower development (Haughn and Somerville 1988) and embryo (Meinke 1985), it seems likely that some would also possess variable vasculatures. In view of the well-documented involvement of auxin levels as limiting factors in some aspects of vascular differentiation (Jacobs 1952; Jacobs and Morrow 1957; Sachs 1981), it would seem particularly appropriate to study the vasculature variation in mutants with genetically modified auxin metabolism such as the "axr1" mutants (Lincoln, Britton and Estelle 1990). The exploration of genetically based research material, coupled with a judicious use of traditional and modern techniques, might provide some novel insights about vascular differentiation and should make a valuable contribution to our understanding of this aspect of flower development.

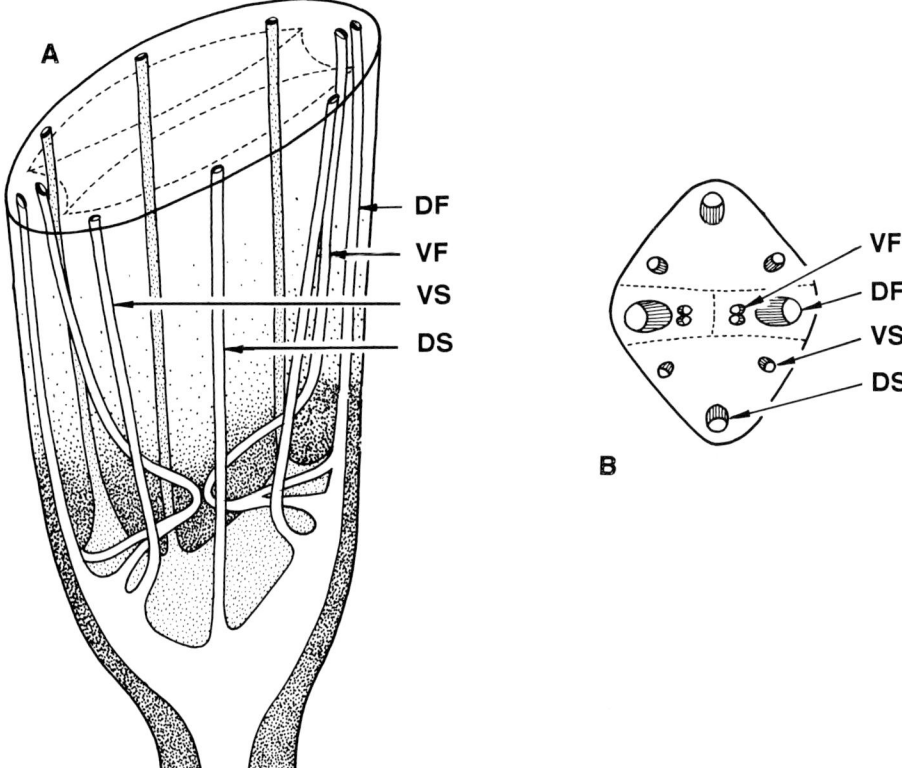

Fig. 6-17A–B. (A) Interpretive diagram of the base of the ovary of a crucifer showing the vasculature, especially the origin of the inversely oriented ventral bundles. VF = ventral bundles and DF = dorsal bundles of the fertile carpel. DS and VS are the dorsal and ventral bundles, respectively, of the sterile carpel. (B) Diagram of cross section of the base of the ovary of *Cardamine Douglasii* showing the vascular supply to carpels. Interpretation of the carpel limits is indicated by dotted line. Labelling is as in A. From Eames and Wilson (1928) with permission of *American Journal of Botany*.

In addition there is a need to describe vasculature development from carefully sampled and timed material as a basis for experimental manipulation. The use of traditional anatomical procedures coupled with computerized recordings and displays should be given some priority. In all studies, care must be taken to detect and measure the variation in any of the characters under study.

Transmitting Tissue

Following the suggestion of Tilton and Horner (1980), and to simplify some of the distinctions in terminology, the cells over or through which pollen tubes grow are all termed **transmitting tissue.** This can extend from the tip of the stigma to the base of the ovarian locule but excludes similar differentiation on the ovule itself. It may be variously modified and may be partly or entirely secretory. It is primarily epidermal but may include sub-epidermal layers. It is a component of angiosperm developmental expression that should possess some distinctive and useful chemical and histochemical markers.

Three types of tissue are recognized: (i) **Stigmatic transmitting tissue** – Highly variable epidermal (+/- hypodermal) tissues of the stigma on which pollen lands and germinates. It is invariably secretory, though variable in the amount of exudate produced. In some flowers it is the site where incompatibility reactions take place. (ii) **Stylar transmitting tissue** – The tissue of the style along which the pollen tubes pass. It can be an epidermal surface lining a stylar canal, or it can consist of several layers filling the stylar canal. Most examples are secretory, providing nutrition and, in some styles, chemotropic or incompatibility cues to the pollen tube; (iii) **Ovarian transmitting tissue.** Transmitting tissue differentiation in the ovarian portion of the gynoecium is assumed to secrete nutrients or chemotropic signals for the pollen tube. Variations of the details are numerous and may include **obdurators.** These are obvious structures in the region of the ovule, placenta or funiculus, such as a pad or swelling of epidermal cells. Tilton and Horner (1980) exclude from this definition of transmitting tissue similar modifications of the ovular integuments or arils that also contribute in some way to pollination.

A classification of stigmas from nearly 1000 species from approximately 250 families, on the basis of whether they are wet or dry and other readily discerned superficial traits of the cells, indicates that useful variation for experimentation with these developmental markers is available (Heslop-Harrison and Shivanna 1977). An alternative classification of stigmas which requires chemical characterization of exudate is also available (Dumas 1978).

Nectar-secreting Tissues and Organs

The inclusion, here, of a discussion of these cells and tissues (**nectaries**) is arbitrary but convenient. Many of those found on flowers (**floral nectaries**) are associated with the gynoecium, but others are part of the perianth, receptacle or stamens (Fahn 1990). In addition, **extra-floral nectaries** develop on vegetative regions of the plant. Many distinctions and classifications are available (Fahn 1979*a;* Schmid 1988) for the numerous variations that have been identified.

The main thrust of the research on reproductive nectaries to date has been the collection and classification of descriptive information on structure and secretion and its application to taxonomic and ecological questions. As with the transmitting tissue, however, an important developmental feature of these cells, tissues and organs is that they are highly localized sites of differential gene expression which exhibit highly distinctive cytology and anatomy and from which significant quantities of molecular secretions can be recovered and monitored. The kinds of molecules in these secretions is large and includes amino acids, sugars, proteins, lipids, antioxidants, alkaloids and many others (Baker and Baker 1983). Their use as developmental markers must, however, await the demonstration that variation in their expression and extent is related to recoverable genetic variability. Ultrastructural studies (Fahn 1979*b;* Durkee, Gaal and Reisner 1981; Durkee 1983) represent some necessary background studies for such an attack.

Ovule Initiation and Differentiation

Placentation

The surfaces on the inside of carpels or ovaries where ovules initiate are called **placentae** (sing. placenta). Of the different mature patterns that develop on the inner ovary walls (Puri 1952), five are illustrated in Fig. 2-4. Taxonomists, phylogenists and ecologists are interested in the variation of ovule number and in pattern variation within and between taxa for their studies on populations and history. Further, these surfaces, as compared to those immediately adjacent, possess a similar but restricted proliferative potential to meristems and therefore represent excellent model material for experiments on the factors that determine pattern and organogenesis. Opportunities, therefore, exist for understanding the developmental factors involved in these features and utilizing the regenerative potential in these tissues for practical applications.

Ovule initiation is most often restricted to what is generally interpreted as carpellary tissue. As mentioned earlier, it is now generally acknowledged that, in some cases, ovules are formed from the floral apex as distinct from the tissues of the gynoecium.

In what are generally interpreted as abnormal or teratogenic situations, ovules can arise in other parts of the flower or vegetative organs. As a special case of carpellody, which is common (Meyer 1966), external ovules have been described in many species and on a variety of organs. These teratomata have been correlated with cytoplasmic male sterility (Meyer and Meyer 1964), nuclear genes (Sawhney and Greyson 1973a) and disease and environment (Meyer 1966). Perhaps a judicious choice of genetically based variation coupled with penetrating experimental protocols could uncover information regarding the factors that underlie ovule initiation, and in particular how the development of ovules can occur apart from the gynoecial tissue.

Ovule Initiation

Cell division sequences, resembling initiation of organ primordia on apical meristems, involving localized periclinal divisions in the L2 [and in most cases the L3 (Bouman 1984)], accompanied by anticlinal divisions in the protoderm, lead to the protuberance that grows into the appropriate species-specific ovule. In *Datura*, Satina (1945) determined through the use of periclinal chimeras that initiation began in the L3.

An illustration of the opportunities for studies of ovule origin and development is provided by Cass and Fabi (1990) on the placentae of *Papaver rhoeas* (the common poppy). The gynoecium is interpreted as arising from a variable number (up to eleven in this study) of carpels from which septa extend toward the center of the single central cavity (unilocular), and the formation of placentae on both surfaces is interpreted as being parietal (Fig. 6-18A–D). Surface views of the initiation events are illustrated in Fig. 6-18A and B. Ovule initiation begins on a

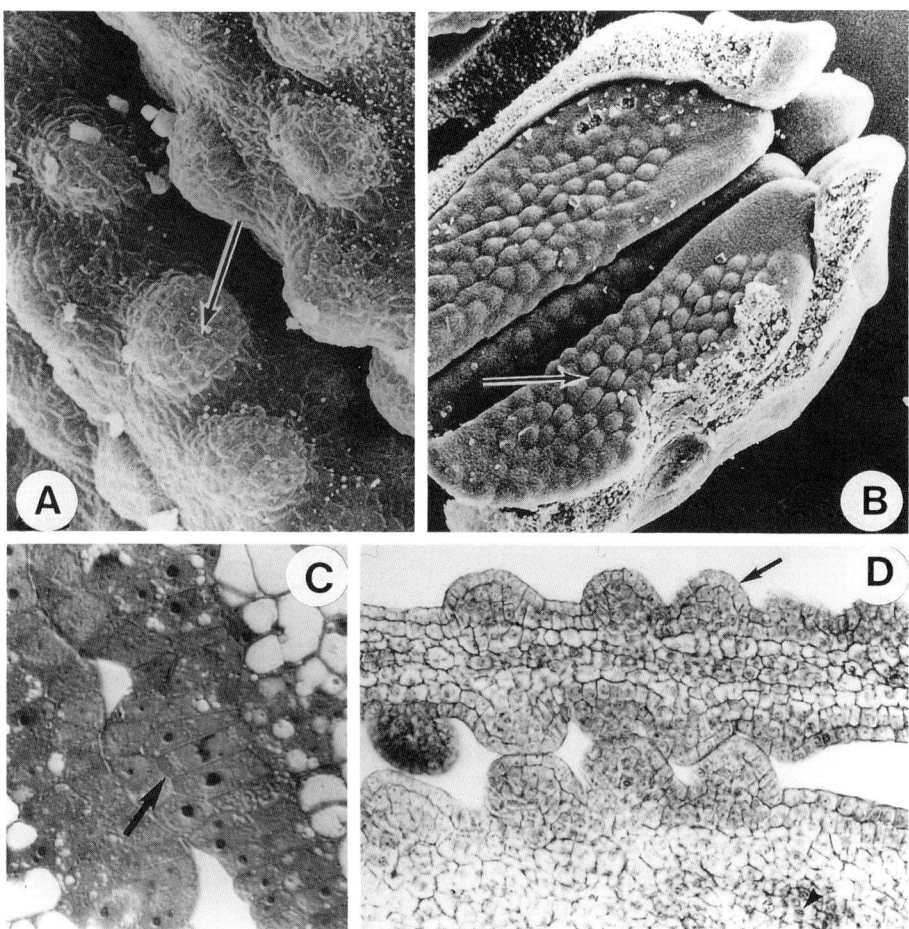

Fig. 6-18A–D. Views of placenta and ovule initiation in *Papaver*. (A) SEM view of ovule initiation showing raised nucelli (arrow) (X360). (B) SEM view of gynoecium that had been opened prior to SEM. Ovule initiation begins near the center of each placenta (arrow) and proceeds both acropetally and basipetally (X75). (C) Histology of placental surfaces using differential interference contrast microscopy, showing a metaphase figure in the second parietal layer (arrow). (D) Several nuclei on two adjacent placental surfaces are illustrated in bright-field microscopy. At this stage each nucellus consists of a protoderm (arrow) and underlying ground tissue. Reproduced from Cass and Fabi (1990) with permission the National Research Council of Canada from the *Canadian Journal of Botany* 68:258–265.

placenta near the center of a septa (note the arrow in Fig. 6-18B) and extends acropetally and basipetally from that point. The cellular make-up of this surface is interpreted as consisting of two layers of anticlinally dividing cells over an underlying mass of placental ground tissue (Fig. 6-18C). Although periclinal divisions in the L2 are involved in ovule initiation, prior elongation of these cells appears also to be involved (Cass and Fabi 1990) (Fig. 6-18D).

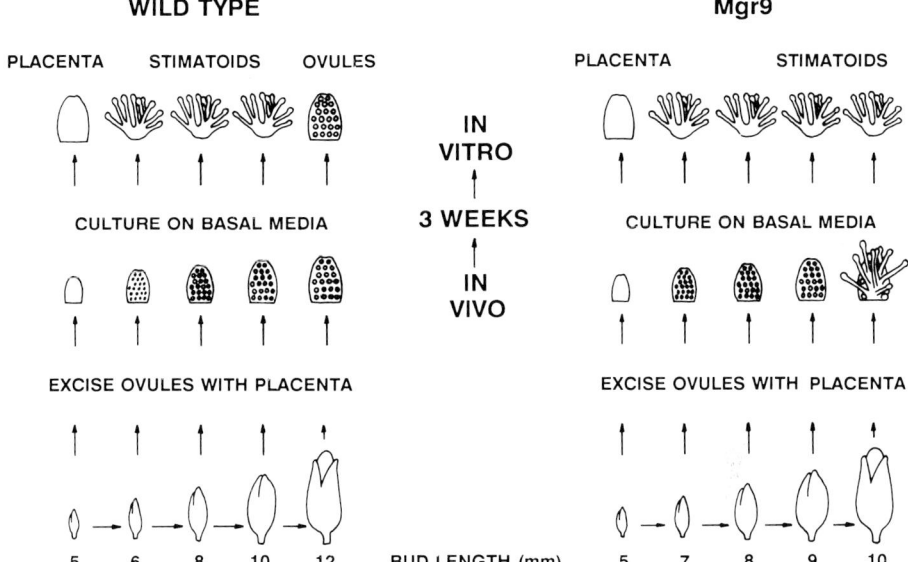

Fig. 6-19. Summary of a series of placenta explant culture studies of wild-type *Nicotiana* and the mutant "Mgr9" as a function of age/size of the bud from which they were dissected. The culture medium consisted of Linsmaier and Skoog (1965) basal medium containing 4% sucrose. From Evans and Malmberg (1989) with permission.

Just how these placental tissues can be used experimentally to explore developmental questions about ovule initiation and differentiation is illustrated in the study of cultured *Nicotiana* placentae of Evans and Malmberg (1989). Cultured placentae of wild-type *Nicotiana* flowers differentiate as stigmatoids or ovules, depending on the stage of development at the time of culture. One can infer from these data just when determination to ovule initiation and formation takes place (Fig. 6-19) and, as well, expose alternative expressions which are normally not expressed. Placenta culture of the polyamine mutant "Mgr9," however, provides a different result which mimics the mutant phenotype and indicates the point in development when the mutant gene is expressed in the tissue. It is possible that histochemical and ultrastructural analyses of uncultured material at different stages might be used to corroborate these observations. The identification and analysis of ovule mutants in *Arabidopsis* (Robinson-Beers, Pruit and Gasser 1992) and other species is an obvious and important research strategy.

Ovule Morphology

A number of different mature ovule forms have been described (Fig. 6-20) of which the **anatropous** form is the most common. Most of the conclusions about the responsible growth patterns are inferred, there being little direct quantitative study of cellular basis of ovule growth. These processes involve the unequal growth that leads to ovule curvatures and the initiation of one or two integuments and the extent of the nucellus and other tissues of the ovule.

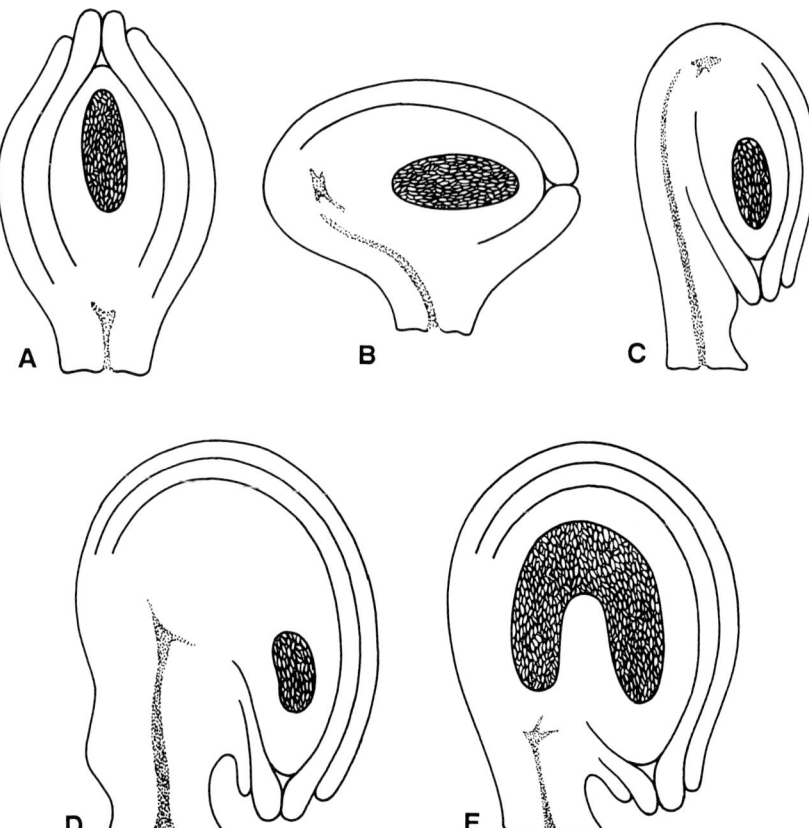

Fig. 6-20A–E. Diagrammatic L/S views of the main types of ovules. Location of the embryo sacs is indicated by hatching. The vascular strands are indicated by stippling. (A) Orthotropous. (B) Hemitropous. (C) Anatropous. (D) Campylotropous. (E) Amphitropous. Redrawn from *Syllabus der Pflanzenfamilien,* 12th ed., vol. 2, by A. Engler. Gebruder Borntreager, Berlin, 1964, with permission.

Bitegmic ovules are the most common, and ategmic ovules are rare. The origin of the **inner integument** is most frequently dermal, while the **outer integument** can be of dermal or hypodermal origin (Bouman 1984). The number of integuments is generally consistent at the level of families, though exceptions are known. While there is considerable scope for the gathering of further detailed information for taxonomic and comparative morphological studies, countless opportunities also exist for the exploration of the factors that influence the cellular processes involved.

The bitegmic mature ovule (Fig. 6-21B) is attached to the placenta by a stalk **(funiculus).** The portion that runs parallel to the nucellus is the **raphe.** The region of the ovule opposite to the micropyle where the integuments are attached is the **chalaza.** The internal region of the ovule **(nucellus)** produces the **embryo sac.**

The cells and tissues of ovules can differentiate in a number of ways. These

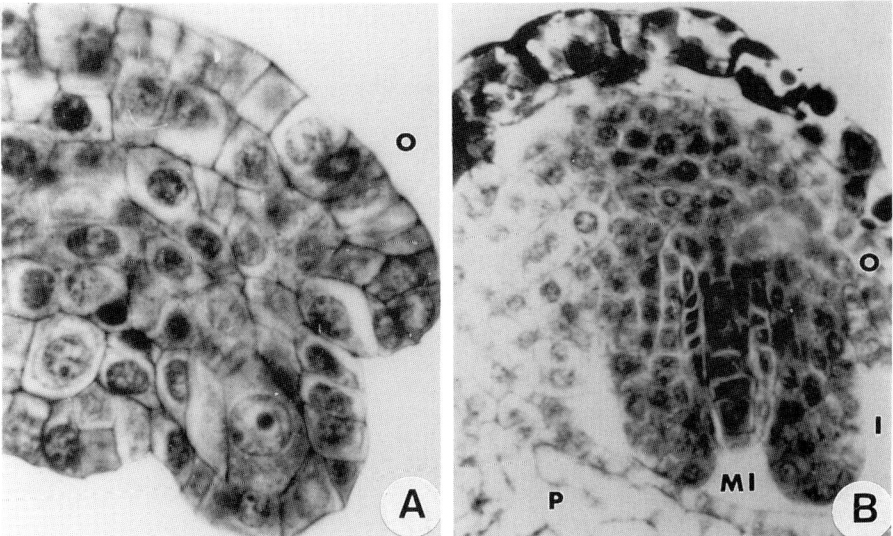

Fig. 6-21A–B. Median L/S of ovules from the California pitcher plant *(Darlingtonia californica)*. (A) Immature ovule at the stage of megaspore development (X300). (B) Ovule with mature embryo sac (X50). O = outer integument; I = inner integument; P = placenta; MI = micropyle.

include the differentiation of vascular bundles in the funiculus and raphe and to the chalazal region. Vascularization also can extend to the integuments and nucellus. The controversial notion that flower vasculature can be "conservative" (Puri 1951), thereby indicating, on occasion, historically earlier organ arrangements of vasculature, has stimulated a number of explorations of vasculature of ovules as well as of the flower as a whole. Because of the relative ease of its exposure, descriptive and experimental explorations to uncover developmental variability seem warranted.

The **ovule endothelium** (integumentary tapetum), superficially resembling the tapetum of the microsporangium, is a hollow cylinder of actively metabolic nucellar cells that surrounds the embryo sac of many, especially dicotyledonous, ovules (Chapman 1987). From the few endothelia that have been studied in detail, their cells appear to be secretory, developing cytoplasmic and nuclear features that resemble the tapeta of microsporangia. They appear to be most variable in structure, and answers to questions about their distribution and role in development must await further study.

Other cellular features that can differentiate within ovules include a **hypostase,** a group of densely cytoplasmic cells that develop at the chalazal end of the nucellus. An **epistase,** derived from nucellar remnants at the micropylar end of the ovule, can form a micropylar plug. Also in the region of the micropyle, some species form an **obdurator** from cells of the funiculus or the placenta. This tissue is thought, because of its transfer cell characteristics, to be secretory in function. Finally, and more generally associated with post-pollination events, are numerous differentiating features in the epidermis, including the production of epidermal hairs.

Megasporogenesis and Embryo Sac Formation

The differentiation of the **megasporocyte** (Fig. 6-21A), from the nucellar cells is a fascinating developmental event complementing the differentiation of microsporocytes in the anther. In the ovule, in contrast, usually only one cell per ovule differentiates. In large ovules consisting of many cells, the megasporocyte frequently develops deep within the nucellus **(crassinucellate)**. In other ovules, with fewer cells, the megasporocyte differentiates from a single hypodermal cell near the tip of the nucellus **(tenuinucellate)**. Other than with regard to location, there seem to be no data available which identify the cues which may be responsible for these crucial cellular events. While the surrounding nucellar cells undergo mitosis, the megasporocyte enlarges and assumes obvious Prophase 1 attributes (Fig. 6-21A).

Megasporogenesis

The number of primary descriptive reports about megasporogenesis is intimidating but has been well summarized. These reviews include Maheshwari (1950), Johri (1963), Jensen (1972), Kapil and Bhatnagar (1981), Bouman (1984) and Willemse and van Went (1984). As with microsporogenesis, part of the complexity arises from the fact that at least three different processes are compressed into the single event labelled megasporogenesis. These are: (i) alternation from diploid to haploid phases of the life cycle through meiosis; (ii) spore production and following megasporogenesis; (iii) production of the female gametophyte and gamete.

The early studies of megasporogenesis and embryo sac formation concentrated on identifying the main cytological events in these processes, which led to a number of different descriptive summaries similar to those depicted in Fig. 6-22. As is indicated, the main differences during megasporogenesis result from the formation or non-formation of walls between the products of meiosis. The most common pattern (monosporic, *Polygonum*-type (Fig. 6-22) accounts for 75% of the 292 species surveyed by Davis (1966). In most cases, the spindles of the divisions are in the long axis of the megasporocyte so that a linear tetrad is formed. In some, a T-shaped tetrad is produced. In most cases, the chalazal megaspore become functional and the remaining three spores degenerate.

While ovules are not favorable material for the study of the cytogenetic features of meiosis, due to the single meiocyte per ovule and relative inaccessibility, considerable interest exists in the cytology of megasporogenesis and embryo sac maturation. The typically large size of embryo sacs has been both a benefit and a problem. Often 100–400μm in length and 30–60μm wide, they are not easily included in a single wax section. Details are therefore compiled from composites of a number of sections. Herr's whole-mount and Nomarski-optics procedures (Herr 1971, 1973) are of considerable value in this regard. Ultrastructural views,

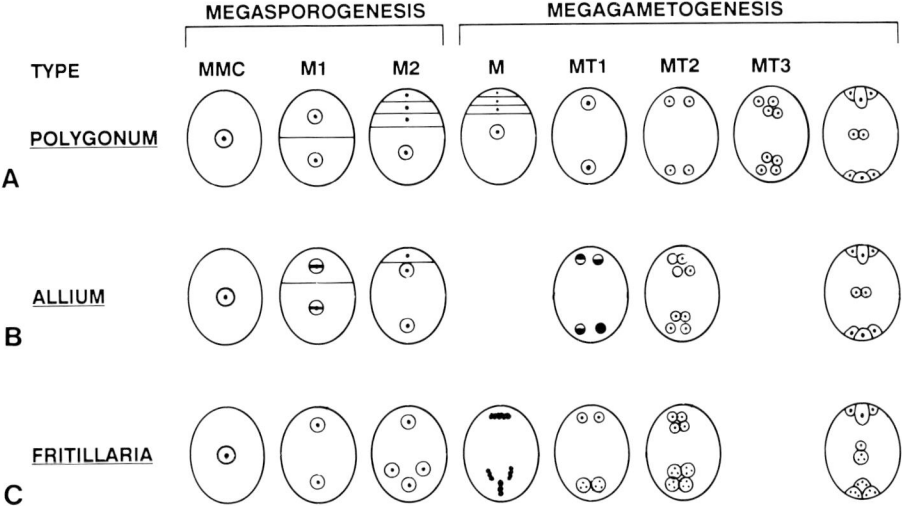

Fig. 6-22A–C. Generalizations of three patterns of embryo sac formation in angiosperms. (A) The "normal," *Polygonum* type (Monosporic). (B) The *Allium* type (Bisporic). (C) The *Fritillaria* type (Tetrasporic). MMC = megaspore mother cell; M1 = meiosis 1; M2 = meiosis 2; M = megaspore; MT1 = mitosis 1; MT2 = mitosis 2; MT3 = mitosis 3. Abstracted and redrawn from Willemse and van Went (1984).

while requiring care, time and patience have been particularly revealing. Recent studies include those of *Brassica* (Sumner and van Caeseele 1990), *Glycine* (Folsum and Cass 1989, 1990) and *Spinacia* (Wilms 1981*a*). Considerable new information can be anticipated from future studies which combine ultrastructural analysis with other technologies.

Female Gametophyte

Whether derived from a monosporic, bisporic or tetrasporic pattern of megasporogenesis (Fig. 6-22), embryo sacs are similar and are composed of three regions: the **egg apparatus,** the **central cell** and the **antipodals**. Because the variations are many, only the most common is discussed here and then only superficially. Other specific examples have been examined in detail (You and Jensen 1984; Wilms 1981*a*).

Prior to fertilization, the mature embryo sac of the *Polygonum*-type consists of two synergid cells, the egg cell, a central cell and antipodals (Fig. 6-23). Normally, three **antipodals** persist and perhaps function in secretion and as a nutrient path to the embryo sac. Some degenerate early in development, while others *(Zea)* proliferate and appear as a multicellular mass, though walls may remain incomplete between many of the cells (Diboll and Larson 1966). The antipodals are highly variable. The **central cell** contains the two polar nuclei, which contribute to the primary endosperm nucleus. The cell is frequently highly vacuolate, with a

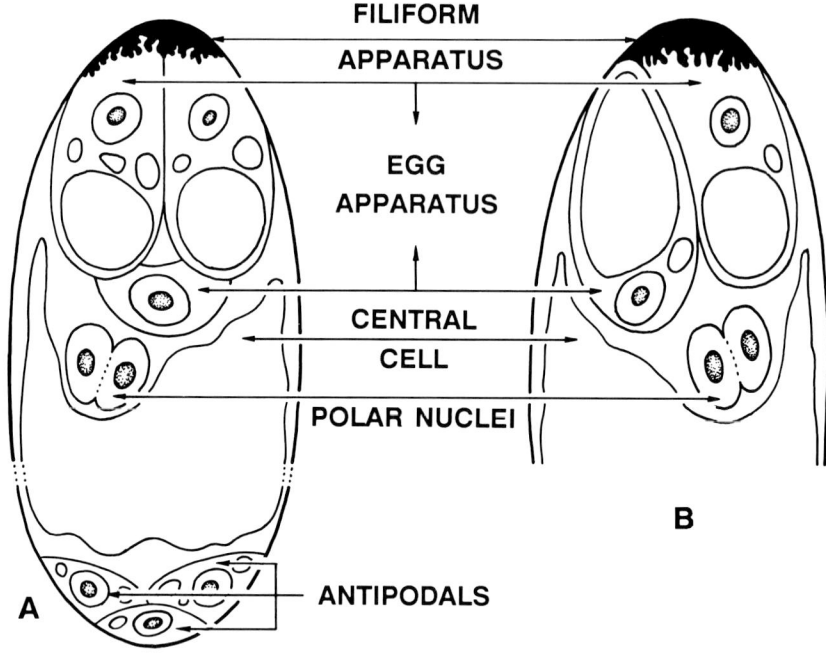

Fig. 6-23. Diagram of a mature embryo sac, plan view (A) and lateral view (B). From Willemse and van Went (1984).

narrow peripheral cytoplasm; but as evidenced by the presence of organelles, it is metabolically active (Willemse and van Went 1984).

The wall of the central cell is variable. Adjacent to the nucellus or integuments, the wall is thick and may have transfer cell wall projections in different regions. Normally no wall is present in the region adjacent to the egg cell. The **synergids**—normally two at maturity, but one frequently degenerates—represent the point of entry of the pollen tube. They are elongate, frequently with extensive wall elaborations, the **filiform apparatus** at the micropylar end, and are often extensively vacuolate at the chalazal end. The **egg** is located near the chalazal end of the synergids, adjacent to the central cell. An added point of emphasis has been provided by identifying the **female germ unit** (Dumas et al. 1984). This encompasses the egg apparatus (the egg plus the synergids) and the central cell (Fig. 6-23).

In reviewing the information available about the developing ovule and the differentiation of the megasporocyte and the embryo sac, two conflicting interpretations are possible. One can be impressed by the mass of descriptive, non-quantitative data obtained primarily from light microscopic images of wax-embedded sections. The record of these treatments and various synthetic summaries are well known. On the other hand, the number of species that have been studied at the level of TEM and cyto-histochemistry is small (<50), and the number of studies that have attempted objective, quantitative, statistically valid morphometric techniques as performed on other tissues (see Weibel 1973; Havelange and Bernier 1974; Mauseth 1980) of the different events along a time-scale is extremely low. One study, which contains some of the elements that begin to reveal the complexity of the activities, is recorded in a series of papers on ovule development in spinach by Wilms (1980*a*, 1980*b*, 1981*b*). These topics will continue to hold the attention of those who have access to these special technologies.

Double Fertilization

While not apparently unique to angiosperms, this phenomenon represents a truly basic feature of angiosperm success, though its significance remains obscure (Friedman 1990). Perhaps the importance of double fertilization lies in the separation of the fertilization of the egg from the fertilization of polar nuclei, which makes possible independent selection for "embryo success traits" as distinct from "early embryo nutrition and environment" as represented by the endosperm.

While earlier studies, both LM and TEM, were essentially descriptive, current research is more sharply focused on solving specific cytological, biochemical and molecular questions (Dumas et al. 1984; Russell 1991) about the fertilization process. They include: (i) What is the significance of the observation that the two sperms of the male germ unit can differ cytoplasmically **(cytoplasmic heterospermy)** and, in *Zea* (Carlson 1986), in their nuclei **(nuclear heterospermy)?** (ii) What is the contribution of sperm cytoplasm to the resultant zygote? (iii) Can isolated sperm nuclei be used in different *in vitro* protocols, particularly in conjunction with the production of transgenic hybrids?

It is possible that answers to these and other questions about the fertilization process may be some of the most important, and ultimately practical, items of information about the angiosperm flower that will be uncovered in the near future.

Morphological Variation of Gynoecium

The Imperfect Flower

Although the hermaphroditic flower, possessing both androecium and gynoecium, is generally accepted as the usual, normal and ancestral condition, imperfect flowers occur naturally and are not uncommon. Indeed, an early view, still held by some, is that the imperfect flower is a primitive condition and the perfect flower is derived. In addition, imperfect flowers are known to arise in some species under abnormal genetic or environmental conditions.

As pointed out earlier (Chapter 2), it is common and perhaps irrevocably entrenched to use sexual terminology (male and female; unisexual and bisexual) when discussing imperfect flowers. Strictly speaking, however, sexual terminology is inappropriate in discussions of the flower and its organs because the mega-

Table 6-1 Percentage of Genera and Species Falling into Five Major Categories in Regard to Sporophyte Sexuality, Based upon Data Published by Yampolsky and Yampolsky (1922)

Sex Type	Monoctyledons		Dicotyledons	
	Genera	Species	Genera	Species
Monomorphic				
Hermaphrodite	74	73	73	71
Monoecius	14	10	7	4
Andromonoecious, gynomonoecious and polygamous	6	7	8	7
Polymorphic				
Dioecious	4	3	6	4
Androdioecious, gynodioecious and trioecious	2	7	1	14

Note: Original sample included 25,145 species (1898 genera) of Monocotyledons and 96,347 species (8,215 genera) of Diecotyledons.
Source: From Heslop-Harrison (1972) with permission.

sporangia and microsporangia produce spores. It is the gametophytes, the embryo sac and the pollen, that produce the cells of sexual fusion, the female and male gametes. This interpretation is discussed more fully by Heslop-Harrison (1958).

Thus, while sexual terminology should be restricted, and **staminate** and **gynoecious** used when discussing imperfect flowers, the tradition is so entrenched that we have little alternative but to continue the practice at certain points in the discussion. A classification of the different possibilities for flower form within the angiosperms is presented in Table 2-1.

This wealth of variability, some explicitly genetically based, some developmentally expressed and some, as well, environmentally sensitive, represents a large and impressive research resource for plant developmental biology. Here, we can only illustrate a small, selected portion of that variation. For more complete treatments, refer to Meyer (1966), Heslop-Harrison (1957, 1972) and Chailakhyan and Khrianin (1987).

Imperfect flowers are found in approximately 25% of the extant flora, but they present a broad range of expression from dioecy to the developmentally late expression of male or female sterility (Heslop-Harrison 1972) (Table 6-1). This continuum of expression can result from the varying interplay of genetic, physiological and environmental inputs.

Cellular Basis of Imperfect Flowers

Unisexual flowers arise in two ways. In some species, both androecium and gynoecium initiate and develop to varying degrees prior to the selective abortion of either stamens or gynoecium. De Mason, Stolte and Tisserat (1982) cite more than ten reports which illustrate this mode of flower development, including *Zea* (Chapter 8). In other cases, only androecia or gynoecia initiate on individual flowers. Sattler (1973a) identifies seven species of this type. For developmental studies, the best example of this form of flower formation is *Cannabis sativa* (Payer 1857;

Heslop-Harrison 1959). Sherry, Eckard and Lord (1993) discuss these alternatives in the development of unisexual flowers in the context of their study of *Spinacia*.

Genetically Based Differentiation of Imperfect Flowers

From a developmental standpoint, it is best to assume that differentiation of imperfect flowers is based in differential gene action, though the cue to its expression can reside at other levels. For example, in *Silene dioica*, dioecy correlates with the segregation of heteromorphic karyotypes (Westergaard 1985), though the genetic basis of expression is complex (van Nigtevecht 1966a, 1966b) [see also *Melandrium alba* (Veuskens et al. 1991)]. Modifications of expected expressions have not been induced through environmental alterations, though some chemical modifications are possible (Lyndon 1985). In other dioecious species, the segregation of male and female plants can be correlated with gene segregation; and in still others, the differences while presumably due to differential gene expression are cued to physiological, environmental, or disease correlates.

Another dioecious species in which male/female segregation is gene-based is spinach (*Spinacia oleracea* L. (2N = 12)). In appropriate seed selections, the development of the flowers on individual plants is in accordance with the segregation of three alleles: Y, Xm and X (Chailakhyan and Khrianin 1987). Gynoecious plants are XX; XmX or XmXm produce intersexes; and XY or XmY plants are staminate. An alternative interpretation invokes sex chromosomes (Metzger and Zeevaart 1985). Suitable experimental cultivars segregate approximately evenly. For example, the segregation ratio in control material used by Chailakhyan and Khrianin (1978) was 48% male and 52% female.

Spinach is a qualitative, rosette-forming LDP that is widely studied in photoperiodicity and flower physiology studies (Metzger and Zeevaart 1985). It has not been widely used in explorations of the developmental biology of flower form, though a number of observations suggest that it could be an excellent candidate. This conclusion is supported by the descriptive and allometric observations of the unisexual flowers of *Spinacia* by Sherry, Eckard and Lord (1993).

For example, the actual proportion of female plants was increased by LD treatments. Further, cytokinin (BA), IAA and ABA applied through root immersion increased the proportion of females plants to 70–86% in treated plants. GA_3 also applied through root immersion produced 79% male plants. (Chailakhyan and Khrianin 1978). Root immersion is an indirect method of PGS application and might itself affect the results.

There are, however, other reasons to assume that these studies may be ambiguous regarding the determination of flower type. The involvement of endogenous GAs in flower differentiation, for example, remains unclear. While LD increased the number of gynoecious plants, the concomitant increases in the form of the GA, generally assumed to be developmentally active, were lowered (Metzger and Zeevaart 1980). One complexity with these experiments, which presumably could be overcome through breeding selections, is that selections are made for differential expression within a heterogenous population of different sensitivities. A

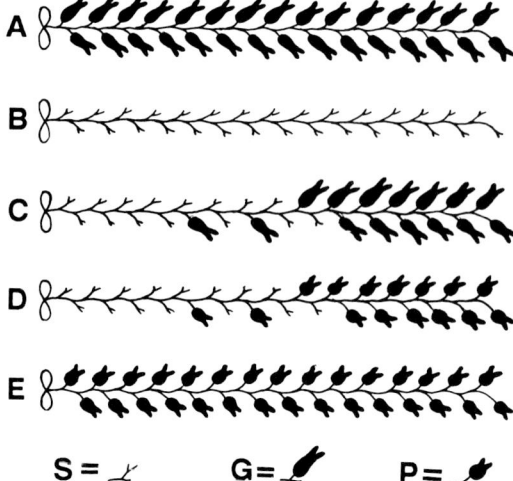

Fig. 6-24A–E. Diagram of different examples of plants available in *Cucumis* from different seed and cultural practices. (A) Gynoecious plant. (B) Androecious plant. (C) Monoecious plant. (D) Andromonoecious plant. (E) Hermaphrodite. Flowers: S = staminate; G = gynoecious; P = perfect. Based upon Shifriss (1961).

judicious selection of a number of specialized lines might improve the quality of the experimental design.

The flowers of spinach are arranged in small non-showy inflorescences. Gynoecious flowers lack petals and stamens and possess a single ovary with four or five styles. Staminate flowers possess four or five stamens in a four- or five-lobed calyx and are borne in clusters on a spike (Metzger and Zeevaart 1985).

In some monoecious species, a fundamental genetic component influencing flower expression has also been identified. For example, the cucumber of horticulture (*Cucumis sativus*) is a predominantly monoecious species, but it can range from lines which produce only staminate flowers to those with only gynoecious flowers. Hermaphrodite lines are also available (Whitaker and Davis 1962) (Fig. 6-24).

In *C. sativus* (2N = 14; Whitaker and Davis (1962)), the kinds and proportions of flowers produced on a plant are reported to be influenced by the action of four independent loci: **A** (influences gynoecious plants); **m; acr** (several alleles which affect the expression of A; and **Tr,** which can influence the appearance of trimonoecious plants (Rudich 1985; Shifriss 1961; Scott and Baker 1975; Kubicki 1969). Interpretations of the genetics involved vary both with the experimenter and the specific lines used.

For experimental purposes, the development of androecious and gynoecious lines is particularly provocative (Scott and Baker 1975). The gynoecious lines were maintained by inducing staminate flowers with applications of GA_{4+7}, and androecious lines were maintained by inducing gynoecious flowers with Ethephon.

The Gynoecium 161

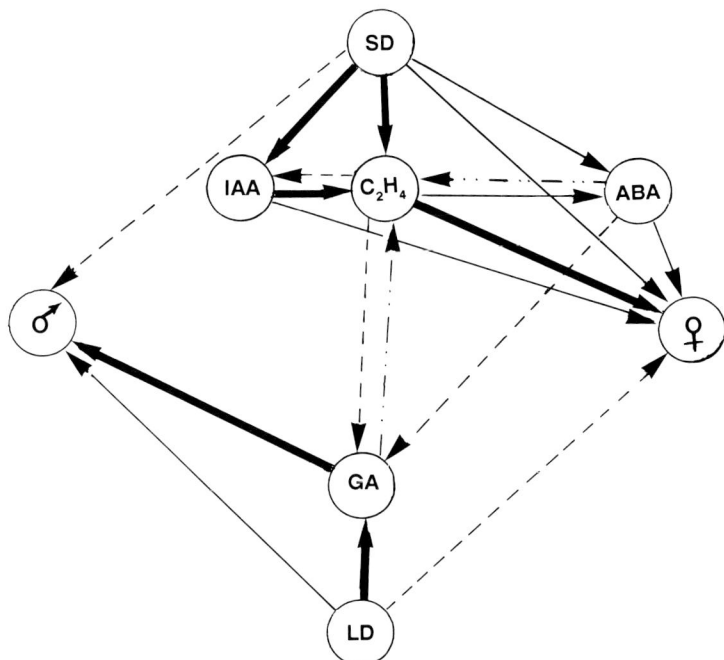

Fig. 6-25. Schematic representation of the interaction of photoperiod conditions and assumed hormonal balance in the regulation of differentiation of staminate and gynoecious flowers in *Cucumis,* based upon a variety of experiments. Key: promotion (——); inhibition (----); strong promotion (■—); no effect (—··—). Reprinted with permission from Rudich, J. 1985. *Cucumis sativus.* In A. H. Halevy [ed.], *CRC Handbook of Flowering,* pp. 365–374. Copyright © CRC Press Inc. Boca Raton, FL.

Thus, four lines (A/A; acrf/acrf) which produced only (or predominantly) gynoecious flowers were produced. Another four lines (a/a; acr$^+$/acr$^+$) produced only staminate flowers.

In terms of flower physiology, *Cucumis* is generally considered to be insensitive to photoperiod, but daylength does affect the degree or proportion of flower types that develop on a stem. More simply, sexuality is affected by daylength. In experimental studies, therefore, the maintenance of appropriate light conditions is critical.

From a developmental perspective, the documentation of the involvement of PGS in the differentiation of staminate and gynoecious flowers of cucumber is most impressive (Fig. 6-25) and is based upon data from experiments involving exogenous application of PGS and their inhibitors, *in vitro* culture, extraction and assay of endogenous levels of PGS and inhibitors and different daylength conditions.

Of all the interactions indicated, those between GAs and the differentiation of staminate flowers are the most convincing support for PGS involvement in regulation. Staminate flower formation is enhanced by applications of GA_3 and GA_{4+7} and inhibited by a variety of growth retardants which, in general, alter GA biochemistry (Pharis and King 1985; Rudich 1985). Methods of application included sprays, rooted cuttings and *in vitro* culture. The analysis of endogenous content in general is consistent with these observations. As indicated in Fig. 6-25, GAs are not the only PGS that can be involved in flower differentiation in *Cucumis*. This complexity, based most frequently on monofactorial experiments and observations, makes any conclusions about the placement of GAs as sole controlling elements in the differentiation of male flowers very difficult.

Begonia is a large and variable genus, consisting of over 1000 tropical species, that possesses many features which make it another desirable research model system for developmental studies on the control of flower sexuality. Individual species are mostly herbaceous, succulent, perennials which have been widely cultivated and hybridized. Despite this variety and the problems it raises for taxonomists, and because of its ease of vegetative propagation coupled with its range of flower traits, *Begonia* deserves more attention from developmental botanists.

All begonias are monoecious, with staminate and gynoecious flowers frequently formed on the same inflorescence (Heide and Runger 1985). Thus, the differential expression for staminate versus gynoecial flowers is expressed developmentally within the same inflorescence and in some cases between adjacent flowers. In some tuberous cultivars, the staminate flowers are large and double; the gynoecious flowers are smaller and less showy. In other cultivars and species, variable patterns of what are termed cymose (Chapter 7) inflorescences are found (Matzke 1938; Kinet, Sachs and Bernier 1985). Staminate and gynoecious flowers from *B. tuberhybrida* are illustrated in Fig. 6-26.

For experimental studies on flower form, it will be advantageous to select and maintain cultivars of *Begonia* which retain fertility, at least under some determined condition and from which both genetically based and environmentally modified variation can be identified. (Note: The common horticultural distinction between fibrous-rooted and tuberous forms of *Begonia* is both misplaced and inaccurate. Tuber formation can be induced in some forms which are originally derived from non-tuberous forms, and the distinction makes an incorrect botanical comparison: tubers are secondarily thickened stems and not roots (Heide and Runger 1985). In *B. tuberhybrida* (Lecocq and Dumas 1975), double flowers produce organs corresponding to petals and stamens which are capped with stigmatoid lobes which resemble normal stigmas in morphology, histology and histochemistry. This condition represents an obvious candidate for inclusion under what is labelled homeosis (Sattler 1988*b*) (see Chapter 9).

The gynoecious flowers of *Begonia* possess a number of additional features that are of interest in their own right. The inferior ovary has been the subject of argument and reinvestigation (Gauthier 1950; Lecocq 1977). Casual personal observations indicate that some horticultural forms of *B. tuberhybrida* can produce intermediate ovary states between inferior and superior. Further, some ambiguity in how best to interpret its formation and placentation has been uncovered: axile, parietal and mixed forms have been described (Charpentier, Brouillet and Barabe 1989). Opportunities for more detailed observations are obvious.

Fig. 6-26A–B. Staminate and gynoecious flowers from a tuberous hybrid specimen of *Begonia.* (A) The terminal "double" male flower. This example had no anthers. (B) An axillary "single" gynoecious flower. The three-winged gynoecium is not visible in this view. (X2/3).

Experimental studies on flowering of *Begonia* indicate that many of the species and cultivars are SDP, a trait that can be affected by temperature and PGS applications. For the purposes of this discussion, however, we are primarily interested in the factors that regulate flower form. Unfortunately, at present their identification appears to be elusive. Low temperature (15°C) and SD (8h) favored staminate flowers (557/1 staminate/gynoecious (S/G)) in experiments on *Begonia* x *cheimantha,* whereas under warmer temperatures (18°C) and LD (24h) smaller inflorescences with fewer flowers were formed but the S/G ratio was 3/1. Interestingly, IAA, GA_3 and the cytokinin benzyl adenine stimulate the formation of gynoecious flowers in this plant (Heide 1969a, 1969b; Heide and Skoog 1967). *In vitro* culture of immature three-flowered inflorescences of *B. franconis* demonstrated growth requirements for nitrate and ammonium and for a cytokinin. The optimum cytokinin level to support the development of female flowers was 30X that required for the staminate flowers (Berghoef and Bruinsma 1979a, 1979b, 1979c).

These observations all place the determinative events governing staminate versus gynoecial flower development as taking place early in the development of the inflorescence and within a small clone of cells that ultimately form each of the flowers. Uncovering the nature of this developmental decision will be difficult, but there appear to be many opportunities for selecting desirable cultivars of *Begonia* for in-depth developmental studies.

PGS Involvement in Development of Imperfect Flowers

It is now well accepted that alterations in the expression of flower development are often accompanied by alterations in the PGS status of the flower meristem and associated tissues. Indeed, a number of research projects have been based on the assumption that PGS status is pivotal for flower differentiation (Heslop-Harrison 1964; Dauphin-Guerin, Teller and Durand 1980; Rood, Pharis and Major 1980; Chailakhyan and Khrianin 1987). Differences in the PGS status of staminate versus gynoecious flowers could take the form of: (a) differences in the rates of synthesis of particular PGS forms; or (b) differences in the rates of metabolism and destruction of particular PGS forms. Either of these might lead to detectable differences in the levels of particular extractable, assayable components. As well, exogenous application of appropriate PGS and their inhibitors would be expected to bring about predicted changes in expression.

Alternatively, differences in PGS status could arise from a change in the manner in which PGS interact within the cell with secondary messengers, PGS binding and acceptor sites (Venis 1985; Libbenga and Mennes 1987; Napier and Venis 1991) and other cellular components of staminate versus gynoecial flowers. In this case, measures of levels of PGS, regardless of the sensitivity or accuracy of the method, would be unlikely to correlate closely with flower differentiation. A further level of complexity arises from the fact that nearly all the identified categories of PGS (auxins, GAs, ABA, ethylene, cytokinins, polyamines) have been implicated to some degree and in some manner in flower differentiation (Metzger 1987). Further, there are numerous examples of developmentally opposite response to particular PGS forms. A further complication lies in the difficulty of knowing the mode of action of PGS or the active form of a particular PGS class.

Considering these complications, it is perhaps surprising that so much information about PGS and flower differentiation has been accumulated, some of which retains its credibility. Notwithstanding, while many of these putative relationships may be valid, tight causal linkages lack the potential for confirmation. Reviews of the large data set and speculation around this topic are extensive (Durand and Durand 1984; Pharis and King 1985; Kinet, Sachs and Bernier 1985; Chailakhyan and Khrianin 1987). No attempt is made to review it here; rather, some individual cases are highlighted.

As summarized above, the involvement of GAs with the determination of maleness in cucumber is supported from a number of different lines of research: endogenous levels (Rudich, Halevy and Kedar 1972; Hemphill, Baker and Sell 1972), exogenous application of available GAs and application of inhibitors (Rudich 1985). GAs are also implicated to varying degrees in the flower differentiation in other species. For example, staminate flowers are promoted in *Bryophyllum*, *Cannabis*, *Humulus* and *Lycopersicon* by applications of GA_3, but pistillate flowers are promoted by the same means in *Ricinus*, *Zea* and *Hyoscymus* (Pharis and King 1985).

A similarly impressive documentation of the involvement of cytokinins with flower differentiation has been documented in *Mercurialis annua* ($2N = 16$). This dioecious strain, from a large, taxonomically complex European weed, has yielded considerable information about the involvement of cytokinins in flower differentiation (Durand and Durand 1985). Cytokinins feminize the genetic males of this line, and the measurement of endogenous cytokinin levels is consistent with this interpretation (Durand and Durand 1984). Other correlative changes in other PGS and differences at the level of enzyme and nucleic acid biochemistry have also been documented (Durand and Durand 1990).

In earlier discussions it seemed reasonable that direct cause-and-effect relationships might ultimately be established between flower differentiation and PGS status. In view of the complexities that have been uncovered, some of which are mentioned above, and in spite of the impressive body of data assembled, it appears that a more accurate assessment should include the following points:

1. While there are some general trends (e.g., GAs often influence staminate expression, and auxins often influence gynoecious development), there are notable exceptions, contradictions and inconclusive correlations.
2. In most examples, flower differentiation simultaneously accompanies variations in a number of PGS. The PGS involvement is probably, therefore, a multi-factor rather than single-factor interaction. There is no single combination that correlates with a particular pattern of differentiation.
3. It is technically exceedingly difficult to devise an experiment which unequivocally demonstrates the primacy of PGS in the differentiation of staminate and gynoecious flowers. Thus, most experiments and analyses can do no more than demonstrate correlations or non-correlations and as such represent the beginnings of more sophisticated physiological analysis, rather than final proofs. One experimental situation which approaches the desired level of control is available in the *in-vitro* culture of *Zea* ears (Bommineni and Greyson 1990), in which cytokinins selectively stimulate the development of staminate flowers and GA_3 stimulates gynoecia (see Fig. 8-6).
4. Determination to flower type may occur early and be localized to specific tissues. Detection of PGS changes may only be available from crude extractions which include large amounts of tissue extraneous to the process and which may be taken from material after determination has occurred.

In view of these realities, future studies of the factors underlying differences between staminate and gynoecious flowers should be thoroughly based in molecular genetics, using the traditional technologies of PGS analysis to confirm and extend insights gained, rather than attempting to prove causal relationships (see additional discussion in Chapter 9). Future research strategies will probably include the insertion of selected genes known to control specific steps in PGS metabolism, accompanied by appropriate tissue-specific promoter sequences and appropriate markers (King 1988; Klee and Estelle 1991), into model plant systems that possess well-defined genetically based modification of flower development. Attempts to connect genes that are known to confer staminate or gynoecious expression with the genetic control of PGS metabolism should also be attempted. An essential precondition for this type of study is that genetic stocks of suitable model material be prepared and maintained.

Environment-based Variability of Perfect/Imperfect Flowers

As noted above in discussing *Cucumis, Spinacia, Begonia* and *Mercurialis*, daylength can elicit different flower expressions. This observation underlines a more general reality that flower differentiation occurs in an environmental context, and in some cases the direction of development is strongly sensitive to the particular set of environmental parameters at that time. These could include: (a) Light (daylength, intensity, wavelength distribution); (b) Temperature (maximum, minimum, regime); (c) Humidity; (d) Soil conditions. In view of the potential of any of these to affect expression, it is imperative, in experiments, to control these parameters and to record any variations.

Genetically Based Variability of Gynoecium

In addition to the presence/absence feature of the male/female flowers discussed above, genetically based variability affecting developmental aspects of gynoecia is also recorded. While the number of gynoecium genes, as demonstrated by *in situ* hybridization studies, is low (Edwards and Coruzzi 1990), there is considerable documentation of genetically based gynoecium variation from breeding experiments. Some examples are discussed below as meristic and heterostylous variation.

Carpellody, which is frequently genetically based, places ovules and stigmatic and gynoecial tissue expressions in unusual locations throughout the flower. Recent reports include: *Lycopersicon* and *Brassica* (Sawhney et al. 1989), *Nicotiana* (Evans and Malmberg 1989) and *Arabidopsis* (Bowman, Smyth and Meyerowitz 1991). Meyer (1966) is a rich source of other examples and references.

Unlike stamen maturation, where limited, tissue-specific mutants called male-steriles are common (Chapter 5), genetically based female steriles, presumably as a function of the relative lack of study, are less commonly described. Male sterility is frequently expressed independently of female sterility, though in maize, for example, some meiotic mutants display female as well as male sterility (Carlson 1988).

Meristic Variation

As discussed in Chapters 4 and 5, organ number per flower can vary significantly from that expressed in floral formulae. Variation in the number of parts is also a feature of gynoecium development and can be seen in the number of individual carpels, the number of locules and septa of the ovary or in the number of styles or stigmas. The causal factors can, in some cases, be traced to genes, or the phenomenon can correlate with plant vigor or flower position on the inflorescence. Frequently, reduced numbers of gynoecial components correlate with smaller-sized apical meristems.

The tomato ovary, during its development, is open to a number of cues which affect the size and the number of its components. The gene Lc in tomato affects the number of locules in the ovary (Young and MacArthur 1947). Alternatively, the gene "fasciated" (Table 5-2) increases the normal locule number. Tomato

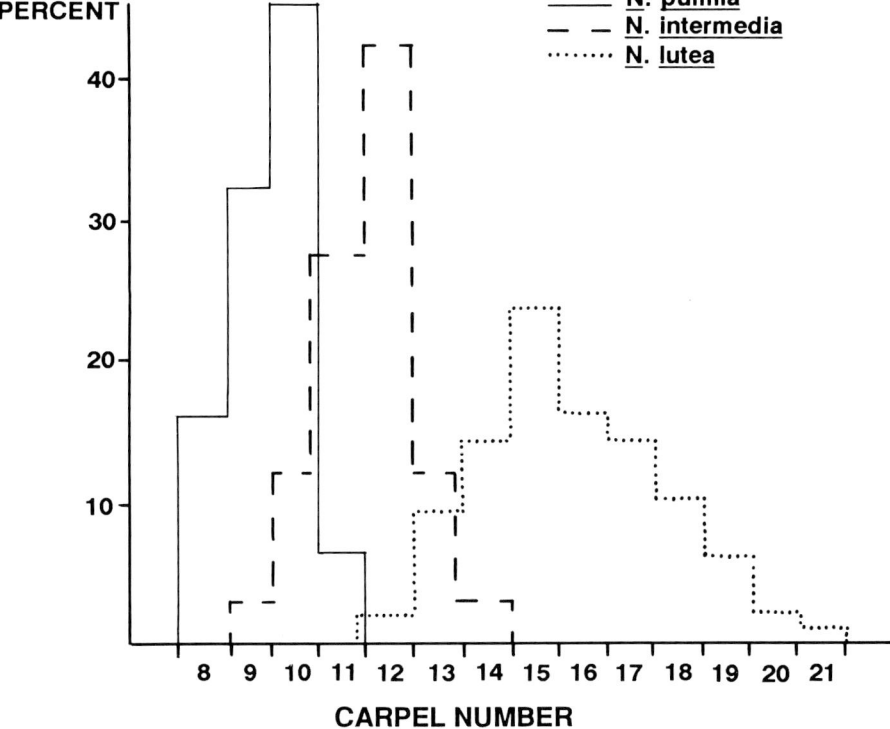

Fig. 6-27. A study of variation in carpel number. Frequency histograms of carpel number in three British taxa of *Nuphar*. From Heslop-Harrison (1953) with permission of the author and the Botanical Society of the British Isles.

locule number has also been shown to vary in response to temperature changes (Sawhney and Polowick 1985) and applications of GA_3 (Sawhney and Dabbs 1978). The general correlation between locule number and fruit size, regardless of cause, is well documented (Houghtaling 1935; Sawhney and Dabbs 1978), though a comprehensive understanding of the sequence of causal events is unavailable.

In a study of the flowers of three English species of *Nuphar* (Heslop-Harrison 1953), carpel number was shown to vary from 8 to 21, with overlapping frequency distributions for the three species (Fig. 6-27). While no attempt was made to record the flower position nor to provide experimental conditions for the plant growth, this study indicates the types of variation that are available for study. Meristic variation in gynoecial characters obviously affects the number of seeds produced and therefore is an appropriate starting position for various studies of "reproductive strategies" (see Lovett Doust and Lovett Doust 1988).

Heterostyly

This genetically based floral polymorphism, usually linked to a sporophytic self-incompatibility system (Chapter 5), has been identified in a small percentage of the extant flowering plants (Ganders 1979). Examples of systems involving two

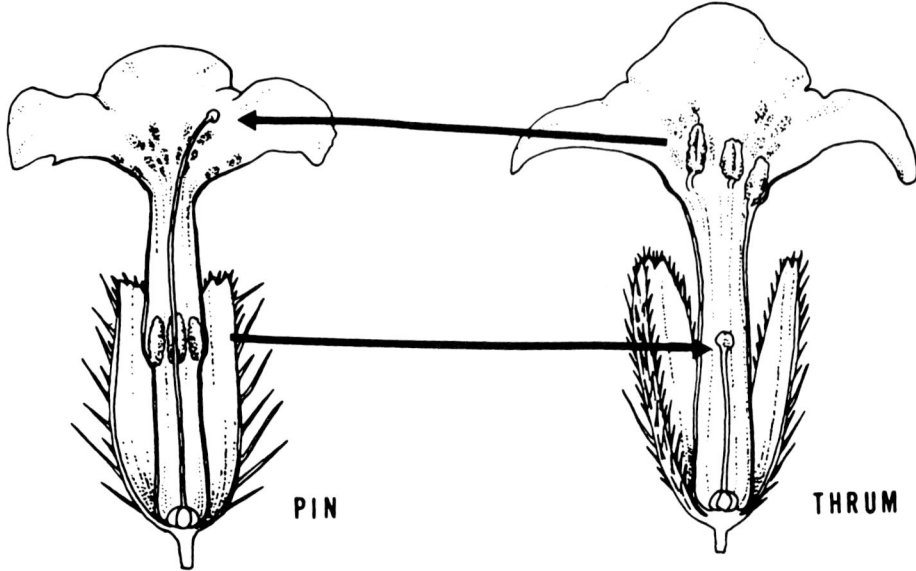

Fig. 6-28. Diagrams of the flowers of a representative distylous species. The reciprocal locations of stamens and stigmas in the two flowers (pin and thrum) are indicated. The arrows indicate the directions of compatible pollinations. From Ganders (1979) with permission of *New Zealand Journal of Botany*.

morphological forms (morphs), **distyly,** and three morphs, **tristyly,** are described (Shivanna and Johri 1985). Botanical interest in heterostyly has been keen, despite its low incidence, and an extensive literature is available.

The most prominent morphological expression of a heterostylous species is the differences in the relative position of the anthers and the stigmas in the different flower forms. In typical distylous examples, these are described as the **pin** and **thrum** forms (Fig. 6-28). Other differences between the flower forms can include corolla shape and size, pubescence, anther size, pollen size and shape and stigma traits. In *Primula,* a much-studied genus since Darwin (1862), documented anatomical features from the style and stigma of the two flower morphs (Table 6–2) provide a useful example (Heslop-Harrison, Heslop-Harrison and Shivanna 1981).

The genetics of heterostyly has been elucidated in only a few species (Ganders 1979). In *Primula,* distyly is regulated by a single locus with two alleles. The allele determining the thrum form is dominant. Pin plants are therefore homozygous recessive, and the thrum plants heterozygous. Homozygous thrums do not normally occur, because of incompatibility.

It seems obvious that many further opportunities exist for careful developmental studies in these heterostylous species, though a need seems to exist for the preparation of lines of seed and plants which can be maintained for dedicated research. Because of the considerable diversity of mechanisms between the different heterostylous species, many different opportunities probably exist for uncovering different developmental features. In particular, the molecular-genetic connections between heterostyly and self-incompatibility warrant exploration.

Table 6-2 Observations on the Stigma and Style of the Pin and Thrum Morphs of *Primula vulgaris* (Sample from a Natural Population in West Wales)

Character	Morph		Ratio (Pin/Thrum)
	Pin	Thrum	
Length of style + stigma (mm)	15.03 ± 0.34	7.29 ± 0.12	2.06[a]
Epidermal cell length (μm)	157.60 ± 3.99	80.78 ± 1.71	1.95[a]
Stigma papilla length (μm)	110.06 ± 0.95	44.26 ± 0.63	2.49[b]

[a] These two ratios are not significantly different.
[b] The mean papilla lengths from the stigmas of pin and thrum morphs are significantly different ($P < 0.05$).

INTERPRETATION: The ratios of the lengths of the pin and thrum styles and the stylar epidermal cells are closely similar, suggesting that the greater length of the pin style is mainly attributable to greater cell lengths. Presumeably, differential cell division levels are not responsible.

Source: From Heslop-Harrison, Heslop-Harrison and Shivanna (1981) with permission.

Apomixis

Deviations from the normal life cycle (Fig. 2-7) whereby seeds are produced asexually should be as provocative to developmentalists as they have been to taxonomists and population biologists. Unravelling the cellular and molecular mechanisms that underlie the different deviations is difficult, however, and data remains primarily at the level of cytological description.

Some definitions are in order! With angiosperms, "apomixis" normally refers to asexual reproduction by seeds and excludes those vegetative propagules such as runners, stolons and buds. Two general modes and some special cases are recognized and described below. There is an extensive literature, primarily in taxonomic discussions:

1. Adventitious embryony—The new sporophytic embryo arises directly from a somatic cell of the ovule, usually from the nucellus. Frequently the process produces a number of embryos (i.e., polyembryony) (Lakshmanan and Ambegaokar 1984).
2. Gametophytic apomixis—The embryo sac arises from an unreduced embryo sac initial.
    A. Diplospory—The unreduced embryo sac originates from the female archesporial cell (megasporocyte).
    B. Aposprory—Embryo sacs originate from somatic cells of the ovule (usually the nucellus).

Complicating factors in the products of these processes include the possibility of fertilization and the source of the endosperm. Since, from all indications, the phenomena have arisen independently in many different families, the individual developmental mechanisms are undoubtedly varied. Further information is available in more complete treatments (Nogler 1984; Lakshmanan and Ambegaokar 1984).

Notwithstanding the difficulty in identifying cellular and molecular mechanisms involved in apomixis, the potential of its eventual exploitation in plant breeding should justify exploration of this phenomenon more fully. There appears to be enough variability, especially in some of the grasses (Knox 1967; Bashaw 1980) that apomixis might eventually be stabilized in desired lines or hybrids (Kim and Rutger 1988) and thereby maintained. Failing this possibility, the induction of apomixis through chemicals, irradiation and, more likely, transgenesis, appears to warrant further exploration.

7

Inflorescences

There is general agreement that knowledge and understanding of the ways in which flowers are arranged into inflorescences lags well behind that of other aspects of angiosperm structure (Weberling 1965, 1989; Wyatt 1982; Gifford and Foster 1989). This ignorance extends beyond traditional morphological and systematic concerns to the current interest in inflorescence architecture as it relates to reproductive ecology and population dynamics. It arises, at least in part, from the variety, complexity and plasticity of inflorescences themselves but has been compounded by the not inconsequential confusion of terminology and conflicting interpretations about the objectives of inflorescence classification (Rickett 1944). The following presentation introduces a few themes from these discussions, as they relate to development. It is not intended to be a definitive, critical review of inflorescence structure.

Primarily because inflorescence terminology accumulated to fill taxonomic and morphological objectives, developmental insights are frequently shallow or missing. Distinctions between patterns of initiation, patterns of flower opening on the inflorescence and the mature flower display are not always maintained. With few exceptions, analyses of different cellular patterns and relative growth dynamics of time- or plastochron-based studies of different parts of inflorescences are rare, as are estimates of variability of expression of different traits within a population. Analysis of the genetic components that are involved in inflorescence development is in its infancy and involves only a few species.

In the foreseeable future, except in selected cases, developmentalists will not change this situation appreciably. Too many unanswered questions about flower development remain unexplored for them to become embroiled in the morphological complexity of inflorescences, though some beginnings of a developmental interpretation of the inflorescence are evident in the analysis of the "inflorescence" genes of *Arabidopsis* (Schultz and Haughn 1991; Shannon and Meeks-Wagner 1991; Weigel et al. 1992). Ultimately, however, developmentalists will have to join forces with reproductive ecologists, taxonomists and phylogenists in developing a comprehensive view of inflorescence structure and its variability. Since most views of angiosperm flower phylogeny already assume some degree of heritable coalescence, and since growth alterations of complex branch systems took place to form present-day flowers, coalescence should, as well, be considered in any developmental interpretation of extant inflorescences. In the process, insights may well be uncovered which shed light on both the phylogeny and the ontogeny of the flower itself. The tendency of some inflorescences to present an impression of individual flowers—for example, the superficial resemblance of the head of *Cosmos* to the individual *Anemone* flower (Good 1956)—should be enough of an impetus to stimulate speculation and collaborative investigation.

Different Interpretations of Inflorescence Structure

The Traditional Classification

A traditional classification of inflorescences, to which North American botanists most frequently refer, based upon 200+ years of observation and speculation, was summarized by Rickett (1944, 1955). Despite its general utility for purposes of identification, taxonomy and comparative morphology, this classification is generally recognized, even by Rickett, as arbitrary, unnatural and incomplete. In his summary, Rickett (1955) dispenses with the traditional distinction between **determinate** (terminal axis forms a flower) and **indeterminate** (terminal apex continues to generate bracts and axillary flowers) because he feels that these terms fail to reflect the complexity and variability that actually exist in nature. This decision leads, therefore, to a list of somewhat independent inflorescence types and definitions. Some of the more commonly recognized types are illustrated and described in Fig. 7-1 and in the boxed text on page 174 (see also Bell 1991).

Developmentalists should use these terms for casual descriptive purposes but should be alert to novel patterns and examples which deviate, sometimes seriously, from classic definitions due to inadequate observation as well as to environmentally and genetically based variation.

Troll's Typology of Inflorescences

An alternative typology of inflorescences, originally developed by Troll (1964, 1969), has been summarized and interpreted by Weberling (1965, 1983, 1989). It is based upon an examination of over 10,000 specimens and appears to possess greater scope for developmental biology than the traditional scheme. It will require considerably more analysis and wider discussion before it is readily available to developmental biologists. Its use and interpretation is more common in Europe than in North America.

In this classification, an initial dichotomy distinguishes **polytelic** and **monotelic** inflorescences (Fig. 7-2). This distinction is comparable to the traditional distinction of determinate versus indeterminate, which Rickett dropped. In polytelic inflorescences, the cluster of flowers which terminates the main axis is called the **florescence** (HF in Fig. 7-2A). The main axis does not terminate with a flower. The lateral branches similarly terminate in co-florescences (CoF in Fig. 7-2A) and repeat, to some degree, the structure of the main axis; these are therefore termed **paracladia.** Second-order paracladia can also be identified.

In monotelic inflorescences, the main and lateral axes terminate with flowers

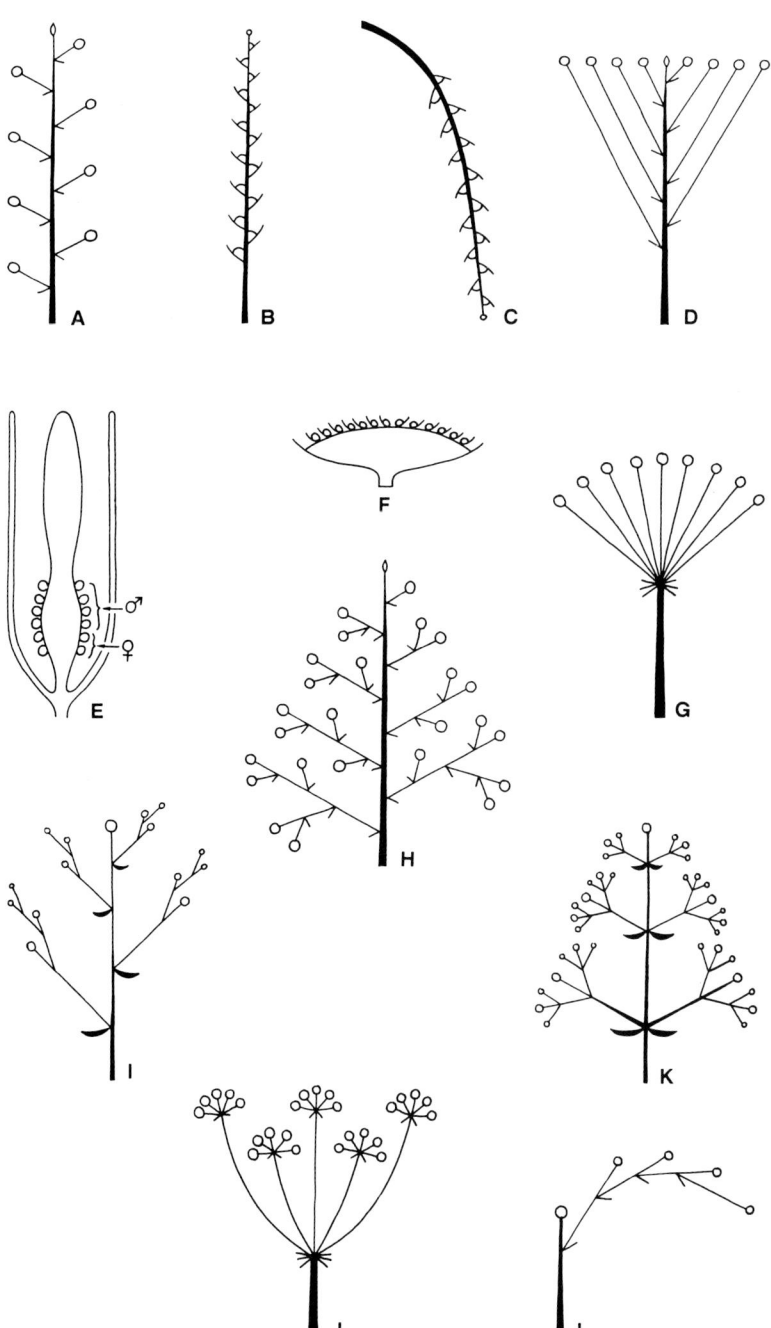

Fig. 7-1A–L. Illustrations of some of the main mature inflorescence types traditionally identified. (A) Raceme. (B) Spike. (C) Catkin. (D) Corymb. (E) Spadix. (F) Head. (G) Umbel. (H) Panicle. (I) Dichasium. (J) Compound umbel. (K) Monochasium. (L) Scorpioid cyme. Parts A–H redrawn from Hill et al. (1960) with permission of McGraw-Hill, Inc. Parts I and J redrawn from Gifford and Foster (1989) with permission.

Traditional Definitions for Different Angiosperm Inflorescences, without the Often Used Determinate versus Indeterminate Patterns

Note: Letters refer to the illustrations in Figure 7-1.

Ament (Catkin): A slender, often pendent, usually scaly spike bearing unisexual, apetalous flowers and finally deciduous as a whole. (i.e., *Salix, Populus, Betula*). (C)

Capitulum (Head): A compact inflorecence (monopodial or sympodial), with a short, often discoid axis. (i.e., Compositae, Dipsacaceae). (F)

Corymb: A raceme whose pedicels are proportionally elongated so as to bring all the flowers more or less to the same level. (D)

Cyme: A compound, more or less flat-topped sympodial inflorescence.

Dichasium: A three-flowered cluster composed of a peduncle which bears a terminal flower and, below it, two bracteoles which subtend lateral flowers. The bracteoles may abort and one or two flowers may also fail to develop. A compound dichasium arises from repetitive branching of a simple dichasium. (I).

Monochasium: A sympodial inflorescence composed of a peduncle which bears a terminal flower and, below it, a single fertile bracteole which subtends a lateral flower; the peduncle of the latter in turn gives rise to a single lateral branch; and so on. The bracteoles may abort. (K)

Bostryx: A monochasium in which successive pedicels describe a helix around the sympodial axis.

Cincinnus: A monochasium in which successive pedicels follow a zigzag path along one side of the sympodial axis. A special case, a **drepanium,** is compressed so that all flowers lie in one plane.

Rhipidium: A bostryx compressed so that all flowers lie in one plane.

Panicle: A loosely branched inflorescence whose ultimate units may be of various types. (H).

Pleiochasium: Like a dichasium, but the peduncle gives rise to three or more lateral branches.

Raceme: A monopodial inflorescence composed of a central axis along the sides of which arise pedicellate flowers usually in the axils of bracts. (A)

Spadix: A spike with a thickened, fleshy axis; usually enveloped by a spathe (i.e., Araceae). (E)

Spike: An inflorescence composed of sessile or almost sessile flowers arranged singly or in contracted clusters along a central axis. (B)

Thyrse: A compact panicle of more or less cylindirical form.

Umbel: An inflorescence (monopodial or sympodial) composed of several branches that radiate from almost the same point and which are terminated by single flowers or by secondary umbels. (G) and (J)

Source: Based upon Rickett (1944, 1955).

Fig. 7-2A–B. Outline sketches of idealized examples inflorescences as defined by Troll. (A) Schematic representation of a polytelic inflorescence. (B) Schematic representation of a monotelic inflorescence. Abbreviations: HF = main florescence; COF = coflorescence; PC' = paraclades of second and third orders; E = terminal flower. Redrawn from Troll (1964) with permission of Gustav Fischer Verlag.

(Fig. 7-2B). The flower-bearing lateral axes, produced from the terminal meristem are assumed to be equivalent to each other and are termed "paracladia." An important question for which there seems no answer at present concerns just how consistent—or alternatively, how variable—examples of these categories are in practice. In addition it will be of interest to learn whether genes can be identified which influence the expression of any of these features and what effect different environmental cues might have on inflorescence structure.

It has yet to be determined whether Troll's classification will be of value to developmentalists as they attempt to dissect and characterize the different cellular and molecular events involved in the production of inflorescences. For those who can interpret the typological discussion (Briggs and Johnson 1979), this intensive examination may well expose a number of important features having physiological and developmental significance. For example, the attention paid to the non-flowering portion of the plant through the identification of zones of inhibition and innovation

(Fig. 7-2) may prove to be important to the physiology of heteroblasty. So, too, the attention given to proliferations (see also Chapter 6), cauliflory, epiphyllous inflorescences (Stork 1956; Dickinson and Sattler 1974, Dickinson 1978), accessory buds and other phenomena may ultimately contribute to a better overall understanding of inflorescence variability (Weberling 1989).

In using either of these classifications to describe their material, developmentalists should be aware of the complexities and contradictions involved in the various presentations. The opportunities for uncovering developmental novelty in inflorescences seems high.

Computer Simulations

Recent attempts (Frijters 1978a, 1978b, Lindenmayer 1984; Janssen and Lindenmayer 1987; Prusinkiewicz and Lindenmayer 1990) to prepare computer simulations of inflorescence structure represent significant contributions to developmental interpretations. For computer analysis and simulation, inflorescences must be viewed as being the result of the initiation and growth of a number of individual elements: internodes, nodes, leaves, bracts, peduncles, pedicels, apical and axillary meristems and flowers (or in Troll's terminology, paraclades, co-florescences, florescences, etc.). In addition, data on rates of initiation of these elements and their growth, coupled with analysis of developmental patterns of maturation, are required for these simulations.

Regardless of how closely they mimic the final form of inflorescences, simulations appear to be inadequate substitutes for careful developmental exploration of the cellular, physiological and environmental components that contribute to the variability of final structure of individual inflorescences. Further, the organ-typology used in simulations may be unable to account for the variability and plasticity that is an important feature of many inflorescences. The identification of these biological counterparts of mathematical variables will be neither obvious nor easy.

Is There a Developmental Approach to the Study of Inflorescences?

Attempts to define the limits of an inflorescence, either morphologically or developmentally, are difficult and arbitrary. Morphologically, in many cases, there is a transition of leaf (bract) and stem traits between the obviously flowering regions and the vegetative portions of the plant. Frequently, no clear morphological border between vegetative plant and inflorescence can be identified (Weberling 1989).

Further, a number of different research studies indicate a physiological continuum along the length of florally induced plants. Cultured thin-tissue explants (Cousson and Tran Thanh Van 1981; Tran Thanh Van, Thi Dien and Chlya 1974; Meeks-Wagner et al. 1989), axillary buds (Singer and McDaniel 1987); internode discs (McDaniel, Sangrey and Jegla 1989) and stem segments (Wardell and Skoog 1969) from the terminal regions of florally induced plants exhibit floral traits: explants from basal regions tend to exhibit vegetative traits. An apex-to-base gradient of expression is frequently detected. The general significance of these individual observations remains unexplored, but it further complicates the question raised in Chapter 3 regarding the expression of the flowering response by different tissues and cells, in addition to the terminal meristem of the plant.

The body of literature on developmental aspects of inflorescences is larger than usually presented. Plant physiologists and developmentalists have frequently explored aspects of stem development but have not distinguished the fact that some of the structures studied were inflorescences. The following topics, normally discussed in introductory botany and plant physiology texts, frequently fail to distinguish inflorescences from vegetative shoots or inflorescence meristems from flower meristems.

Initiation and Growth of Inflorescences

The basic literature dealing with descriptions of inflorescence initiation, as distinct from flower initiation, is small and preliminary. The introductory, descriptive observations on heads of the Compositae and Dipsacaceae (Philipson 1946, 1947a, 1947b, 1948a), the cereal grass inflorescences (Bonnett 1936; Barnard 1955), raceme of *Lobelia* and other inflorescences of Campanulaceae (Philipson 1948b) have not, in general, been followed by more thorough quantitative observations of initiation and subsequent growth and maturation of pedicels and peduncles. In addition to qualitative descriptions of changes in inflorescence meristems, quantitative changes in cell number, mitosis patterns and histochemical and *in situ* hybridization studies need to be pursued. As well, comparative data between inflorescence and floral meristems are required.

An example of one type of study that will be required is that by Molder and Owens (1973) on *Cosmos bipinnatus* (Compositae), a quantitative short-day plant. This work: (i) documents an outline summary of the changes that take place in the terminal meristem and during the development of axillary heads (Fig. 7-3) (unfortunately, not time- or plastochron-based); (ii) generates quantitative measures of mitotic activity in the different layers of the terminal meristem; and (iii) documents the changes that take place in these parameters and in the dimensions of the terminal meristem at six stages following photoperiodic induction. Further, flowering of non-induced plants was achieved by applications of GA_3, and effects of this obviously important PGS class on some of the cellular parameters of meristem development were documented. Other descriptive anatomical investigations of initiation of heads of Compositae include the studies on *Chrysanthemum* (Popham and Chan 1950, 1952).

Substantive, quantitative, developmental studies are needed on the initiation and growth of other composite heads and other inflorescences. For example, developmental comparisons between a series of racemose and cymose growth patterns might reveal some major physiological and perhaps genetically based differences which have heretofore gone undetected. For studies such as this, it will be wise to use model systems for which physiological and genetic data are already available [refer to case study collections (Evans 1969d; Halevy 1985)].

Fig. 7-3A–I. Longitudinal sections of apices of *Cosmos* during transition to the inflorescence stage. Leaf pair is indicated by number 7. (A)–(D) Early transitional apex during various phases of a double plastochron. (E) Late transitional apex. (F)–(G) Apices at prefloral I stage of bract initiation. (H) Apex at prefloral II stage of bract initiation. (I) Early inflorescence apex during disc floret initiation. Abbreviations: b = leaf base of opposite leaf pair; BB = bract buttress; BP1 = bract primordium; Br1 = bract of the outer whorl; CH = chaff primordium; FP = disk floret primordium; SM = shell meristem. Reproduced from Molder and Owens (1973) with permission of National Research Council of Canada from *Canadian Journal of Botany* 51:535–551.

Growth Physiology of Inflorescence Stalks

The general term used to identify the stalk of an inflorescence is **peduncle;** the stalk of a single flower is a **pedicel.** Because of their convenient size and significant growth response over short periods of time, the peduncles of different composites have been popular research subjects—for example, *Taraxacum* (Chao 1947), *Bellis* (Uyldert 1928) and *Gerbera* (Sachs 1968) [see Lang (1961) and Sachs (1965) for further discussion and other examples]. In physiological presentations, distinctions between peduncle, pedicel and vegetative internodes are not always maintained. Since there are many different types of peduncles, one might expect the identification of a number of different growth physiologies.

The best-studied case of peduncle growth, though incomplete by present standards, is that of the growth of the scape (stalk) of *Gerbera* (Sachs 1968). The scape grows from 2–3cm to over 50cm in approximately 30 days (Fig. 7-4A). The sigmoidal growth curve corresponds to the first (pre-anthesis) of the biphasic stages described in *Taraxacum* (Chao 1947). In scapes 1–3cm in length, cell division and cell enlargement occur at more or less the same rate throughout the length of the scape. In scapes longer than 3–5cm, cell length of pith and cortical cells increase from apex to base. As indicated in Fig. 7-4B, this apex-to-base trend continues in longer scapes. Cell division in the scape ceases in the apical region in scapes 20^+ cm in length. Using the current insights about the cell kinetics of organ growth (Crone and Lord 1991; Silk 1984), no doubt far greater detail and precision about the cellular events of peduncle growth could be uncovered.

As with *Bellis* and other composites (Lang 1961), removal of immature inflorescences inhibits scape growth (Fig. 7-4C). A difference between decapitation and deflowering was observed in *Gerbera* (Sachs 1968). In some cases, a strong auxin correlation, both response to exogenous application and analysis of endogenous content, has been observed. In *Gerbera,* data from experiments in which IAA and GA were applied in lanolin to immature, decapitated inflorescences provide strong evidence that both of these PGS may be involved in normal scape elongation.

Kaldewey's analysis of the growth of the whole *Fritillaria* stem, including the terminal pedicel (Kaldewey 1957) also deserves mention for the care and precision of its documentation. A number of the growth and bending responses of the plant were correlated to availability of extractable auxin-like materials. Further details of these systems are discussed by Lang (1961), Sachs (1965) and Kinet, Sachs and Bernier (1985).

In many LDP, a different kind of growth response associated with flowering can be related to gibberellin metabolism. In the case of many LD rosette plants and vernalizable species, gibberellins have a marked effect on the different components of the inflorescence, and correlations with extractable forms of endogenous GA-like materials are possible (Pharis and King 1985).

These examples provide only a small view of what must be a much larger and more complex body of information concerning the physiological concomitants of growth in inflorescences. Many aspects of control may yet be uncovered involving

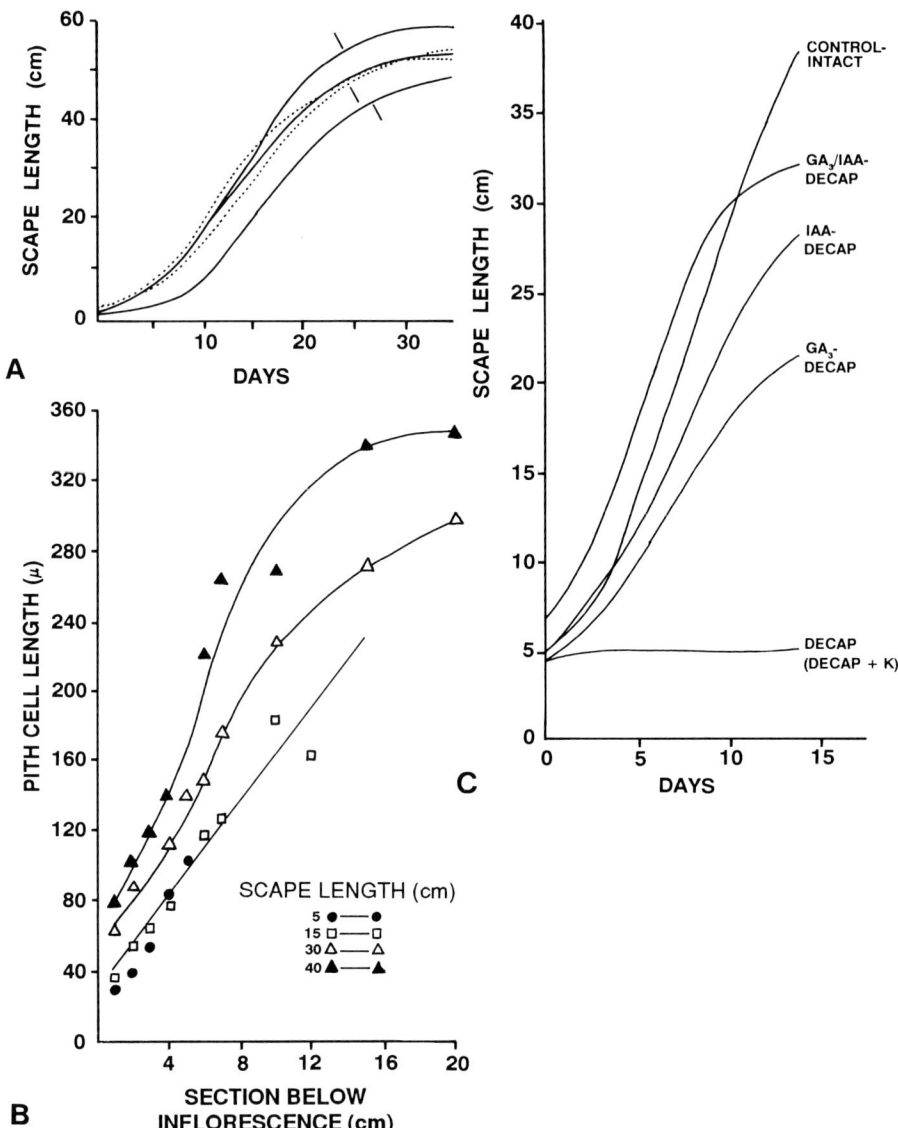

Fig. 7-4A–C. Observations on the elongation of the scape of *Gerbera*. (A) Elongation of several scapes from the 2cm stage to anthesis (arrows). (B) Pith-cell length as a function of scape length and distance below the inflorescence. Each value is the mean for 50 cells of each of two scapes. (C) Elongation response of scapes to GA_3 (1 mg/g lanolin), IAA (0.1 mg/g lanolin), GA_3/IAA (same concentrations) and kinetin (K) (0.1 mg/g). The control is an intact scape. Decap = decapitated head. From Sachs (1968) with permission of *American Journal of Botany*.

additional physiological mechanisms and molecules. Indeed some of the present insights may be shown to be incomplete or inaccurate. Research on other inflorescence examples may suggest different physiological control mechanisms. The area of PGS regulation of many inflorescences is probably complex and subtle but could, if pursued vigorously and in accordance with current knowledge of PGS action and technology, add measurably to our understanding of inflorescence development.

Apical Dominance

One assumes that at least part of the distinction traditionally recognized by morphologists between determinate (monotelic) and indeterminate (polytelic) inflorescences lies in the different types of organs that physiologists have traditionally included under the topic of apical dominance. It is not clear, at present, just how flower and inflorescence development overlap or involve this widely recognized but variable phenomenon.

Although apical dominance is a popular topic for physiological investigation, no single explanation of its mode of action is available (Phillips 1969, 1975; Tamas 1987). In some vegetative (uninduced) specimens, frequently legumes, inhibition of axillary buds appears to be auxin-based. Auxin from the apical tissues is transported basipetally and inhibits axillary bud development. In other experimental materials, cytokinins, ethylene, gibberellins and abscisic acid can be implicated to varying degrees and in different ways. Further, other physiological components such as competition for nutrients, water and gas exchange seem important (McIntyre 1977). One commonly described concomitant of flowering is the loss of apical dominance in the lower parts of many plants. Unfortunately, however, despite the data collected, it is not possible at present to explain how apical dominance is related to the regulation of inflorescence growth.

Meristem Abortion

A dramatic but poorly studied component of inflorescence architecture is the selective but regular abortion of axillary buds. Alternatively, axillary buds may fail to initiate or may be selectively inhibited. These examples would be analogous to the regular abortion of the vegetative terminal apex of *Syringa* (Garrison and Wetmore 1961) and the resultant sympodium branching pattern and the phenomenon of unequal bud development (anisoclady) discussed by Cutter (1967). No doubt many examples of selective growth and/or abortion of meristems exist which affect the final form of the inflorescence. Both the correlative effects on the rest of the plant and the individual cytological mechanisms of meristem abortion are significant developmental questions, though poorly studied. (Refer to *Pisum*, below, and *Zea* in Chapter 8).

Heteroblasty

It seems obvious that many of the changes in organ structure along a stem, changes in leaf shape, size and anatomy and changes in internode length which have been discussed, since Ashby (1948a), as a special case of heteroblasty are related to the onset of flowering. (Note: The original meaning of heteroblasty, as defined by Goebel (1900), has undergone a number of interpretations and redefinitions (see also the section on developmental variation in Chapter 5)). The special case that interested Ashby (1948a) involved the node-to-node changes that occur on many angiosperm vegetative stems. The factors that influence these changes have stimulated a number of studies [see Nojuku (1956, 1957) and Sinnott (1960)]. Inflorescence stalks frequently possess leaf-like bracts and bracteoles of changing shape and size in much the same way that other uninduced plants produce a range of leaves along the stem (Ashby 1948b). Just how heteroblasty along non-induced stems can be compared to analogous changes along the inflorescence stalks is unknown, since no recent studies are available.

Genetically Based Variability

Attempts to identify explicit genetically based aspects of inflorescence structure are rare and relatively recent (Marx 1983). This is perhaps to be expected in view of the modest interest in inflorescence development in general and the imprecision of most treatments. One should expect that future interest will increase, especially in crop species. In general we should expect differential gene action affecting inflorescence structure to express between the time of floral evocation of the terminal meristem and the initiation of individual floral meristems. The expression of some inflorescence-related genes may also persist throughout the growth of the inflorescence and throughout the rest of the plant. The following examples indicate opportunities available for study.

The wild-type inflorescence of *Arabidopsis* is frequently described, according to traditional interpretations, as a compound raceme. More accurate descriptions identify the average number of paracladia arising in axillary positions of lower nodes and the number of individual flowers occurring in the terminal florescence. Bracts are a variable trait (Napp-Zinn 1969, 1985) in this species, depending upon the genetic background and growth conditions. Two illustrations of wild-type inflorescences of isolates of the Columbia ecotype are illustrated in Fig. 7-5A and C.

Two recessive inflorescence-related genes, both derived from ethylmethylsuphonate (EMS) mutagenesis programs, have been described. In "leafy" (lfy) (Schultz and Haughn 1991), the number of axillary coflorescence-like shoots that develop lacking normal flowers is much larger than in the wild type (Fig. 7-5B). In later-forming shoots, some gynoecium-like and sepal-like flower traits develop, though fully normal flowers are not formed. Weigel et al. (1992) describe the phenotype slightly differently. In their description, the "leafy" mutation affects the primary inflorescence in two ways: (1) the number of secondary inflorescence shoots which are subtended by cauline leaves is increased; and (2) the flowers that eventually do develop are abnormal and show some characteristics of secondary inflorescences.

In the "terminal flower" mutant (tfl1-1) (Shannon and Meeks-Wagner 1991) (Fig. 7-5D), the inflorescence of LD-grown homozygotes bears a small cluster of

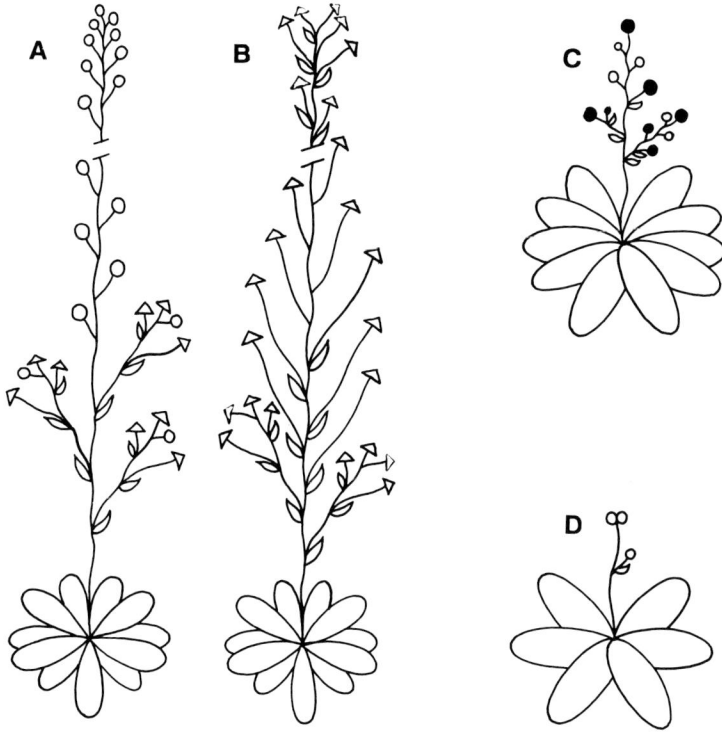

Fig. 7-5A–D. Diagrams of wild-type and mutant inflorescences of *Arabidopsis*. (A) Sketch of wild-type plant. (B) Sketch of leafy (lfy) mutant plant. Symbols: ◯ = flower; ▽ = coflorescence or coflorescence-like structure (on B). Redrawn from Schultz and Haughn (1991) with permission of ASPP. (C) Sketch of wild-type plant grown under LD conditions. (D) Sketch of the mutant "terminal flower" mutant (tfl1-1). Symbols: ◯ = flowers or flower meristems; ● = coflorescence. A joined circles in (D) represent the flowers of the terminal floral structure. Redrawn from Shannon and Meeks-Wagner (1991) with permission of ASPP.

partially fused flowers and occasionally one or two lateral branches that terminate in a single flower. The number of flowers forming at the tip of the inflorescence is variable, and some chimeric flowers were noted.

The inflorescence of the garden pea *(Pisum sativum)* consists of individual axillary, or few-flowered clusters, developing acropetaly along a monoaxial leafy stem (Haupt 1969). Normally, no terminal flower develops. This description is generally assumed to fit the classification of a raceme, though Tucker (1989), from SEM analysis, documents some of the complications. Although environmental conditions (light intensity and temperature (Murfet 1985) affect the number of flowers per raceme and the number of flowers per node (Hole and Hardwick 1976), these traits are generally assumed to be regulated by polygenic systems (Reid 1980; Murfet 1985). The gene Sn has been shown to delay flower initiation, flower development and apical senescence through the production of a graft-transmissible inhibitor during darkness.

Genetically based research materials such as these from *Arabidopsis* and *Pisum* and others such as "ramosa" in *Zea mays* (Chapter 8) and "floricaula" (flo) in *Antirrhinum* (Coen et al. 1990), when studied carefully, should help clarify the cellular dynamics and the developmental biology of inflorescences.

Model Inflorescences for Developmental Study

It is not immediately apparent which model specimens should be selected for basic studies of developmental differences between monotelic and polytelic patterns. Perhaps there are genera or species in which examples of both patterns are expressed. Similarly, there will be many suitable specimens for the exploration of apical dominance, meristem abortion and the other aspects of inflorescence development that were identified above.

The **capitulum** (head) of the Compositae possesses a number of features which make it attractive for developmental studies, and many species possess commendable features for the exploration of many basic problems in flower development. Some species are photoperiodically sensitive, have large inflorescence meristems, produce a variety of flower forms, and are available in a variety of genetically based forms. A short-list of suitable model candidates, each for different reasons, might include: *Cosmos* (Molder and Owens 1973, 1985), *Chrysanthemum* (Cockshull 1985*b*), *Calendula* (Cutter 1971), *Tusilago* (Wardlaw 1961), *Xanthium, Helianthus* and *Microseris*. The last three are discussed here in more detail.

Xanthium strumarium L. *(Cocklebur) (syn.* X. Pennsylvanicum*)*

A major weed in many parts of North America and other parts of the world (Weaver and Lechowicz 1983), *Xanthium* has become a classic research material for studies on photoperiodic control of flowering. The "Chicago Strain," commonly used by physiologists, is highly sensitive to photo-induction and initiates flowering following a single SD inductive cycle. It has been used for more than 50 years in flower initiation studies (see detailed reviews of Salisbury 1969, 1985). Populations from other latitudes can exhibit different daylength sensitivities: some show quantitative rather than qualitative sensitivity to daylength (Ray and Alexander 1966). A comprehensive review of research on *Xanthium* including quantitative studies of leaf growth, phyllotaxy and flowering was prepared by Maksymowych (1990).

Morphological and anatomical investigations of the initiation and development of flowers in *Xanthium* have been selective, concentrating mainly on the terminal staminate (male) inflorescence, and exploratory rather than probing. These include Farr (1915), Naylor (1941), Lance (1957), Wetmore, Gifford and Green (1959) and Kirk, Morrow and Jacobs (1967). Following a single SD inductive cycle and continuous light, only the terminal meristem is transformed into an inflorescence meristem which initiates staminate flowers. A single SD inductive cycle followed by LD cycles stimulates the production of gynoecious inflorescences in the axillary positions. Other qualitative and quantitative responses are possible with different light regimes (Kirk, Morrow and Jacobs 1967).

The attributes that have made this a desirable research plant for photoperiodic and physiological studies of flowering also make it a useful plant for developmental studies. These include: (i) its sensitivity to induction stimuli; (ii) its quantitative response to different numbers of inductive cycles; (iii) the development of a sensitive assay system of flowering response based on the terminal inflorescence (Salisbury 1963) [other assay systems are compared in Salisbury (1969)]; (iv) the ability to study PGS components of the flowering response; (v) the well-defined nature of growth conditions for experiments (Salisbury 1985).

In addition, *Xanthium* is monoecious, producing staminate inflorescences at the tips of the main stem and branches and gynoecious inflorescences in the axillary positions along the major axis and its branches. The combination of a sensitive, controllable research plant which possesses developmentally variable expressions of staminate and gynoecious inflorescences should attract the interest of developmentalists. Seed dormancy and ways to control it for the routine production of research material are reviewed by Salisbury (1985).

More details of the development of staminate and gynoecious inflorescences are illustrated in Fig. 7-6 and Fig. 7-7. Staminate flowers are restricted to the

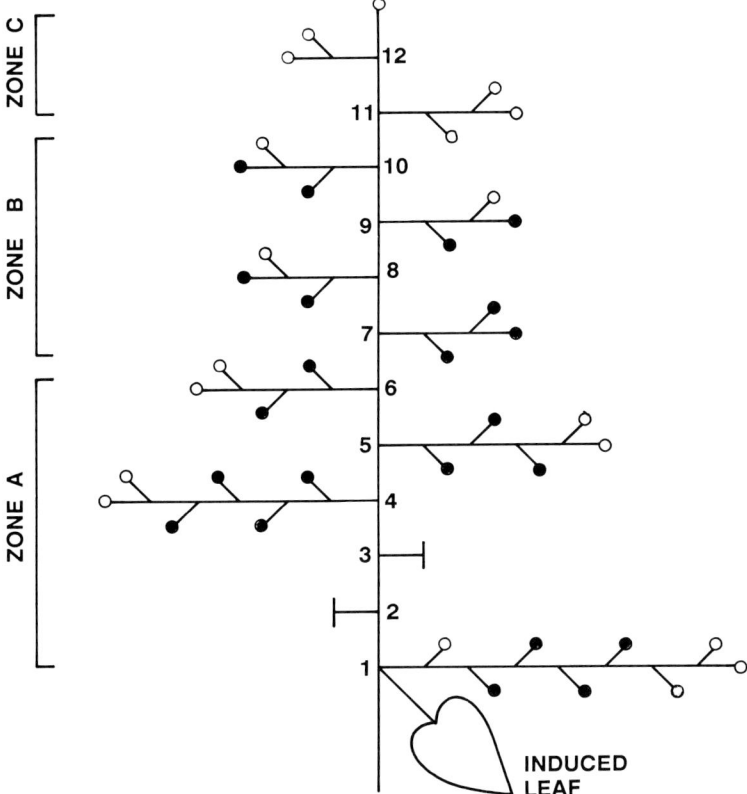

Fig. 7-6. Distribution of staminate (○) and gynoecial (●) heads on a plant of *Xanthium strumarium* induced by a single long night. Observations were restricted to the terminal bud on the main stem and the terminal and axillary buds of the first-order branches. Side axes are numbered acropetally from the induced leaf. From Leonard et al. (1981) with permission of ASPP.

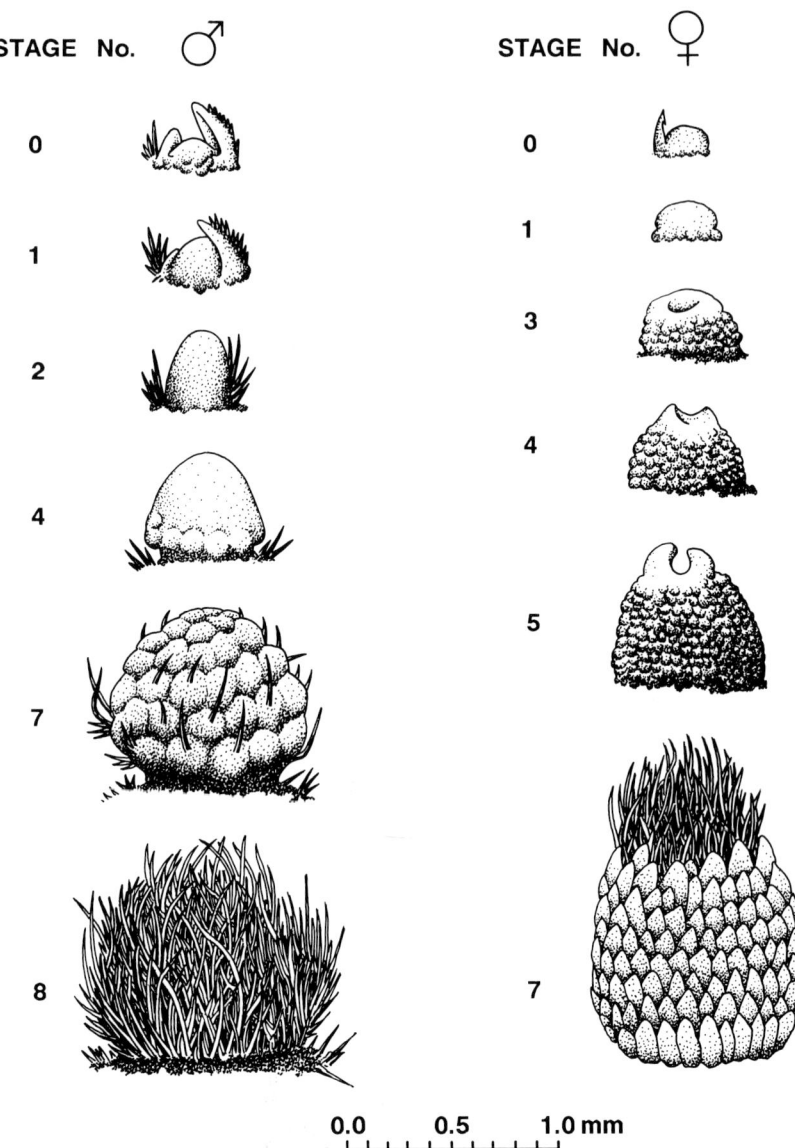

Fig. 7-7. Comparison of the development of the staminate (♂) and gynoecial (♀) heads on *Xanthium strumarium*. Details of the individual stages of both types of heads are presented in Table 7-2. Staminate details redrawn with permission from Salisbury (1985) in A. H. Halevy [ed.], *CRC Handbook of Flowering,* vol. 4, pp. 473-522. Copyright CRC Press Inc. Boca Raton, FL. Carpellate inflorescences redrawn with permission from Leonard et al. (1981) with permission of ASPP.

terminal florescence and, with few exceptions, to the tips of axillary paracladia (Fig. 7-6). The number of florets (150–170) (Kinet, Sachs and Bernier 1985) per staminate inflorescence varies depending upon its position on the stems and the vigor of the plant. Each staminate flower consists of a corolla (normally five-lobed) and five stamens fused by their filaments. A vestigial or aborted stigma is reported in *X. commune* (Farr 1915). Each flower is subtended by a bract.

By contrast, only two flowers develop on each gynoecial inflorescence. A comparison of the development of the gynoecial and staminate inflorescences is illustrated in Fig. 7-7 and Table 7-1. Brief histological descriptions of the early stages of these events in the terminal staminate inflorescence meristem were recorded by Wetmore, Gifford and Green (1959).

The gynoecial flowers developing in the axillary inflorescences are distinctive in a number of features. First, their development lags slightly behind the terminal inflorescence at varying times, depending on the position on the stem. Development of the gynoecial inflorescences can also vary, depending on the level of flowering stimulation (Kirk, Morrow and Jacobs 1967). Second, the gynoecial inflorescence is complex and is not easily homologized to classic composite structures. The most prevalent interpretation is that each gynoecious inflorescence initiates 9–15 involucral bracts, which bear a number of papillae which will develop into the hooked spines of the fruit and two fertile flowers (Farr 1915). The gynoecia of the two flowers of each inflorescence grow at different rates and therefore produce seeds of different sizes at maturity. The mature spiny, two-seeded fruit of *Xanthium*, therefore, represents a matured inflorescence. The opportunities for

Table 7-1 Description of Selected Floral Stages from Staminate and Gynoecial Heads of *Xanthium strumarium* Illustrated in Figure 7-7

Stage	Description
Staminate Heads	
0	Vegetative. Shoot apex relatively flat.
1	First clearly visible swelling of the shoot apex.
2	Floral apex at least as high as broad, but not yet constricted at base.
4	First visible staminate flowers covering up to 1/4 of the inflorescence meristem.
7	Inflorescence completely covered by staminate flower primordia. Bracts slightly visible.
8	Hairy bracts cover flowers. Some differentiation of the flower parts. Basal diameter at least 1mm.
Gynoecial Heads	
0	Vegetative. Meristem small and flat.
1	First visible bract primordia at the base of big, round inflorescence primordium. Wider than tall.
3	Inflorescence primordium with an apical depression, completely covered by spine primordia.
4	Development of two beaks through rapid growth of the receptacle margins.
5	Elongating and bending of the two beaks.
7	Elongation of the spines. Inflorescence primordium taller than wide.

Source: Staminate heads extracted from Salisbury (1985); gynoecial heads extracted from Leonard et al. (1981).

more detailed and careful, but primarily traditional, studies which extend these initial observations (Farr 1915; Naylor 1941; Kirk, Morrow and Jacobs 1967) are substantial. In addition, because of its superb status as a research plant, allowing many opportunities for exploring developmental questions about the qualitative control of the alternative developmental programs of monoecious plants, *Xanthium* should be seriously considered.

It seems obvious that considerable basic information on the pattern of initiation, the number of flowers in the staminate inflorescence under different environmental conditions, has yet to be uncovered. Similar details seem also to be lacking for the axillary, gynoecial inflorescences. More interesting from a developmental standpoint are the factors that influence the staminate versus gynoecial expression. Other than some crude exogenous application experiments, little research has been done in this direction.

Helianthus annuus L. (Sunflower)

Because of the large size of its inflorescence, and despite the large size that plants attain, the commercial sunflower possesses a number of traits which recommend it as a plant for the exploration of certain developmental questions. Questions having to do with organ spacing on meristematic surfaces and the expression of ligulate versus disc-flower traits are particularly appropriate. Aspects of flower initiation and growth in *Erigeron* (Harris, Tucker and Urbatsch 1991) provide an introduction to these processes in the Compositae. The selection here of *Helianthus* over other composites such as *Cosmos, Chrysanthemum,* and *Dahlia* is arbitrary.

Shoot apical meristems of both vegetative and florally induced meristems of *Helianthus* are amenable to observation and experimentation. Autoradiography of vegetative shoot apices with thymidine-H^3 suggests that the cells of the central zone possess lower rates of DNA synthesis and mitosis than do the peripheral cells (Steeves et al. 1969; Davis, Rennie and Steeves 1979). These differences between central and peripheral zones diminish as the inflorescence develops. This zonation, reminiscent of the *meristeme d'attente* (Buvat 1952; Nougarède 1967; Gifford and Corson 1971), has been independently demonstrated through cell-cycle studies in many species (Lyndon 1990) for meristems which give rise to flowers and inflorescences. Further insight into the organization and dynamics of the meristem was obtained through the production of chimeric sectors by sublethal doses of gamma radiation (Jegla and Sussex 1989) (see Fig. 3-3B).

Depending upon the cultivar, sunflower is reported to be a quantitative SDP or a DNP. Application of GA_3 tends to shorten the vegetative period and the time to flowering (Schuster 1985). Based upon a variety of different observations, vegetative growth also possesses a significant gibberellin involvement (Jones and Phillips 1966) and may well be a significant factor in aspects of flower development. There are many different cultivars available commercially.

Sunflower is not as well known genetically as many research plants. The basic haploid chromosome number is N = 17; and diploid (2N = 34), tetraploid (2N = 68) and hexaploid (2N = 102) species are known (Whelan 1978). Genetic variation within the species includes isolates for incompatibility, for genic (g-mst) and cytoplasmic (c-mst) male sterility and for flower color. Unfortunately, beyond the recessive "tubular ray-flowers" and the more complex genetic condition "chrysanthemum flower," in which all flowers express the ray flower trait (Fick 1976), few floral or inflorescence morphological genes have been identified and selected, and a linkage map is not available (Fick 1978).

At maturity, a typical head of an oilseed cultivar can contain 700 to 3000 flowers; the number is even higher in non-oilseed cultivars (Knowles 1978). The outer appendages on the head are called **involucral bracts** (phyllaries) and the first flowers initiated mature into **ray flowers** which are sterile and showy, possessing five-lobed dorsiventrally elongated ray petals (Figs. 7-8 and 7-9). The **disc flowers** cover the remainder of the head. They appear in arcs **(parastichies)** radiating from the center of the head to the margin. At least two sets, major and minor, can be observed. (Note: Parastichies are frequently discontinuous, reflecting the changed relationships as initiation nears the center of the meristem.) Each disc flower is subtended by a short, sharp, chaffy bract and two scaly sepals (sometimes called the pappus), five anthers and an inferior ovary (Fig. 7-9B).

Depending upon the cultivar and conditions, the apical meristem following induction broadens and becomes dome-shaped. This is followed by the initiation of involucral bracts, ray florets and disc flowers. A developmental timetable for the cv. Sunfola 68-2 is given in Table 7-2.

The rate of initiation of primordia on the sunflower head is spectacular. Assuming a duration of 19 days from FS 3 to FS 8 (Table 7-2), an inflorescence producing 2000 disc flowers and the same rate of initiation over the period, this would mean more than 100 primordia/day or approximately 5/h. That means a new primordium is initiated every 10–15 minutes. This astonishing value is made more credible by acknowledging that initiation, at least in the early stages, takes place at the end of 30–50 parastichies and, therefore, assuming single initiations and a helical sequence of initiation, a number of hours would separate initiations along a single parastichy.

Because of its large size, the sunflower inflorescence lends itself admirably to the exploration of a number of developmental questions. These can involve the cellular properties of meristematic tissue in general, changes in the meristem associated with the onset of flowering, pattern formation on meristems and organ differentiation. Although we should not necessarily assume that similar developmental mechanisms govern the initiation and growth of individual flowers on the capitulum with the initiation and growth of organs on the flower meristem, many analogous situations can be explored and mechanisms revealed.

For example, Palmer and colleagues have uncovered a number of intriguing features of the sunflower meristem as it is transformed to flowering and as it generates flowers. There are a number of parallels between the dynamics of transformed sunflower meristems and those of induced flower meristems (Marc and Palmer 1982). Their direct measure of mitotic counts supports the interpretation demonstrated by others (Steeves et al. 1969; Davis, Rennie and Steeves 1979) that a central quiescent zone is present in the vegetative meristem. Following induction, which occurred (in cv. Sunfola 68-2) within 16 days from germination, mi-

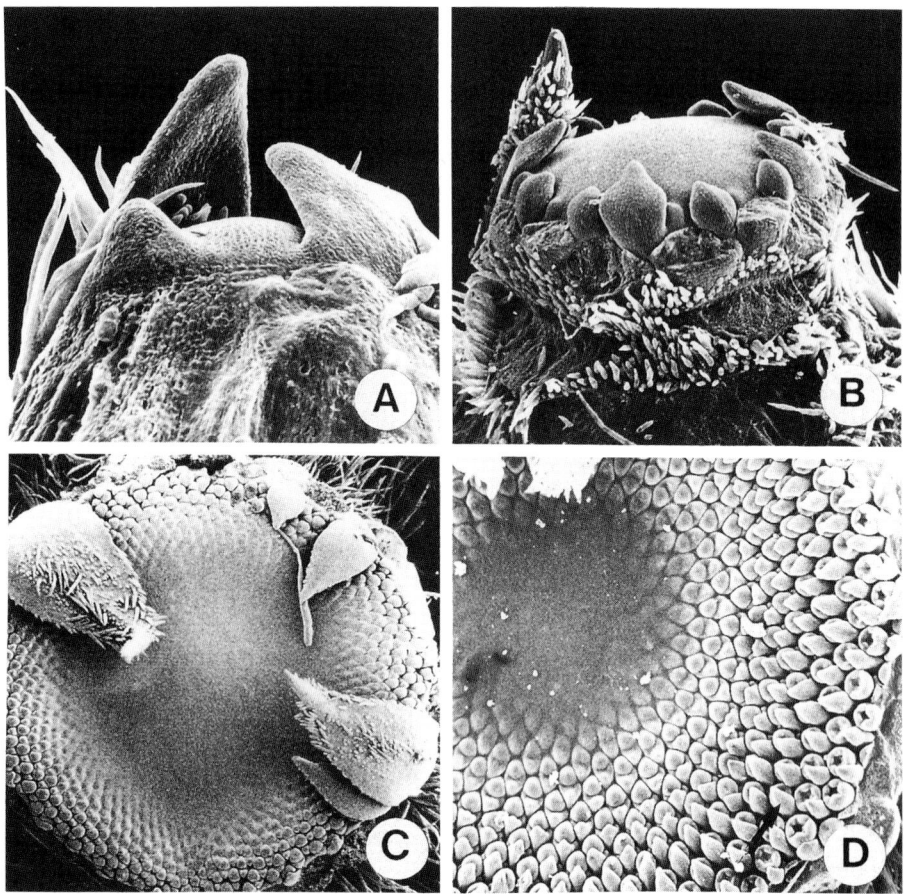

Fig. 7-8A–D. SEM views of the development of the inflorescence of *Helianthus annuus*. (A) The apical meristem prior to floral initiation (FS1). (B) The apical meristem increases in size and becomes flatter, and involucral bracts initiate at the periphery of the dome (FS3). (C) Further expansion of the apex continues, the periphery of the rim becomes elevated and the diameter of the apex exceeds the diameter of the stem. Floret primordia (F) are first initiated at the periphery of the apex (FS5). (D) Florets first appear as a single ridge that later develops into the floret bract (arrow) and then the floret primordium. Initiation of the florets continues acropetally. Parastichies become obvious (FS6). From Moncur (1981) with permission of CSIRO Division of Forestry. Floral stages (FS) adapted from Marc and Palmer (1981).

totic activity was distributed throughout the meristem, which had taken on a domed appearance (Fig. 7-10).

Because of its size, the number and kinds of primordia initiated on its surface and their predictable position, the sunflower inflorescence meristem is a most appropriate object for the study of pattern of initiation. Questions regarding the time to initiate a primordium, the timing and nature of organ differentiation and the

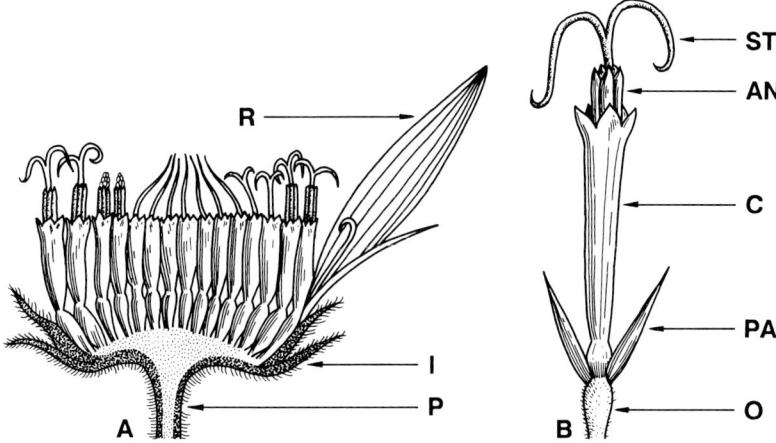

Fig. 7-9A–B. (A) Sketch of the head of *Helianthus* at early anthesis. Abbreviations: P = peduncle; I = involucral bracts; R = corolla of a ray flower. (B) A disc flower prior to anthesis. Abbreviations: O = ovary; C = corolla; AN = anther; ST = style; PA = pappus. Redrawn from Benson (1957) with permission.

conditions and factors that influence differentiation and maturation of the flowers can be investigated.

In experiments on the meristem of *H. annuus* cv. Sunfola, Palmer and Marc (1982) made surgical cuts on the surface of the meristem at different times during development. In one experiment, following a linear bisection at an early devel-

Table 7-2 Time-Base for Development of Different Inflorescence Characters of the Sunflower cv. "Sunfola 68-2" (an Early Oilseed Cultivar)

Floral Stage (FS)	Age From Sowing (Days)	Principal Features
1	18	Meristem begins to swell and flanks become round.
2	20	Meristem is clearly dome-shaped.
3	23	Dome broadens and flattens. Appearance of involucral bract primordia.
4	28	Receptacle disc appears. Margin overhangs stem.
5	32	Florets appear at the rim of the receptacle. Up to 1/3 of the receptacle radius occupied.
6	36	Florets occupy 1/3–2/3 of the receptacle radius. Appearance of the 5-lobed corolla in peripheral floret primordia. Floret bract is smooth and bent towards the center.
7	39	2/3 or more of the receptacle radius is covered by florets.
8	42	Floret bracts visibly differentiated at receptacle center.
9	46	Central florets developing 5-lobed corolla.
10	50	Floret bracts at receptacle center becoming hairy.

Note: The controlled growth conditions included: temperature, 28°C; light, fluorescent and incandescent lamps providing a total quantum flux at plant height of 420 $\mu Em^{-2}\ s^{-1}$; photoperiod, 18h.

Source: From Marc and Palmer (1981).

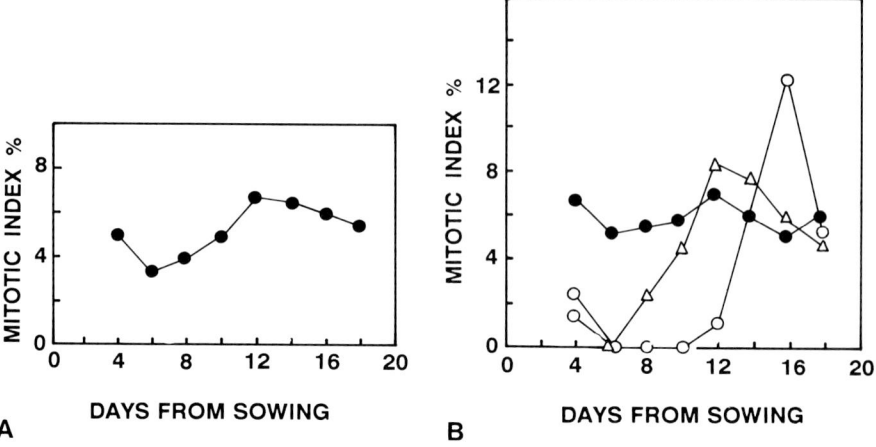

Fig. 7-10A-B. Mitotic index (MI) in the apical meristem of *Helianthus*. (A) MI in the whole apical meristem within the youngest leaf primordia and the transverse axis. (B) MI in individual segments of the meristem. Symbols: ● = peripheral zone; △ = distal cells; ○ = central tunica. Values are means of four apices from the median section and one displaced by 16μm from the median. From Palmer and Marc (1982) with permission of *American Journal of Botany*.

Fig. 7-11A-B. Observations on the response of the *Helianthus* inflorescence to surgical wounding. (A) Twin inflorescences formed following receptacle wounding at FS2 (slightly younger than Fig. 7-8B). From Palmer and Marc (1982) with permission of *Plant and Cell Physiology*. (B) Circular incision in the receptacle surface (FS4 or 5) to form a plug of undifferentiated tissue (Bar = 300μm). Within 3–6 days, floral initials appeared on the flanks of both rims and later on the top of the plug and the surrounding receptacle surface. The first formed initials developed into involucral bracts or ray florets and the later ones into disc florets. From Hernandez and Palmer (1988) with permission of *American Journal of Botany*.

opmental stage, twin inflorescences developed with the initiation of involucral bracts and ray flowers along the margins of the wound and disc flowers initiating centrifugally (Fig. 7-11A). In another study (Hernandez and Palmer 1988), a 1mm cylinder was cut into the center of the meristem. After 3–6 days, primordia initiation took place on the original meristem rim and that of the new cylinder. Involucral bracts, ray flowers and disc flowers were initiated on both portions of the meristem (Fig. 7-11B). These experiments point to some opportunities for considerably more focussed explorations of the questions of organ determination and maturation and are akin to those performed on single-flower meristems (Cusick 1956; Hicks and Sussex 1971).

Microseris

Another example of the opportunities that inflorescences, as distinct from flowers, provide for developmental studies is found in the variation in the number of pappus parts on the inflorescence of different species of the western North American composite *Microseris* (Bachmann 1983). (Note: The **pappus** is the collective term used to define the frequently scarious and hairy set of appendages at the base of and external to the corolla. They occur in the region normally referred to as the calyx). In these studies the pappus parts were observed on the mature fruits, the **achenes.** In *M. laciniata,* a perennial, pappus number ranges freely between 5 and 10 (Fig. 7-12). Besides varying from plant to plant, the number of pappus parts varies on each head: flowers at the margin of the head have generally more

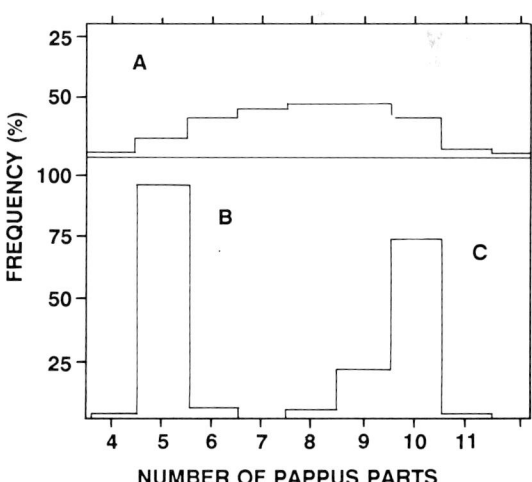

Fig. 7-12A–C. A frequency distribution of achenes bearing various numbers of pappus parts in three species of *Microseris.* (A) *M. laciniata* (perennial). (B) *M. bigelovii* (annual). (C) *M. pygmaea* (annual). All populations show five and ten as limiting numbers, with very few achenes bearing fewer or more pappus parts. In the annuals, in addition, there is a clear canalization towards one or the other of these limits. From Bachman and Chambers (1978) with permission of Springer-Verlag Wein.

pappus parts than those near the center of the head. In this case, expression of the trait is seen to be plastic or epigenetic.

In two annual species, however, mean pappus part number/head is canalized to either 5 *(M. bigelovii)* or 10 *(M. pygmaea)* (Bachmann and Chambers 1978) (Fig. 7-12). The insight which makes these studies of interest developmentally comes from the subsequent breeding studies on the progeny from hybrid lines (Bachmann, Chambers and Price 1981; Bachmann et al. 1982). Based on their data, these researchers conclude that pappus part number is canalized between 5 and 10 through the action of four segregating unlinked genes. Thus, in these very preliminary studies, we are able to see some of the complexity involved in the interaction of genetic and epigenetic components which determine development. Other examples and further discussion of the control of meristic variation are found in Chapters 4 and 9.

Summary

Future studies of developmental aspects of inflorescences will obviously be more sharply focussed than in the past, though comprehensive insights will be slow to accumulate. Specific topics that should be explored include:

1. Descriptive and experimental studies of the developmental distinctions between racemose (polytelic) and cymose (monotelic) patterns of initiation and maturation.
2. Cellular, physiological and molecular biological aspects of meristem abortion, apical dominance and unequal development.
3. Identification of inflorescence-specific genes and the patterns of their expression from a variety of different genera and families. One might then be able to comment upon any homologies between different species and homologies to other flower-specific genes.
4. Characterization of the PGS relationships to developmental patterns and different stages of maturation.
5. Exploration of selected developmental mechanisms such as site selection during flower initiation in composite heads.
6. Uncovering the factors that regulate patterns of maturation and flower opening (aestivation) as separate from patterns of initiation.

8

Flowers and Inflorescences of Grasses

Four aspects of flower and inflorescence development of grasses dictate the choice of a separate chapter. First, the Poaceae (Gramineae) is a large and variable family (Watson, Clifford and Dallwitz 1985; Chapman and Peat 1992), but its flowers, often small and scarious, are less well studied developmentally. Second, many features of grass flower development and mature architecture differ from those (primarily dicotyledonous) examples discussed earlier. Third, as model research material for studies on flower development, the four examples selected—maize, wheat, rice and *Lolium*—exhibit many desirable attributes despite the limited basic descriptive developmental information. Finally, rice, wheat and maize, the preeminent grains of human nutrition, deserve emphasis for the potentially practical applications that knowledge of their flowers can provide. Readers who wish more information about the flowers and inflorescences of grasses other than the four examples discussed here are directed to Arber (1934), Williams (1966a), Gould and Shaw (1968), Clifford (1987) and Shewry (1991).

As discussed in Chapter 7, the most common form of grass inflorescence is defined as a **panicle.** Paniculate growth and maturation are normally determinate, with the terminal spikelets maturing first on each branch and the basal ones last (Gould and Shaw 1968). The rice inflorescence is a typical panicle, though (as described below) the maturation of non-terminal spikelets deviates from a strictly basipetal maturation sequence.

Descriptions of the inflorescences of other grasses are less consistent. In the tassel and ear of maize, the initiation of spikelets is acropetal, with the first primordium at the base and the last near the tip. In traditional terminology, this is the manner in which racemes are initiated. If the flowers or spikelets are sessile, the inflorescence is called a **spike;** if the flowers or spikelets are pedicellate, the inflorescence is called a **raceme** (see Chapter 7).

The ultimate unit of the grass inflorescence is the **spikelet,** which in its most generalized form is interpreted as consisting of a series of overlapping bracts **(glumes** and **lemmas),** arising distichously from the apex of a short axis **(pedicel** or **rachilla)** (Clifford 1987) (Fig. 8-1). Short, lateral reproductive shoots arise in the axils of some of the lemmas. These reproductive shoots, including their subtending lemmas, are termed **florets.** It should be appreciated that the descriptions of individual cases may differ as a consequence of real differences but also from different interpretations and from the variable depth of knowledge about each case.

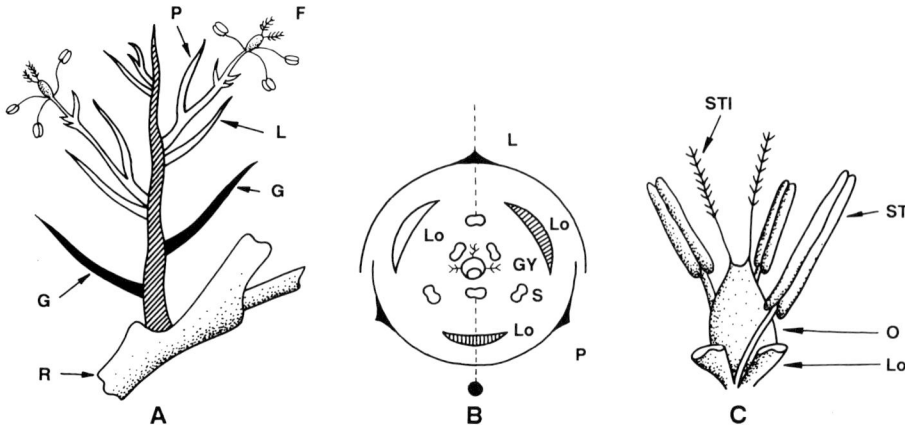

Fig. 8-1A–C. Outline sketches of a generalized grass spikelet. (A) Schematic diagram of a grass spikelet. Abbreviations: G = glume; R = rachis; L = lemma; P = palea; F = flower. (B) Floral diagram of a generalized grass flower; the median plane is indicated by the dashed line. Abbreviations: L = lemma; P = palea; LO = lodicule; GY = gynoecium; S = stamen. (C) Typical grass flower (minus glumes, lemma and palea) containing two lodicules, three stamens and a gynoecium. Abbreviations: LO = lodicule; O = ovary; ST = stamen; STI = stigma. Redrawn from "Spikelet and Floral Morphology" by H. Trevor Clifford in *Grass Systematics and Evolution,* edited by Thomas R. Soderstrom, Khidid W. Hilu, Christopher S. Campbell and Mary E. Barkworth. Published by Smithsonian Institution Press. Copyright 1987, Smithsonian Institution, p. 23.

Maize (*Zea mays* L.)

At first glance, maize is an unpromising candidate as a model system for studying flower development. It is physically large, requiring space and care in its culture. It has a relatively long life cycle (50–100 days); inflorescences are hidden throughout most of their development; and its developmental program appears to be inflexible.

Fortunately, most of these deficiencies can be circumvented (Sheridan 1982). The main justification for using maize in flower research remains its developmental control of monoecy and the impressive array of genetically based flower variability (Neuffer, Jones and Zuber 1968: Coe, Neuffer and Hoisington 1988). More recently, the ability to study the molecular biology of individual maize genes and their expression makes the traditional genetic information even more attractive. For example, the first demonstration of a homeobox in angiosperms was in the gene for knotted (Kn1), which was cloned and sequenced (Vollbrecht et al. 1991). (See Chapter 9 for discussion of homeotic mutants.)

For this discussion, the selected topics center on important developmental features about the origin and development of inflorescences and flowers of maize. Many other details about morphology and physiology in maize flowers have been reviewed previously (Weatherwax 1916, 1955; Bonnett 1940, 1948, 1953, 1966; Kiesselbach 1949; Sass 1977; Cheng, Greyson and Walden 1983; Hanway and Ritchie 1985).

Origin of the Inflorescences and Flowers

Transition to Flowering

Despite considerable genotypic variation between and within races and maturity groups, as well as considerable variation in response to temperature/photoperiod interactions, most studies of high-latitude cultivars conclude that maize is a quantitative SDP: long photoperiods delay tassel differentiation, but *Zea* will flower under 24h light conditions (Hanway and Ritchie 1985). Tropical cultivars are more adapted to daylengths of 12h to 13h (Goodman and Brown 1988)

The number of leaves that initiate prior to the transformation of the apical meristem into a tassel primordium is relatively constant for individual races (Duncan and Hesketh 1968) and single-cross hybrids (Hesketh, Chase and Nanda 1969), though it varies slightly in response to changes in daylength and temperature.

This relative consistency inspired a provocative study of cultured shoot apices from plants with different numbers of leaves (Irish and Nelson 1988). Following the initiation of roots, the explants were transferred to pot and field. The number of leaves (and by inference the onset of flowering) was shown to be determined in these seventeen-leaf plants sometime between initiation of leaf 10 and initiation of the last leaf. A later study (Irish and Nelson 1991) suggested that meristems were not determined to form a tassel until after all vegetative leaves had been initiated. Determination to form flowers was shown to be a separate and still later process.

Morphology of Ear and Tassel

Both inflorescences of maize are racemose rather than paniculate (Fig. 8-2). The ear can legitimately be described as a spike, though the classification of the tassel, because of its branches and the short pedicels on one-half of its spikelets, is less clear-cut. Maturation of flowers on the tassel and ear, despite the pattern of initiation, which is consistently acropetal, begins near the center and proceeds both acropetally and basipetally. Despite these discrepancies, however, maize inflorescences are also referred to as panicles (e.g., Nickerson 1954).

An extensive and disputatious literature has accumulated on how best to interpret the morphology of the ear and tassel of corn. The reviews of Mangelsdorf (1974), Beadle (1980), Iltis (1983) and Galinat (1985) represent different approaches to the problem. While some data from these studies are relevant to understanding the ontogeny of these two complex inflorescences, most of these discussions relate to the proposed origins of maize and not to ontogeny.

Initiation and Growth of Spikelet-Branches, Spikelets and Flowers on the Tassel

As the distichous pattern of leaf initiation proceeds, the shoot apical meristem enlarges in both width and height; dimensions vary with cultivar and growing conditions. Approximate dimensions for a commercial single-cross hybrid at this

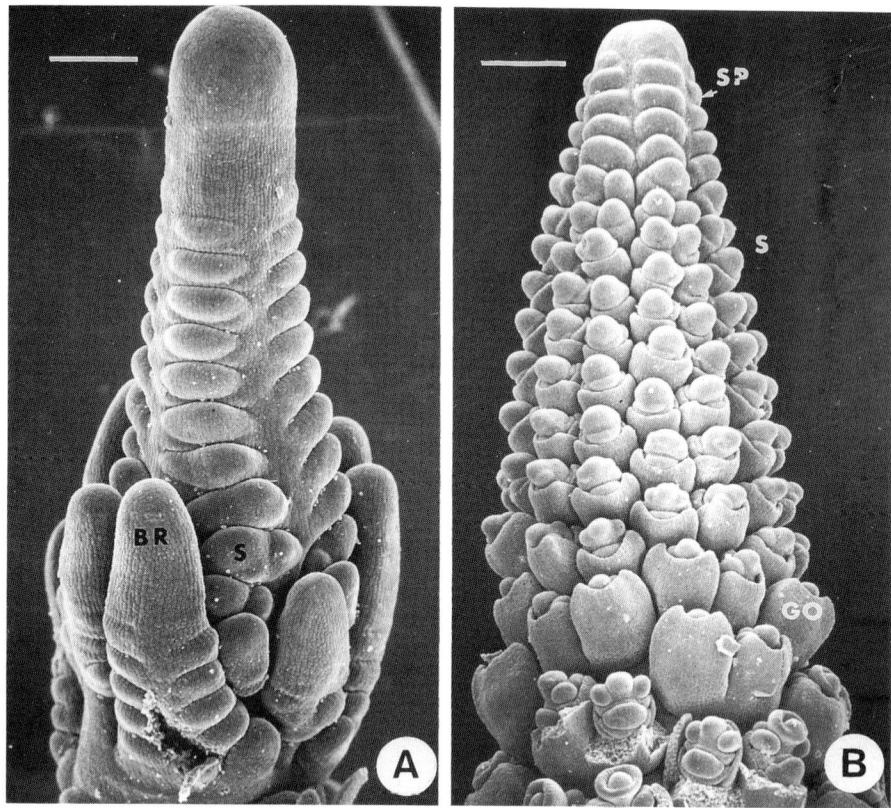

Fig. 8-2A–B. SEM views of immature inflorescences of *Zea mays* cv. Se60. (A) Immature tassel inflorescence approximately 25–30 days after germination (DAG). (B) The tip of a developing ear from a plant approximately 35–40 DAG. Some of the outer glumes of the older spikelets, near the base, were removed. Abbreviations: GO = outer glume; S = spikelet primordium; SP = spikelet branch. Bars = 150μm. From Cheng, Greyson and Walden (1983) with permission of *American Journal of Botany*.

stage, 26 days after seeding, were 160μm height × 180μm diameter (Greyson, Walden and Smith 1982). After the last leaf is initiated, the transition apex elongates prior to the initiation of **branch primordia** (Fig. 8-2). To help in collecting staged specimens for research, stages in inflorescence development can be correlated with the number of visible leaves. In "Illinois High Oil," for example, meristem elongation began when the seventh leaf was visible in the whorl of the sixth leaf (Leng 1951).

Branch primordia (their number varies among cultivars) initiate acropetally in an apparently helical sequence. Some reports identify ridges (rachis-flaps) initiating below each branch primordium (Bonnett 1948, 1953; Cutler and Cutler 1948; Sundberg and Orr 1986). These ridges are of interest in discussions of the homologies of organs but do not develop beyond the ridge stage in most cultivars (see discussion of double ridges in the section on wheat, this chapter).

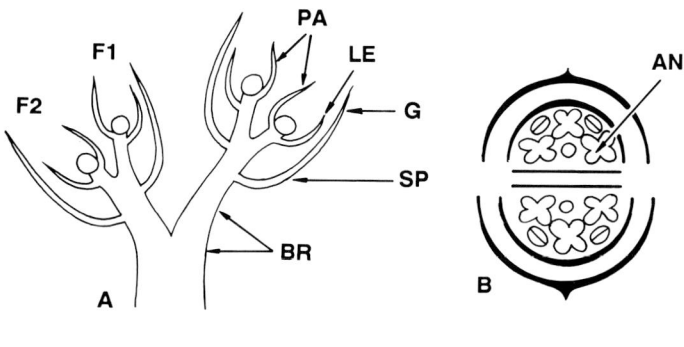

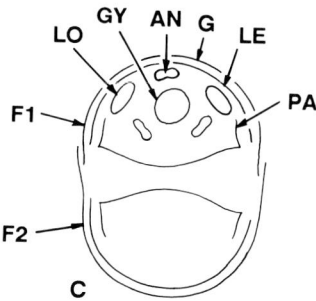

Fig. 8-3A–C. Sketches of the arrangements of organs on spikelets of *Z. mays*. (A) A pair of spikelets each composed of a pair of flowers. The branching and the position of organs are presented diagramatically. The orientation of the spikelets with respect to each other is not correct. The circles represent the lodicules, stamens and gynoecium. (B) Cross-sectional diagram of a two-flowered tassel spikelet at maturity. (C) Cross section of a spikelet from an ear. The upper flower of the spikelet bears three degenerating stamen primordia and a well-developed gynoecium. As indicated, the organs of the lower flower abort. Abbreviations: AN = anther; BR = branches; F1 = upper flower; F2 = lower flower; G = glume; GY = gynoecium; LE = lemma; LO = lodicule; PA = palea; SP = spikelet. A and C redrawn from Bonnett (1953) with permission of University of Illinois Agriculture Experiment Station. B redrawn from Cheng, Greyson and Walden (1983) with permission of *American Journal of Botany*.

Spikelet-branch primordia initiate acropetally on the main rachis and the branches. In the sweet corn hybrid Se60 (Fig. 8-2), two rows initiate along the branches, whereas four rows form along the main rachis. Two spikelets, one sessile and one pedicellate, develop from each spikelet-branch primordium. Each spikelet initiates two glumes and two lemmas. Flowers initiate in the axils of each lemma and initiate a palea (Figs. 8-3 and 8-4). Subsequently, each flower initiates two lodicules, three stamens and a gynoecium with its elongate silk (style and stigma).

The cellular basis of these different initiations has been assessed qualitatively. Bonnett (1953) concluded that glumes, lemmas, and paleas initiate, as do leaves, distichously, from periclinal cell divisions in L1 and L2 of the shoot apex. The lower flower was initiated by divisions in L3, and no periclinal divisions were detected in the L1 or L2. Lodicules and stamens initiated by periclinal divisions in the L3, whereas carpels initiated by periclinal divisions in the L1 and L2.

Following organ initiation in both flowers of each tassel spikelet, the gynoecia abort and the remaining organs grow and mature. At anthesis, the stamen filaments elongate rapidly over a 3–4h period, extruding anthers. Simultaneously, the lodicules enlarge and push the glumes apart. The second flower of the spikelet undergoes anthesis within a few days.

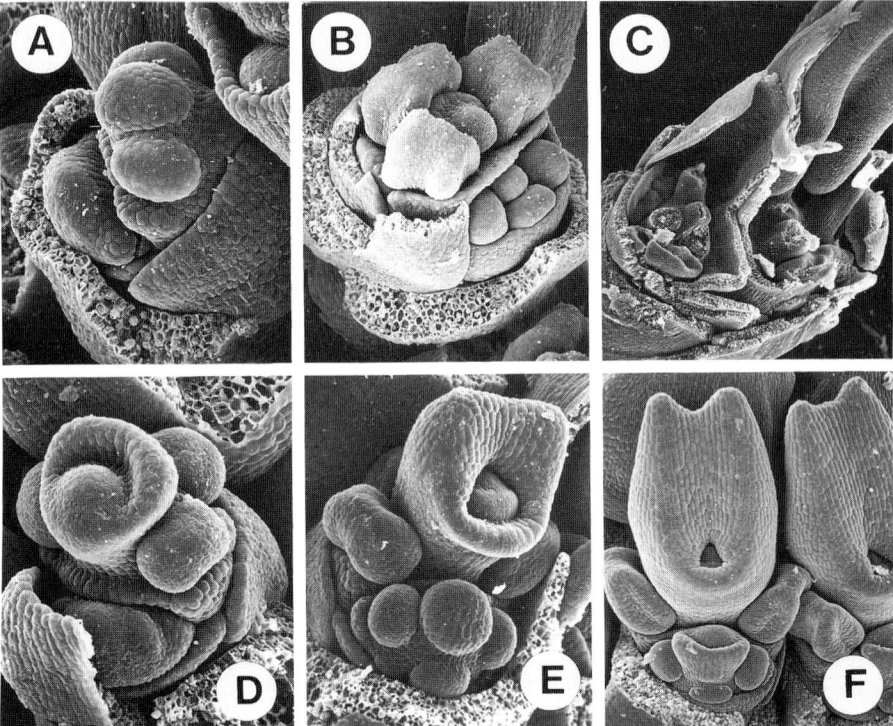

Fig. 8-4A–F. SEM view of androecium and gynoecium initiation in *Z. mays*. (A) Stamen initiation on the upper flower. The third stamen initiates later and slightly below the level of the first two (X160). (B) Later stage in tassel flower development. Development of the upper flower precedes that of the lower flower. The degenerating gynoecium is still visible (X120). (C) Anthers of the upper and lower flowers prior to anthesis (some anthers removed); filaments. Degenerating gynoecia of upper and lower flowers indicated (X25). (D) Upper and lower flowers of ear spikelet. The gynoecial ridge is forming on the gynoecium of the upper flower (140). (E) Developing gynoecium of the upper flower. The stamen primordia are initiated. The lower flower is initiating stamens (X120). (F) Early silk development on the gynoecium of the upper flower with stylar canal. The Lower flower is considerably delayed (X75). Originals courtesy of P. C. Cheng.

Initiation and Growth of Spikelet-Branches, Spikelets and Flowers on the Ear

The early initiation stages of spikelet-branches, spikelets and flowers on the ear, though more compact, are comparable to the same processes in the tassel (Fig. 8-4). Flower development, however, differs significantly.

As in the tassel, both flowers of the ear spikelet initiate, and the lower flower aborts in most cultivars (Fig. 8-4). This abortion results in the development of two rows of flowers, one from each spikelet. In the traditional sweet corn cultivar, "Country Gentleman," the lower flower of each pair fails to abort, and the fruits, therefore, are irregularly arranged on the ear (Weatherwax 1916; Bonnett 1953). The stamen primordia of ear flowers also abort after initiation (Fig. 8-4). In ear flowers, the gynoecial ridge overgrows the flower apex, leaving a small stylar canal (Figs. 8-4 and 8-5), and continues to form the elongate silk. In some lines, the gynoecial ring is triangular in outline and develops three vascular traces. This observation appears to be the basis of the interpretation that the maize gynoecium is phylogenetically derived from three carpels. Initially, the silk primordium is bifurcated and develops two vascular strands. Detailed developmental studies of the silk, which frequently exceeds 30cm at maturity, are urgently needed.

Functional ear shoots initiate in leaf axils near the mid-point of the stem. The position of the upper ear shoot, which is the one that normally develops, coincides with an inflection point in a base-to-tip curved plot of length/width ratios. The factors which determine the axillary buds of these particular nodes to develop as ears, rather than as vegetative buds (tillers), are as unstudied as other heteroblastic features of the maize plant—and for that matter, other grasses (Greyson, Walden and Smith 1982).

The first primordium that initiates on the axillary apical meristem of an ear shoot primordium is called the **prophyll**. It is distinguished from the other leaves (husks) on the ear shoot by possessing two major vascular strands along its flanks. The vascularization of other husks resembles more closely that of the vegetative leaves. In some cultivars, axillary buds associated with husks develop to varying degrees (Fig. 8-5B). In corn, the prophyll shares this two-vascular-strand feature with other first-primordium situations: the coleoptile of the embryo, and the palea of the flower meristem. It is also a common feature of other grasses (Arber 1934).

The Developmental Basis of Monoecy

From a developmental perspective, many aspects of maize inflorescence development deserve further research attention. Foremost are the selective organ abortions that result in male flowers in the tassel and female flowers in the ear. Following initiation, the cells of the aborting tissue become vacuolate (Cheng, Greyson and Walden 1983). Further observations on the cell cycle, the hydrolytic and degradative enzymes of these tissues, are indicated. In view of the observation of Bommineni and Greyson (1987, 1990) that kinetin stimulates stamen development in cultured ears of Se60, whereas GAs stimulate gynoecium development, the pos-

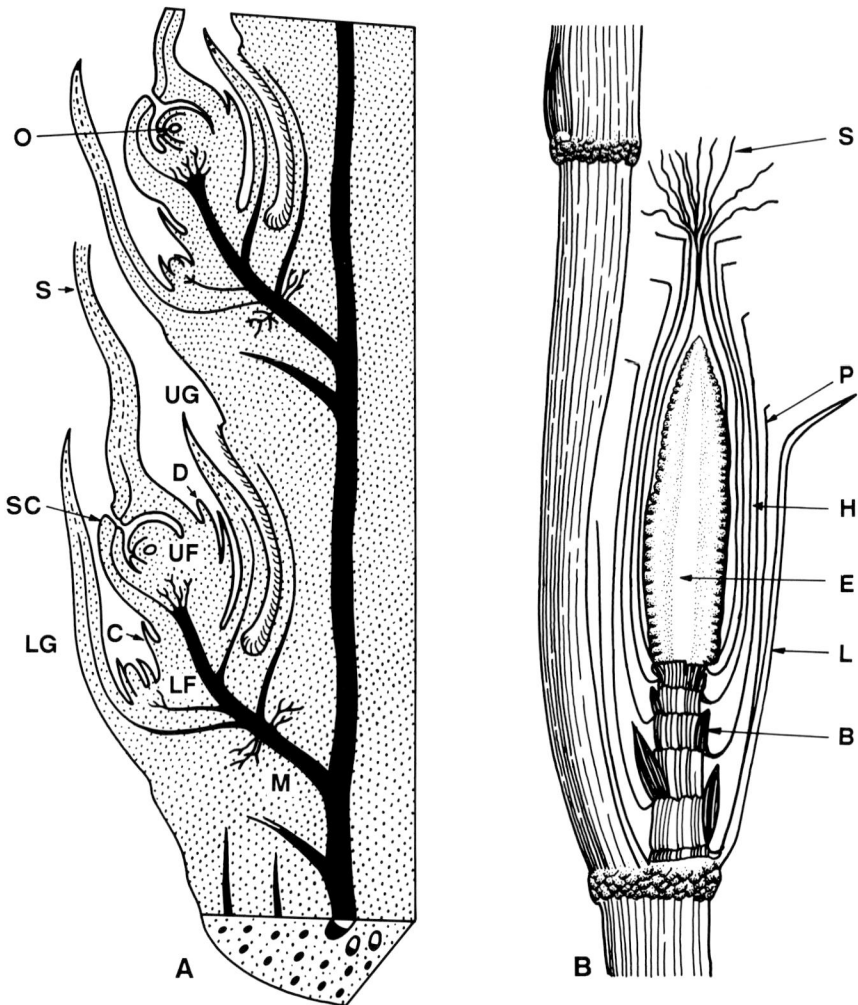

Fig. 8-5A–B. (A) Semi-diagrammatic longitudinal view of an immature spikelet on an ear of *Z. mays*. One flower of the spikelet has aborted. The ovule of the fertile flower (O) is surrounded by the gynoecial wall and the base of the silk (S): the stylar canal (SC) is indicated. Adaxial to the fertile flower is a structure interpreted as a cupule-forming adnate prophyll. Abbreviations: LG = Lower glume; UG = Upper glume; D = lemma of fertile flower; C = palea of the fertile flower; UF = Upper fertile flower; LF = lower non-fertile flower; M = main vascular bundle of spikelet. Redrawn from Nickerson (1954) with permission of American Journal of Botany. (B) Sketch of a longitudinal section of a mature ear branch attached to the main stem. Abbreviations: B = buds in the axils of husks; E = ear; L = leaf; H = husks; P = prophyll; S = silks. From Weatherwax (1955).

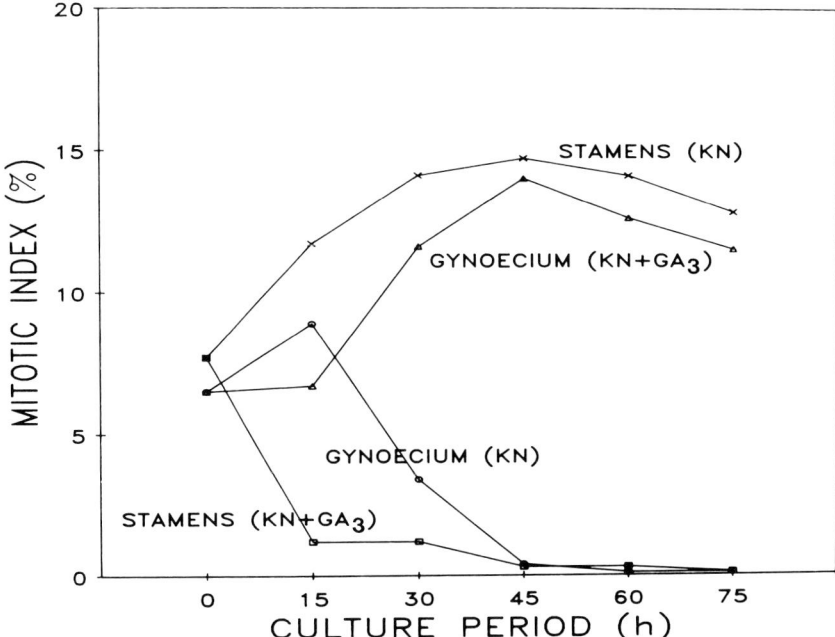

Fig. 8-6. Summary of the mitotic activity associated with developing and degenerating flower organs from the spikelets of cultured maize ear inflorescences after different periods of culture in kinetin or kinetin + GA_3. Because the accumulated counts of cells in mitosis were enhanced by treatment with 8–hydroxy quinoline, the mitotic index (MI) is overstated. From Bommineni and Greyson (1990).

sibility that some of these developmental decisions involve the localized and selective actions of these two classes of PGS requires detailed examination. The selective stimulation and repression of mitosis in these developmental events is illustrated in Fig. 8-6.

Genetic Variability in Flower Development

The documentation of genetically based variability of flower traits in maize represents a substantial opportunity for developmental study and is one of the main reasons for considering maize as a serious model system for flower research. The availability of genetic stocks in known genetic backgounds is also exemplary. Only a portion of this resource is discussed here. Three types of genetically based flower variability (Table 8-1) are discussed. Additional information about the genetic resources is available elsewhere (Coe, Neuffer and Hoisington 1988), particularly in the annual *Maize Genetics Newsletter* (Appendix 2).

Cytoplasmic Male Sterility (c-mst)

In maize, three major types of cytoplasms (T, S and C; each of which has been identified in a number of different sources) which confer male sterility on plants

Table 8-1 Mutants of Maize *(Zea) mays)* (2N = 20) Affecting Flower and Inflorescence Structure (Excluding Male Sterile Examples)

Symbol	Name	Map Location	Phenotype
an_1	anther-ear	1L-104	Andromonoecious dwarf; responds to gibberellins.
d_1	dwarf-1	3S-44	Andromonoecious dwarf; responds to gibberellins.
d_2	dwarf-2	3	Dwarf; like d_1.
d_3	dwarf-3	9S-59	Dwarf; like d_1.
d_5	dwarf-5	2S-34	Dwarf; like d_1.
D_8	Dwarf-8	1L-133	Andromonoecious dwarf; unresponsive to gibberellins (Harberd and Freeling 1989).
Mpl1	Miniplant	1L-near Adh1	Miniplant; andromonoecious; unresponsive to gibberellins (Harberd and Freeling 1989).
ms	male sterile		More than 20 male steriles are listed and available for research.
ra	ramosa	7-32	Ear branched; tassel conical.
ts_1	tassel-seed-1	2S-74	Tassel pistillate and pendant.
ts_2	tassel-seed-2	1S-24	Like ts-1.
ts_4	tassel-seed-4	3L-73	Tassel compact and upright with gynoecious and staminate flowers.
Ts_5	Tassel-seed-5	4S-53	Nearly normal tassel; scattered gynoeciuos flowers with short silks.
Ts_6	Tassel-seed-6	1L-158	Tassel compact, gynoecious to mixed; short silks.
Tu1	Tunicate	4L-101	Tassel with few granches, glumes large, and some gynoecious flowers with silks; ears with enlarged glumes enclosing kernals.

Source: Taken from Coe, Neuffer and Hoisington (1988)

containing them have been recognized, in addition to the normal (N) cytoplasm. For each c-mst, a selection of nuclear fertility restorer (Rf) genes has been isolated (Kaul 1988).

The expression of these traits is restricted spatially and temporally to the anther, with little pleiotropic expression in other tissues, at least under normal conditions; c-mst T plants are, however, susceptible to toxins from the pathogen *Helminthosporium maydis* (Race T). In c-mst T plants, anthers form but fail to extrude, the mitochondria of tapetal cells become altered and pollen aborts during microsporogenesis (Warmke and Lee 1977). Restoration is conferred by the genetic constitution of the sporophytic tissue (sporophytic reversion) (Lee, Earle and Gracen 1980). In c-mst S plants, expression is restricted to the abortion of nearly mature pollen (Lee, Earle and Gracen 1980), and restoration is conferred by the nuclear restorer factors in the pollen (gametophytic restoration). Anthers from c-mst C plants develop abnormal tapetal cells at the early tetrad stage and the pollen aborts (Lee, Gracen and Earle 1979). Since it was first tied to a mitochondrial lesion, T strain c-mst has been studied extensively (Leaver and Gray 1982; Laughnan and Gabay-Laughnan 1983). The characterization of the molecular features of this phenomenon proceeds at a reasonable pace, though many features remain unexplained (Leavings and Brown 1989).

Genic Male Sterility (g-mst)

Because its cytogenetics and genetics have been studied so intensively, maize must be considered one of the better model systems for the developmental study of g-mst. Technically, any genetic aberration which leads to a reduction in the amount of viable pollen can be said to be a genic male sterile (See discussion in Chapter 5).

In practice, however, g-mst is restricted to those situations where the action of a single nuclear gene (frequently recessive) leads to the failure of microsporogenesis and/or the failure of viable pollen production. Few pleiotropic traits in other tissues are observed. Description of the cellular events from light and TEM images is the most common study technique, though attempts to uncover the physiological and molecular biological concomitants are being made. Of the more than fifty genes currently known to affect flower development in maize, nearly thirty are listed as being g-mst (Coe, Neuffer and Hoisington 1988; Kaul 1988), and sample-seed for more than twenty of these g-mst lines is available from the Maize Genetics Stock Center (Appendix 2). Descriptive studies, to date, of individual g-mst strains include the observations from anther squash preparations representing thirteen different g-mst loci (Albertson and Phillips 1981) and TEM and SEM observation on ms-10 (Cheng, Greyson and Walden 1979) and ms-9 (Greyson, Cheng and Walden 1980).

Because of the restricted expression of so many individual genetic sequences, this collection in maize represents a very special resource for developmental studies of gene expression and cell differentiation. One assumes that these spatially and temporally localized developmental failures arise because of separate individual aberrant syntheses at different stages and at perhaps different cytoplasmic sites during microsporogenesis. For example, ms8 and ms9 break down in early meiosis, whereas ms15, ms11, ms17 and ms24 express during or after microspore mitosis (Coe, Neuffer and Hoisington 1988). In some cases, the time and site of gene expression can be identified with precision. In a number of cases, the gene has been mapped to a chromosome locus. The exploitation of this genetic information in the context of *in situ* localization of protein and mRNA technology should be most revealing.

Because most g-mst mutants are recessive and few pleiotropic traits can be used to identify g-mst plants, serious future research with g-mst mutants of maize will require the production of homozygous stocks. A number of schemes have been suggested to accomplish this objective:

1. Patterson's duplicate-deficient chromosomal procedure (Patterson 1975). This has been the most successful method of generating nearly 100% ms seed of a few male steriles. The preparation of the seed, however, requires a thorough familiarity with the use of the appropriate cytogenetic manipulations in breeding.
2. Exploitation of close linkage between a genic male sterile gene and an easily scored vegetative marker trait. At present only one example is readily available: the male sterile-silky trait (si1) (position 20 on chromosome

6), which is listed as being 3 centimorgans from yellow endosperm (y1) (Coe, Neuffer and Hoisington 1988). Dales (personal communication) used this linkage to sort and use homozygous plants for use in manipulative experiments.
3. Chemical reversion (see chemical complementation in Chapter 5). To date, attempts to revert g-mst mutants of maize through the application of chemicals have not succeeded (Pareddy 1990), suggesting that the genetic blocks to development in the lines tested are not open to chemical complementation (i.e., are cell autonomous); alternatively, the appropriate metabolite has not yet been tested. Records of variable expression of some g-mst mutants of maize are occasionally mentioned, though no well-documented citations are available. Obviously, demonstrations that some features of g-mst in maize are developmentally variable are most enticing. So, too, are the report of differential expression of male sterility on a tiller of a male sterile plant and the production of fertile pollen on tillers of ms-14 plants (Albertson and Phillips 1978).
4. The establishment of a proliferative axillary bud culture (Raman, Walden and Greyson 1980; Walden et al. 1989) of g-mst lines so that a continuous supply of male sterile plants is available for study. The initial establishment of the culture conditions, depending on background, will require some effort, as will the eventual selection of clones from which a steady supply of homozygous g-mst plants can be derived. Parallel +/ms and +/+ clones should also be maintained as control material. This technique is a variation of traditional clonal propagation by standard procedures of isolated g-mst plants for regenerative species such as *Oryza, Lycopersicon* and *Glycine*.
5. Molecular approaches. It seems inevitable that the recently developed knowledge and facility in manipulating genes and their expression will ultimately result in a variety of new methods to control microsporogenesis and pollen fertility. The induction of male sterility in tobacco and rapeseed through the incorporation of a chimeric ribonuclease gene linked to a tapetum-specific expression factor is an example of one such opportunity (Mariani et al. 1990). The isolation and utilization of the g-mst genes themselves will no doubt also be attempted.

Tassel-Seed Flower Traits

In addition to environmentally-based tassel variability, a number of genetic variants have been identified and preliminary morphological analyses have begun (Emerson 1920; Nickerson and Dale 1955; Irish and Nelson 1989). Five different tassel-seed mutants are listed in Table 8-1.

The tassels of tassel-seed 1 (ts_1) and tassel-seed 2 (ts_2) plants are similar in that they are heavily feminized with well-developed silks. The upper stem internodes are shortened, and the tassel (or branches) is pendant. The main difference between the two phenotypes of these mutants lies in the proportion of the flowers that are female with well-developed silks. In ts_1, tassel flowers are entirely female with well-developed silks; both florets of a spikelet frequently develop; and occasional staminate flowers appear at the tip of the inflorescence (Nickerson and Dale 1955). In ts_2, both female and perfect flowers are produced in some material, though male flowers are not recorded (Nickerson and Dale 1955). These mutants differ morphologically from each other in the number of gynoecia with silks that develop and in the severity of the pendant tassel trait.

The obvious potential of ts_1 and ts_2 for experimental exploration of the genetic and physiological relationships in flower development is unfortunately limited by the difficulty of obtaining quantities of pure homozygous seed. Irish and Nelson (1993) document the development of ts_2 inflorescences through SEM observations. Tassel-seed 4 (ts_4), by contrast, is readily maintained in pure lines because most tassels produce some flowers with functional stamens and viable pollen (Fig. 8-7). Because of the considerable variability in flower and spikelet structure, this material deserves extensive and detailed morphological observation. The report of anomalous silks associated with glumes of ts4/ts4 (Phipps 1928), though questioned by Nickerson and Dale (1955), deserves reexamination in view of the report of adventitious silks in the mutant mssi (Dales, personal communication). In fact, all these tassel-seed mutants, if properly studied, might yield information on flower structure not available in standard-pedigreed wild-type lines.

The dominant mutants, tassel-seed 5 and 6 (Ts_5 and Ts_6), while easily maintained in 1:1 segregating lines, are less easily maintained as homozygous lines, presumably due to a degree of lethality or poor results with self-pollinations. The

Fig. 8-7A–C. Mutant inflorescences in maize. (A) Tassel of tassel-seed 4 (ts4). (B) Tassel of tassel-seed 5 (Ts5). (C) Tassel of tassel-seed 6 (Ts6).

tassels of Ts_5 plants (Fig. 8-7) bear staminate, gynoecious or perfect flowers, though this tassel-bears fewer silks than the other tassel seed mutants. Many tassels of Ts_6 are profusely covered with silks, some with few or no stamens. In other Ts_6 tassels, staminate and perfect flowers provide some viable pollen. The ears of all these plants are reduced in size and number of fertile ovaries, compared to appropriate wild-type plants. Ears of tassel-seed plants develop better if tassels are removed early in development. While heterozygous and homozygous examples of these two dominant mutants segregate from crosses (unpublished data), no careful morphological comparison has been made between their tassels. Preliminary tests of applications of GA_3 on some of these tassel-seed mutants gave equivocal results (Nickerson 1960*a*).

The dominant mutant, tunicate (Tu) also deserves detailed attention as a developmentally useful inflorescence and flower trait. Its phenotype includes enlarged glumes with feminization of some flowers of the tassel and greatly enlarged glumes with enlarged ear rachis and proliferation of flower primordia. The Tu locus is complex, the parts of which can be separated and recombined in different proportions (Mangelsdorf and Galinat 1964).

Anecdotal data (Greyson, unpublished) indicate that homozygous (Tu/Tu) plants are occasionally produced. However, the most consistent method of preparing research material is from a 1:1 segregation of a cross between Tu/+ and +/+. Further, immature Tu/+ tassels were shown to contain measurably higher levels of extractable cytokinin-like materials as measured by bioassay and spectrophotometry. Presumably, this fact correlates with the cauliflory of some ear specimens.

Poethig (1985) considers that three dominant mutations (Corngrass (Cg); Teopod 1 (Tp1) and Teopod 2 (Tp2)) can be interpreted as examples of homeosis in that they involve transformation of reproductive structures into leaves. The expression of these traits is variable depending on background and different conditions. (Chapter 9 has an additional discussion of homeosis.) Nickerson (1960*b*) obtained near normal phenotypes of plants of all three mutants by exogenous application of GA_3; however, Poethig found no such effect of GA_3 on Tp1 and Tp2. These preliminary observations, coupled with additional ones on other flower- and inflorescence-variable material (Coe, Neuffer and Hoisington 1988), represent the beginning of what could become an extensive body of basic information on plant developmental genetics.

Andromonoecious Dwarfs

In addition to being dwarf, due to shortened internodes, the maize mutants d_1, d_2, d_3, d_5, an_1, D_8 and Mpl1, to varying degrees, develop stamens on their ears (Abbe and Phinney 1940, 1942). The degree of expression of stamen formation in individual flowers and the resultant patterns of perfect flowers on the ear and the degree of maturation of individual stamens vary and are presumably subject to background and developmental and/or environment-based inputs. Most frequently, stamens in these ears do not produce viable pollen and do not extrude their anthers (Coe, Neuffer and Hoisington 1988). On ears of an_1, the flowers at the tip of the ear are primarily male; those near the center of the ear primarily contain stamens and gynoecia; and those near the base are primarily female (Karpoff, personal communication). Tassel development is generally normal, though compact, and pollen fertility can be variable.

With the exception of D_8 and Mpl1, near-normal plant height of mutant specimens can be achieved through appropriately timed applications of GAs. These mutant conditions in d1, d2, d3, d5 and an1 are assumed to be based on genetic blockages at different steps in the biosynthesis of GA_1 (Phinney and Spray 1983). In D_8 and Mpl1, which are not GA-revertible, the mutant conditions are thought to arise by some failure in the GA_1 receptor mechanism (Harberd and Freeling 1989).

Environmental and Experimental Modification of Flower and Inflorescence Development

The development of silks, ovaries and mature fruits in the tassel of maize can be induced in wild-type populations by short days, cool temperatures and disease (Heslop-Harrison 1961; Heslop-Harrison 1972; Rood, Pharis and Major 1980). This plasticity has provided a continuing incentive to uncover the physiological basis of this developmental expression, an important aspect of which is the involvement of PGS.

PGS and Flower Development in Maize

Various observations implicate PGS as naturally occurring and normal components of the regulation of male and female flower development in the tassel and ear, respectively, of maize. Serious questions, however, remain about the modes of action and the levels of causality involved. The most obvious PGS involvement in maize is that of the GAs.

The conclusion that GAs are normally involved in flower development follows from three different kinds of data: (a) estimation of endogenous content under different flowering conditions; (b) exogenous application; and (c) the correlation between flower variation and GA metabolism in andromonoecious dwarf mutants.

The best estimates of total levels of GA-like substances in developing tassels and ears are those of Rood, Pharis and Major (1980). These data (Table 8-2), document that low-light-induced feminization of tassels is associated with total levels of GA-like substances above 100ng/kg between 17d and 39d. Comparable estimates from tassel shoots under high-intensity light fell off to <1ng/kg by 39d after emergence. The total GA levels of 39d ears under these conditions were intermediate. Thus, the development of silks and ovules in tassels and also in ears takes place in association with relatively high tissue levels of GA-like activity. Differences between the levels of polar and non-polar GA-like substances were also monitored.

These data appear to be consistent with a number of reports of exogenous application (Nickerson 1959; Nickerson and Embler 1960; Hansen, Bellman and Sacher 1976; Krishnamoorthy and Talukdar 1976) and with the *in vitro* study on immature maize ears (Bommineni and Greyson 1990). The coincidence of stamen

Table 8-2 Estimates of GA-like Activity (μg GA$_3$ Equivalents per kg of Meristematic Tissue) in Terminal (Tassel) and Axillary (Ear) Normal and Feminized Shoot Meristems of Maize (Inbred 66A4-2)

Tissue	10 Days	17 Days	24 Days	39 Days
Normal	25.5	109	1.8	0.2
Feminized	33.6	182	162	122
Normal ears	ND[1]	ND[1]	ND[1]	27.6

Note: Normal flowering plants were grown under natural light plus supplemental light (8h/968 μEm^{-2} s^{-1}).
Feminized plants developed under natural light plus supplemental light (8h/323 μEm^{-2} s^{-1}). Estimates determined by bioassay in serial dilution on dwarf rice (cv. Tan-ginbozu) seedlings.

[1] ND = Not determined

Source: From Rood, Pharis and Major (1980) with permission of American Society of Plant Physiology.

development in ears of andromonoecious dwarfs and the demonstrated lowered levels of GA-like substances in some tissues of some dwarfs are also consistent with this general assumption (Phinney 1984; Phinney and Spray 1983, 1985). An opposite interpretation (Sladky 1986)—that enhanced levels of GAs are associated with maize stamen development, based on extraction and bioassay techniques—is obviously at variance with the above interpretation. So, too, is the generalization about PGS involvement in monoecy, derived from other species (Chailakhyan and Khrianin 1987).

That GA physiology (including total levels of GA-like materials and conjugates, levels of individual GAs and the status of binding sites) is only one component of a complex PGS involvement in inflorescence and flower development is suggested by additional observations involving auxins, ethylene and cytokinins. Exogenous applications of auxins and ethylene stimulate the feminization of tassels (Heslop-Harrison 1964; Krishnamoorthy and Talukdar 1976). A suggestion that cytokinins are also involved arises from the demonstration that stamen development is stimulated and gynoecium growth is inhibited by kinetin in *in vitro* cultured ears (Bommineni and Greyson 1990).

Wheat (*Triticum aestivum* L.)

The selection of wheat as a model research plant for studies of inflorescence and flower development, over other grains, is arbitrary. For some research objectives, barley, oats or rye could serve equally well or better. However, its agricultural preeminence throughout the world and the diversity of sites where it is cultivated (Briggle and Curtis 1987) argue heavily for its selection. The following discussion relates to the "spring wheats," which do not respond to vernalization, as opposed to the "distinct winter" and "intermediate" varieties (Pinthus 1985). The relative ease with which many varieties will flower under a variety of conditions should encourage a number of different basic studies. Cultivated wheat includes four species, of which the most common is *T. aestivum*. It represents the world's bread and biscuit wheats (Pinthus 1985).

Origin of the Inflorescence and Flowers

Vegetative stems (culms, tillers) arise from axillary positions on a short rhizome (Lersten 1987). The number of tillers is variable in response to cultivar and cultural differences. Each tiller bears 5–7 nodes and terminates in an inflorescence (Lersten 1987). The winter cultivar "Chlumecka 12" (Opatrna, Seidlova and Benes 1964) produced 10–11 vegetative leaves before initiating an inflorescence. Other qualitative and quantitative analyses of the leaf- and internode-generating activities of wheat are summarized by Williams (1975).

The wheat inflorescence, frequently referred to as an ear or head, is technically a spike with two rows of lateral branches (spikelets), each of which has two sterile glumes and 2–5 florets (Lersten 1987). It arises when the terminal shoot apex stops producing leaves, elongates and begins to form bracts and axillary spikelet primordia (Bonnett 1936; Barnard 1955; Moncur 1981; Lersten 1987). The paired bract and spikelet primordia form the **double ridges,** which are the first positive criterion of inflorescence initiation. The double ridges are interpreted as equivalent (homologous) to the leaf/axillary bud relationship of the vegetative portion of the plant. The number of double ridges, which indicates the number of spikelets that will form, is variable depending on the cultivar and growing conditions.

Transition to Flowering

Considerable cultivar- and environment-based variation in the number of days to floral initiation has been documented. In a semi-dwarf spring wheat sown in Israel, apex elongation (AE) occurred at 20 days after seedling emergence, and floral initiation (FI) occurred at 30 days (Fig. 8-10A) (Pinthus 1985). FI is identified by the appearance of "double ridges" on the meristem. Another experiment documenting the total number of organs, leaves and spikelets at different times after planting is illustrated in Fig. 8-10B. Considerable variation in these different parameters is produced with different cvs. and different conditions.

Spikelet and Flower Initiation

The distichously arranged "double ridges" are interpreted as the initiation of a leaf primordium and its associated axillary bud (spikelet primordium) (Fig. 8-8). The pattern of initiation of these ridges is distichous and acropetal.

The cellular details of initiation of this leaf and spikelet primordium were followed by Barnard (1955). His observations are both striking and provocative from a developmental point of view. The leaf primordia of the inflorescence primordium—more correctly, bracts, as with vegetative leaves—arise through periclinal divisions in L1 and L2 (Fig. 8-11). Spikelet primordia, however, are initi-

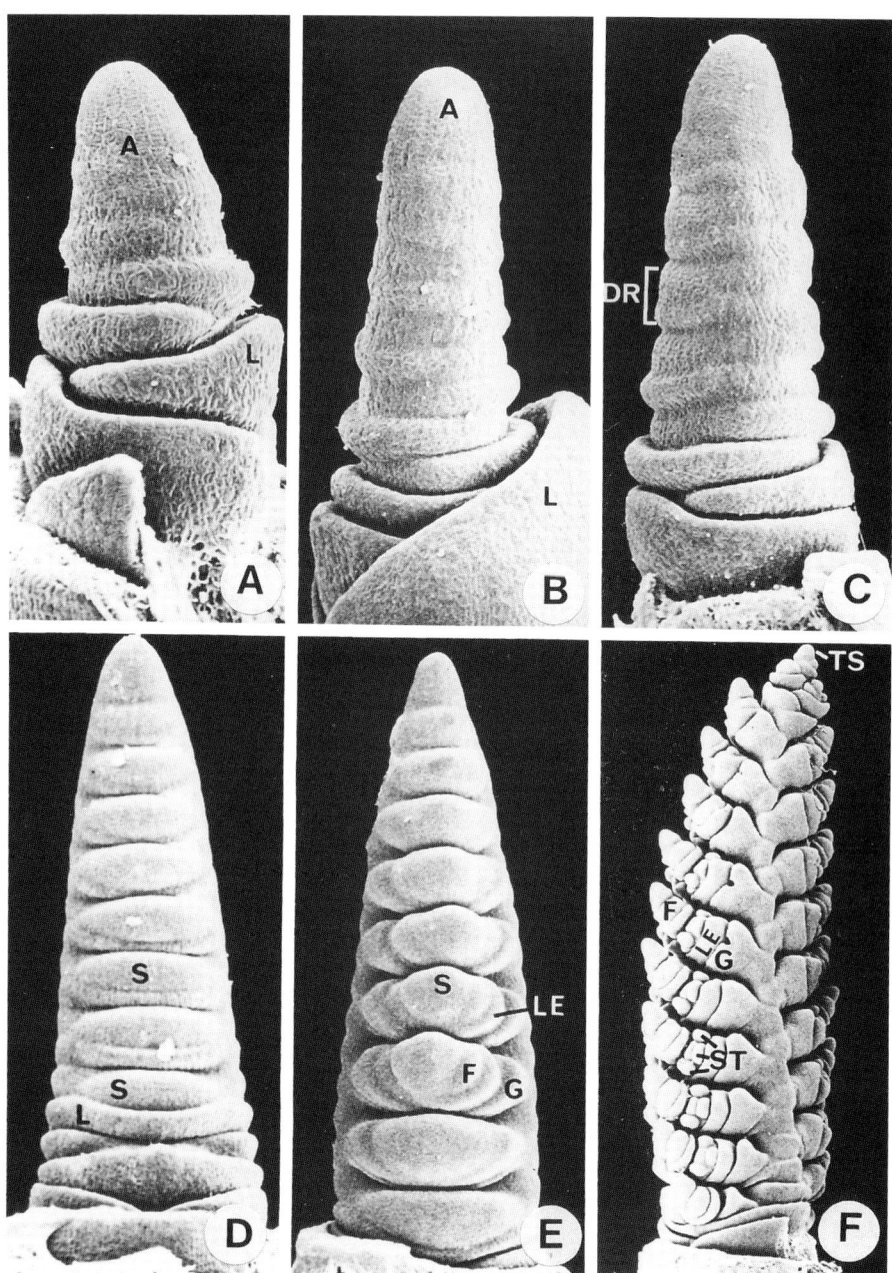

Fig. 8-8A–F. SEM views of the shoot apex of *Triticum aestivum* during the transition from leaf formation to flowering. (A) Vegetative phase. (B) Pre-initiation phase with elongation of the apex. (C) Flower initiation: double ridges (DR) initiated. (D) Spikelet primordia enlargement. (E) Glume initiation. (F) Floret initiation. Abbreviations: A = apex; L = leaf primordium; DR = double ridges; S = spikelet primordium; F = flower primordium; G = glume; LE = lemma; ST = stamen; TS = terminal spikelet. From Moncur (1981) with permission of CSIRO Division of Forestry.

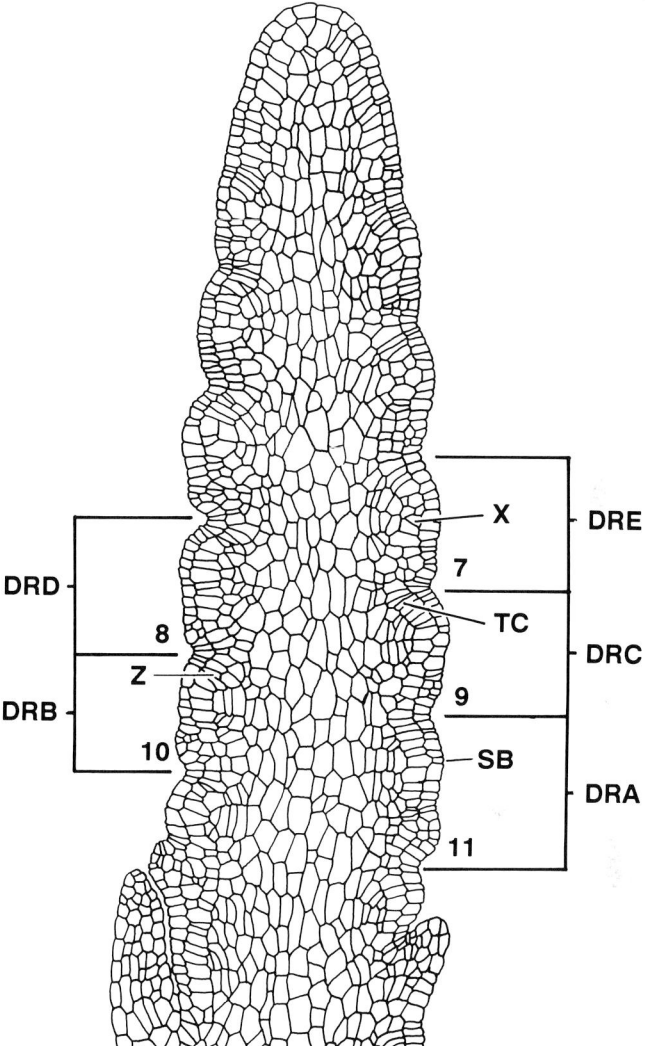

Fig. 8-9. A "mosaic" reconstruction of a median longitudinal section of an immature spring wheat (cv. "Thatcher") inflorescence primordium at the "double ridge" stage. Abbreviations: SB = spikelet buds; TC = thin cells; X = periclinal divisions in the hypodermal layer; Z = periclinal divisions in the thin cells; DRA–E = five double ridges A–E. Numbers refer to leaf primordia subtending axillary buds of the double ridges. From Sharman (1983) with permission of *Annals of Botany*.

ated through periclinal divisions of the cells of the outer layer of the corpus (subhypodermal).

This interpretation of leaf and bud initiation is corroborated by Sharman (1983), who studied three different cultivars: a British winter variety, a British spring variety and a Canadian spring wheat. His conclusions are as follows:

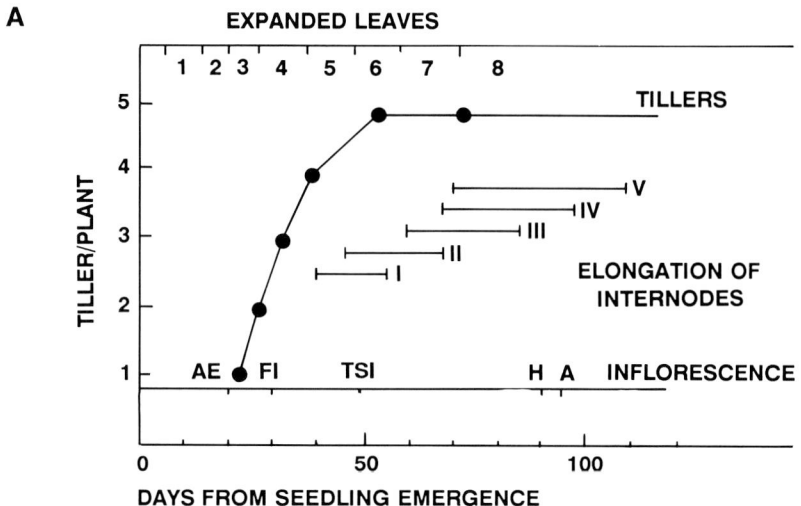

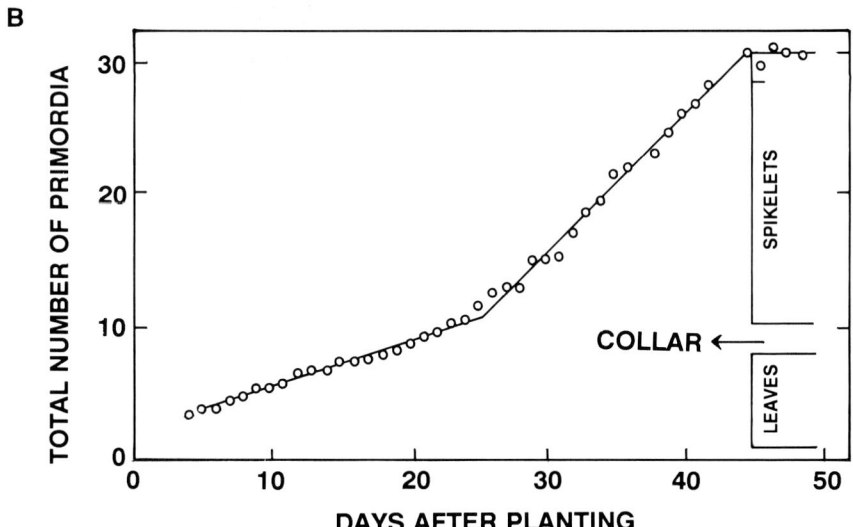

Fig. 8-10A–B. (A) Phenological presentation of vegetative and reproductive development of an early semi-dwarf spring wheat sown in Israel in November. Abbreviations: AE = apex elongation; FI = floral initiation; TSI = terminal spikelet initiation; H = heading; A = anthesis. Reproduced with permission from Pinthus (1985) in A. H. Halevy [ed.], *CRC Handbook of Flowering*, vol. 4, pp. 418–443. Copyright CRC Press Inc. Boca Raton, FL. (B) Observations of the total number of leaf and spikelet primordia at various times after planting in a spring wheat (cv. Kolibri). From Kirby (1974) with permission of Cambridge University Press.

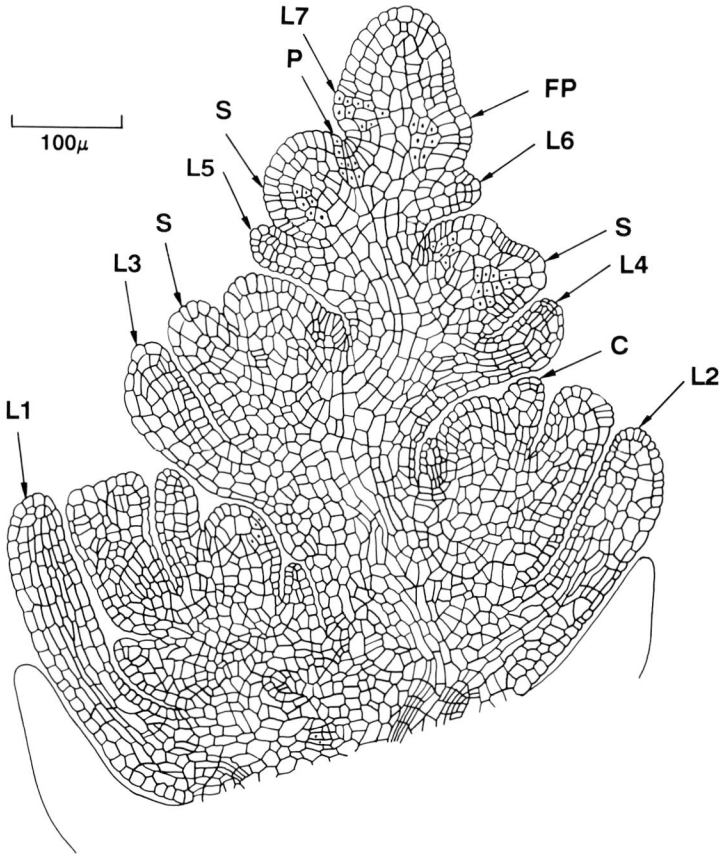

Fig. 8-11. Longitudinal section through a wheat spikelet in which the seventh lemma (L7) is arising; the origin of the youngest flower primordium is shown at FP. Spikelet from spike 6 mm long. Abbreviations: P = palea; S = anterior stamen; C = carpel; L1–L6 = first to sixth lemmas. From Barnard (1955) with permission of *Australian Journal of Botany*.

> The tip of the shoot apex normally consists of a core of irregularly arranged cells covered by two uniseriate self perpetuating layers—no third, inner layer is present. Leaf initiation involves periclinals in the cells of the dermatogen and hypodermal layers but not in the core. Buds involve many periclinals in the outer cells of the core, a few occasionally in the hypodermal layer but never any in the dermatogen.

Some of these observations are summarized in Fig. 8-9. In the initiation of the spikelet primordia, no periclinal divisions were observed in the dermatogen. Unfortunately, the anatomical investigation of wheat inflorescence development of another winter wheat (Opatrna, Seidlova and Benes 1964) did not examine this question, nor is it possible to confirm or reject the conclusion from the latter's illustrations.

After initiation, the apical meristem of the spikelet primordium resembles, in its organization, that of the vegetative apex (Barnard 1955). The spikelet meristem forms, distichously, a pair of glumes (Barnard 1955) followed by a variable number of lemmas (Fig. 8-12 and Fig. 8-13A). Florets (flowers) develop in the axils of some (perhaps two to five) of the lemmas. Organ initiation on the floret meristem includes a two-keeled palea, two lodicules, three stamens and a central gynoecium (Fig. 8-13B) (Percival 1921; Lersten 1987). Stamen initiation, as with the spikelet and flower, begins with periclinal divisions in the subhypodermal cells (Barnard 1955). As indicated by Percival (1921), a gynoecial ring overgrows the apex, developing two pointed styles (Fig. 8-14). This gynoecial ring, as with leaf initiation, involves periclinal divisions in dermatogen and hypodermal layers. The flower apex develops directly into the ovule (Figs. 8-12 and 8-14). While there has been limited careful study of the early growth of stamen and gynoecium, microsporogenesis and megasporogenesis were described by Bennett et al. (1973). The vasculature of the wheat spikelet was explored by O'Bricn et al. (1985), and the flower vasculature of grasses has been described by Pizzolato in a series of papers (see Pizzolato 1993 for references).

In spite of the work already accomplished (Barnard 1955; Williams 1960, 1966b; Percival 1921), considerable scope for further, carefully detailed descriptive analysis of normal development seems warranted. These studies could serve as a basis for examination of aberrant development, whether naturally occurring or from experiments. Sharman's (1967, 1978, 1983) studies of abnormal flowers from both untreated populations and plants treated with 2,4-dichlorophenoxyacetic acid (2,4-D) represent preliminary explorations of this interesting study in wheat.

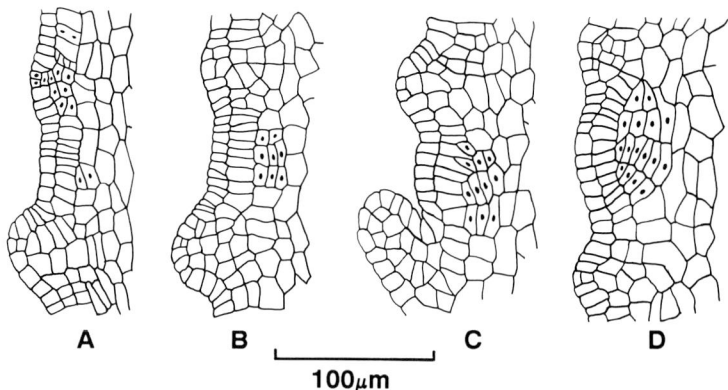

Fig. 8-12A–D. Longitudinal microscopic views through a spike of wheat to illustrate cell division patterns during bract and spikelet initiation. (A) Periclinal divisions in the protodermal layer of bract initial, and first periclinal division in the sub-hypodermal layer. (B) Two periclinal divisions in three sub-hypodermal cells of spikelet initial. (C) Inclined walls in hypodermal cells in later spikelet initial. (D) Spikelet initial is an obvious bulge with files of cells derived from the original sub-hypodermal cells and the first division of the fourth sub-hypodermal cell. From Barnard (1955) with permission of *Australian Journal of Botany.*

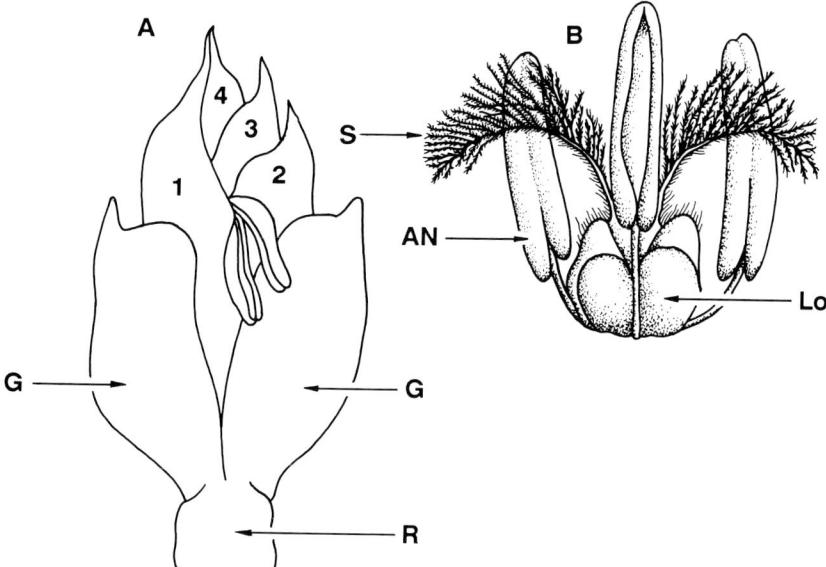

Fig. 8-13A–B. Views of the spikelet and flower of *Triticum aestivum*. (A) A spikelet with two basal glumes (G) and four numbered lemmas (1–4). Two anthers dehiscing. R = rachis. Reproduced with permission from Pinthus (1985) in A. H. Halevy [ed.], *CRC Handbook of Flowering*, vol. 4, pp. 418–443. Copyright CRC Press Inc. Boca Raton, FL. (B) Wheat flower, lemma and palea removed. Abbreviations: AN = anther; Lo = lodicule; S = style. Reproduced from Percival (1921) with permission of Gerald Duckworth and Co. Ltd.

Opportunities for Experimental Studies

Genetic Studies

For a number of reasons, opportunities to exploit genetic aspects of flower development are limited. Genetic linkage maps in wheat are sparse (McIntosh and Cusick 1987), and the chromosomal complement is complex. Nonetheless, some opportunities are available.

Frankel (Frankel, Shineberg and Munday 1969; Frankel 1976) describes a series of "base sterile" (speltoid) wheats in which the lowest one, two or more florets of a spikelet are missing or defective. The genetic features are complex. They result from a deletion or an inactivation of the Q factor of *vulgare*, a complex gene or supergene responsible for the *vulgare* syndrome of soft glumes, tough rachis and short internodes in the spike (Frankel, Knox and Considine 1981). Genetically controlled gradients govern the pattern of sterility between the spikelets in a spike and within spikelets. In addition, some strains exhibit temperature and daylength sensitivity.

The timing of expression of two of these speltoid genes, St_1 and St_3 (Fig. 8-15), was investigated by stereological analysis and quantitative histochemical analysis for nucleic acids (Considine, Knox and Frankel 1982). RNA content of the lem-

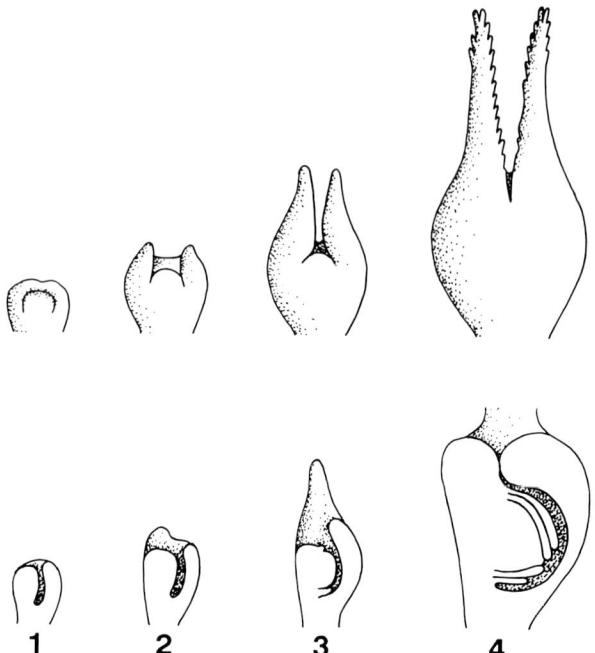

Fig. 8-14. Lateral and sectioned views of four stages in the development of the wheat gynoecium. An ovule arises from the terminal portion of the flower meristem. A gynoecial ridge overgrows the ovule and forms the ovary wall and the styles. Reproduced from Percival (1921) with permission of Gerald Duckworth and Co. Ltd.

mas subtending presumptive fertile or sterile flower primordia was strikingly different in the floral meristems of the two types. There was a linear increase in the nucleic acid content during fertile flower development, but no significant increase in the RNA content in sterile St_3 flower primordia. A stereological study revealed that rates of cell number increase in the wild-type and the speltoid plants were significantly different. The authors conclude that these fertility-controlling alleles exert their effect prior to the appearance of a visible floral primordium and probably after the initiation of the lemma.

This system of speltoid genes appears, therefore, to be another example of a genetically regulated and environmentally sensitive developmental situation in the grasses which affects organ growth and maturation. It appears to be comparable to the developmental control of stamen and gynoecium growth in maize, where genetic, environmental and PGS variability are documented. These localized regions of selective growth inhibition represent a significant opportunity for future developmental exploration.

There are many opportunities for the study and manipulation of male sterility (Kaul 1988), though the complexity of the genome and the lack of linkage data vitiate the usefulness of this feature in wheat in developmental studies. The ultimate objective, comparable to that for barley (Hockett, Baenziger and Steffens

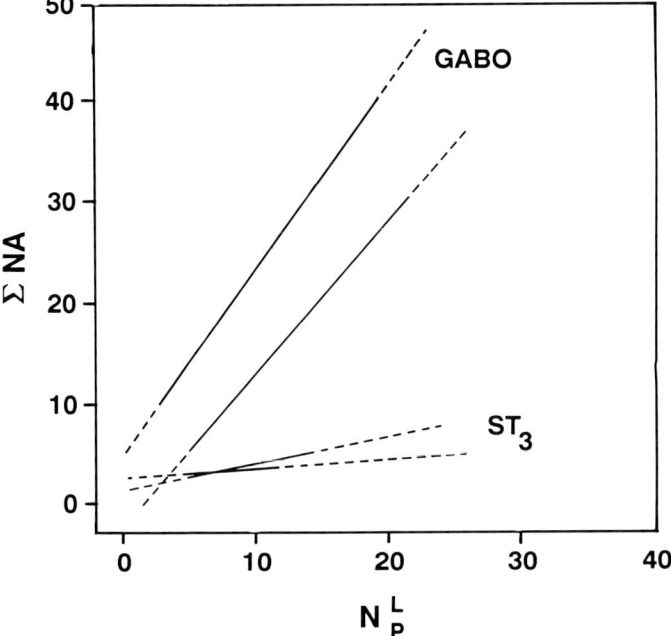

Fig. 8-15. A comparison between the total nucleic acid content (Σ NA, arbitrary units) measured in florets of a fertile wheat (var. Gabo) and sterile (St_3) spikelets of wheat plotted as a function of the number of epidermal cells transected in the perimeter of the subtending lemma primordium (N L/P). The regression equations were computed from the data and used to plot the lines of best fit. Analysis of covariance demonstrated that the regressions fitted for the two Gabo spikelets are statistically different from the two St_3A spikelets at $P < 0.01$. From Considine, Knox and Frankel (1982) with permission of *Annals of Botany*.

1978), however, of a revertible sterility to facilitate the production of hybrid wheat, is being explored primarily by commercial seed and chemical companies.

In Vitro *Culture*

The ease with which immature wheat inflorescences can be dissected and isolated makes them most suitable candidates for a variety of *in vitro* culture studies. A number of unpublished anecdotal reports suggest that considerable success with developmental studies might be possible. Wheat has been successfully employed in a number of different tissue culture protocols including anther haploid production (Kudirka, Schaeffer and Baenziger 1986) and callus and suspension cultures (Schaeffer, Lazar and Baenziger 1984; Sharma, Gill and Sears 1984). Various opportunities for anther and ovary culture in wheat and other cereals were reviewed by Dunwell (1985a). A quite different objective, the culture of wheat inflorescences in order to study the factors responsible for spikelet maturation and grain-filling, has been well-tested (Donovan and Lee 1977, 1978; Singh and Jenner 1983; Nicholls 1986; Armstrong, Soong and Hinchee 1987; Trione and Stockwell 1989).

Rice (*Oryza sativa* L.)

Rice has been cultivated, initially and primarily in Asia, for over 6000 years, and varieties have been selected to succeed in the most diverse environmental situations of any major food crop (Vergara 1985). The common cultivated rice of Asia is derived from *O. sativa* with three generally recognized subspecies, *Indica*, *Javanica* and *Japonica*. These are distinguished by a number of morphological and physiological characters (Takehashi 1984; Oka 1991). Cultivars are available for cultivation from sea level to 3000m altitude and from 30°S to 49°N latitude and are adapted to a variety of sites ranging from dry land to submerged (Yoshida 1981). Varying sensitivities to temperature, soil water status, photoperiod and other conditions are associated with these differences.

Most of the cultivated and primitive isolates are classified as SDP, though many modern forms have been selected for photoperiod insensitivity (Vergara 1985; Vergara and Chang 1985). Descriptions of the vegetative morphology and anatomy are included in Kaufman (1959a, 1959b, 1959c) and Yoshida (1981).

The annual, vegetative plant of rice arises following germination of the fruit, producing the main stem (culm) and axillary buds from which the primary tillers are formed. Secondary and tertiary tillers subsequently develop. Stems with internodes, increasing in length from base to tip, bear leaves in a distichous phyllotactic pattern. The main stem bears more leaves than are found on the secondary and tertiary tillers. The terminal meristem of each stem ultimately is transformed into an inflorescence.

Discussions of the morphology and development of flowers and inflorescences of rice are more difficult to locate and are less complete than for other suggested model systems. There is an obvious need to review and translate insights from the Asian sources in non-developmental disciplines such as crop science and plant breeding into more widely circulated developmental publications. The beginnings of this process are evident in Khush and Toenniessen (1991) and Takeoka et al. (1992). In addition to its importance as a crop, rice warrants consideration because of its modest list of genetic mutants of flower and inflorescence, its ease of culture and the ease of vegetative clonal propagation.

Flowering of Rice

Inflorescence Initiation from the Induced Meristem

As with other research plants, the growing conditions need to be standardized to produce essentially uniform material. A protocol for raising a cultivar common on Hokkaido is presented in Appendix 1. The apical meristems of the main stem and tillers are transformed and develop into inflorescence meristems (Fig. 8-16). Prior to floral initiation, the apical meristems of the main stem and the tillers enlarge,

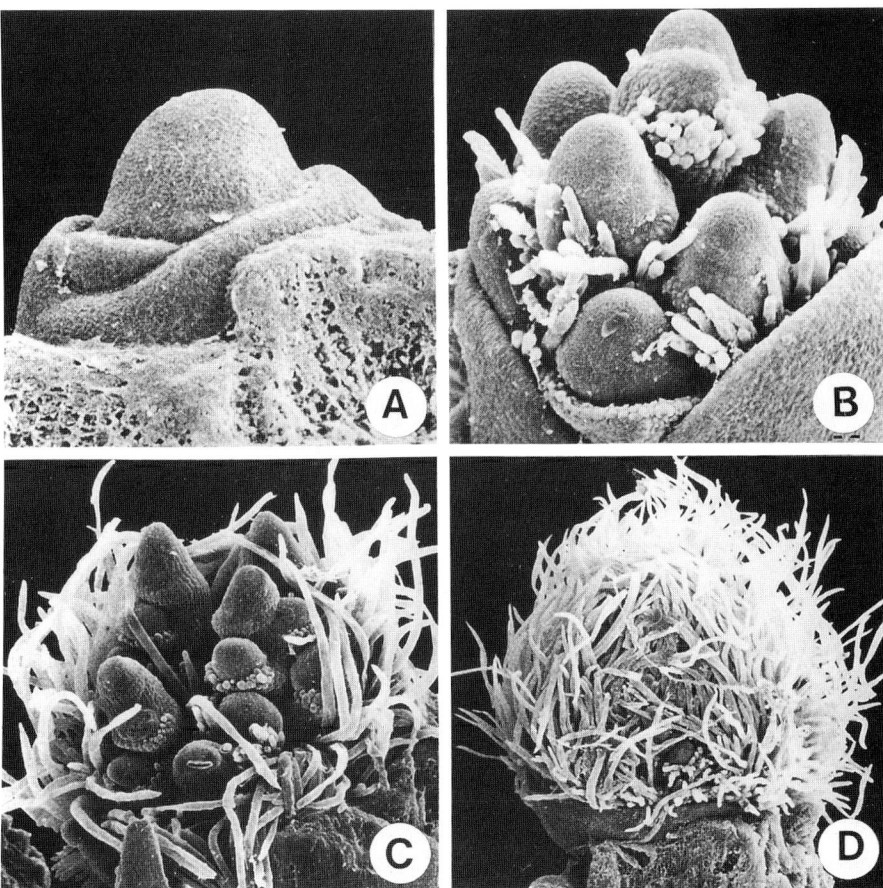

Fig. 8-16A–D. SEM view of the development of the inflorescence meristem of *Oryza sativa*. (A) View of the vegetative meristem. (B) During the change from vegetative apex to inflorescence meristem, bract primordia initiate and primary branch primordia arise in their axils. (C) Hairs develop on the bracts and grow rapidly. Secondary branch primordia develop in the axils of the bracts that form on the branch primordia. "Spikelet differentiation begins with the initiation of both empty and rudimentary glumes at the top of each primary branch. Next to initiate are the lemma and palea which, when grown, completely cover the developing stamen. The lemma is larger than and partly encloses the palea." (Moncur 1981). (D) Before the branches elongate, the inflorescence is completely covered with hairs. From Moncur (1981) with permission of CSIRO Division of Forestry.

and bract primordia (B) initiate distichously. Branch primordia (BP) arise in the axils of these bracts. Branch primordia proceeds acropetally. According to Evans (personal communication) the plastochron is greatly reduced immediatly after floral evocation, and leaf and spikelet primordia appear in rapid succession. Thus, the apex grows rapidly in size, but its dome remains much the same size until it aborts. After a generally fixed number of primary branch primordia are initiated, the residual meristem is assumed to abort. It leaves a scar at the base of the most apical primary branch (Xu and Vergara 1986). Secondary branch primordia (SBP) develop in the axils of bracts, and hairs (H) initiate and grow, eventually obscuring the inflorescence primordium (Fig. 8-16D or E). It is obvious that these inflorescence hairs complicate developmental studies on the immature inflorescence. Developmental studies with immature inflorescences would, therefore, benefit from the isolation and characterization of a "hairless" mutant: none, however, are listed (Kinoshita 1984). Spikelet differentiation begins with the initiation of two or more empty and rudimentary glumes at the top of each primary branch, followed by lemma and palea initiation. The mature rice inflorescence is illustrated in Fig. 8-17. A more detailed discussion of spikelet morphogenesis is provided by Takeoka et al. (1992).

The opening of flowers begins with the terminal flower of the main rachis (Fig. 8-18). The terminal flowers of primary branches open in a generally basipetal pattern. Other flowers on each branch, however, open acropetally (Fig. 8-18), though cultivar differences are recorded (Xu and Vergara 1986).

The Mature Flower

At maturity, the lemma is larger than the palea and partially encloses it (Fig. 8-19). The lemma keel can develop an awn. Unlike other cereal flowers, the rice flower has six stamens. The development of anthers and the details of microsporogenesis were studied by Raghavan (1988) (see Chapter 5). The single gynoecium bears a short, bifurcated and plumose stigma (Fig. 8-19). Two scale-like, transparent lodicules are located at the base of the flower attached to the palea (Chang and Bardenas 1965).

Opportunities for Experimental Studies

The present state of developmental knowledge about rice flowers and inflorescences is rudimentary, and the relevant literature is scattered. Efforts to improve this situation can be justified on a number of grounds: (1) the knowledge about genetic variability of flower and inflorescence is substantial, and samples of mutant stocks are available in a number of centers (2) ecotypic variability is also striking; (3) normal culture of experimental material is as convenient as for other grains; (4) the potential for vegetative propagation is important for the maintenance of clones, especially of infertile mutant material; (5) the preeminent place of rice in human nutrition requires that basic developmental information about flower development be as complete as possible.

Genetic Variability

The chromosome complement of *O. sativa,* though variable, is generally accepted as $2N = 24$ (Kurata and Omura 1984; Khush and Kinoshita 1991). Opportunities for genetical analysis are available in the many genes that have been identified and assigned to twelve linkage groups (Kinoshita 1984; Khush and Kinoshita 1991).

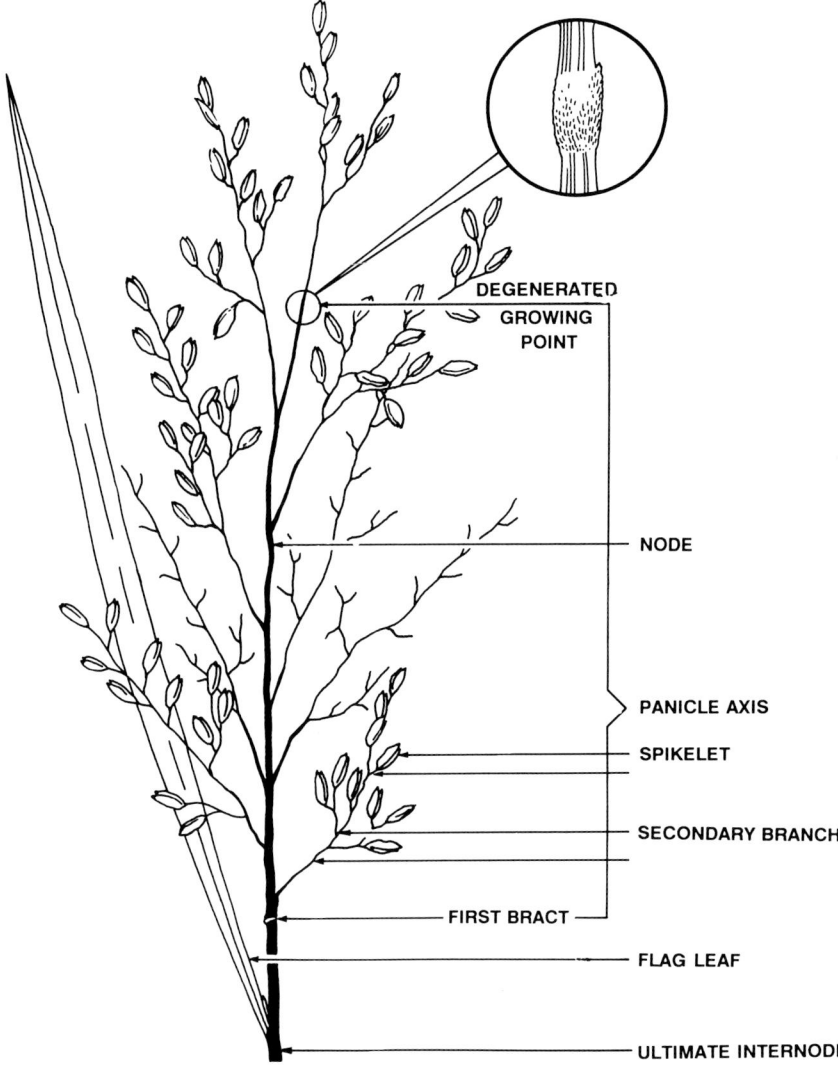

Fig. 8-17. Component parts of the mature rice panicle. Redrawn from Xu and Vergara (1986) with permission of IRRI Publications Office.

Genes which are involved in the development of inflorescence and flower include those listed in Table 8-3.

In addition to anthocyanin and fragrance traits which express in the flower, a number of morphological mutants have been identified. Kinoshita (1984) distinguishes those that affect the awn and spikelet (twenty-five listed), the grain (fourteen listed) or the panicle (twenty-two listed). He identifies fifty-nine reproductive characters, which include male sterility (g-mst and c-mst), fertility restorers, ga-

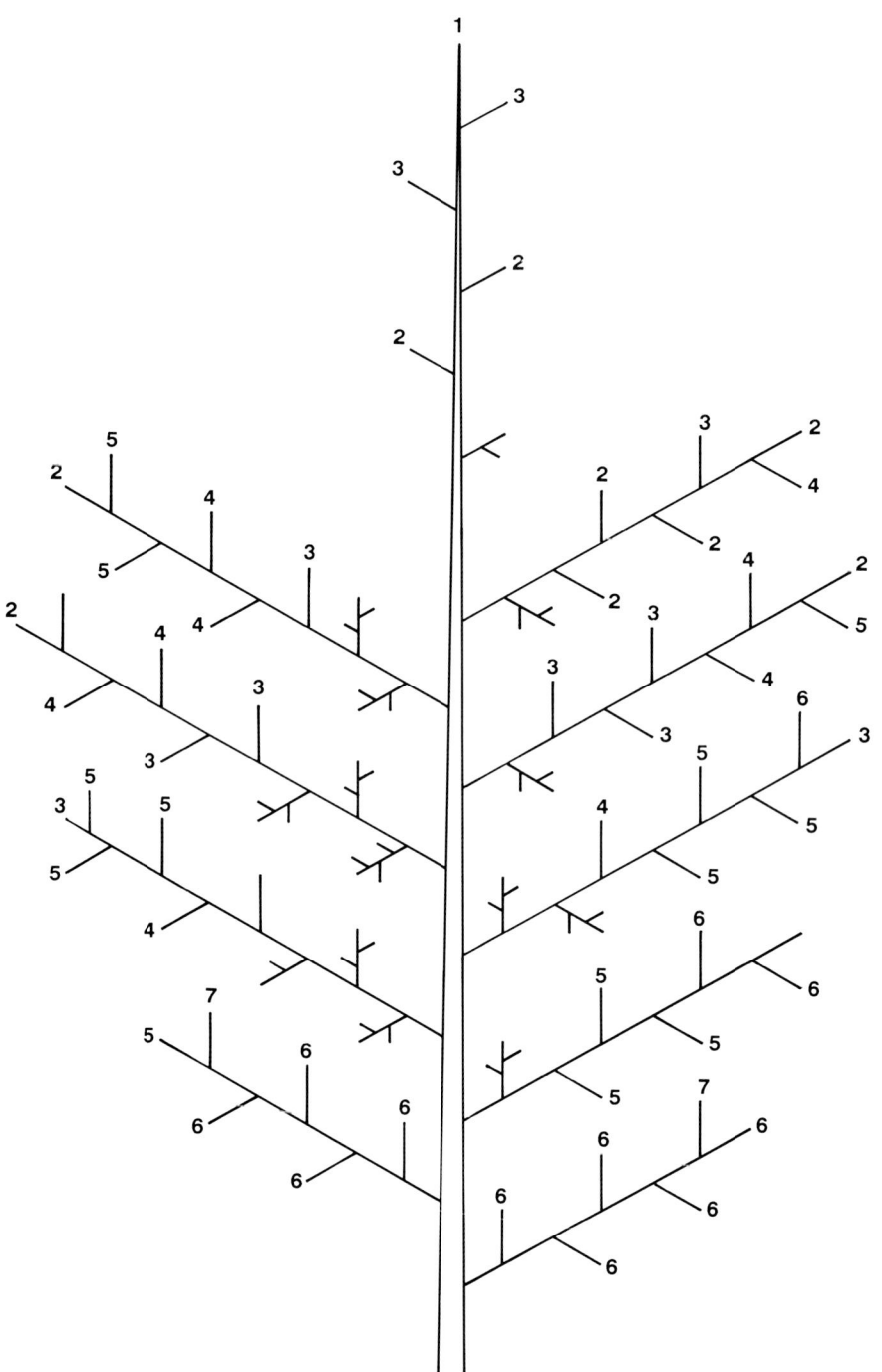

Fig. 8-18. Sequence of flower opening (flowering dates) of individual spikelets on the primary branches of a rice panicle (Nongken 58 variety). Redrawn from Xu and Vergara (1986) with permission of IRRI Publications Office.

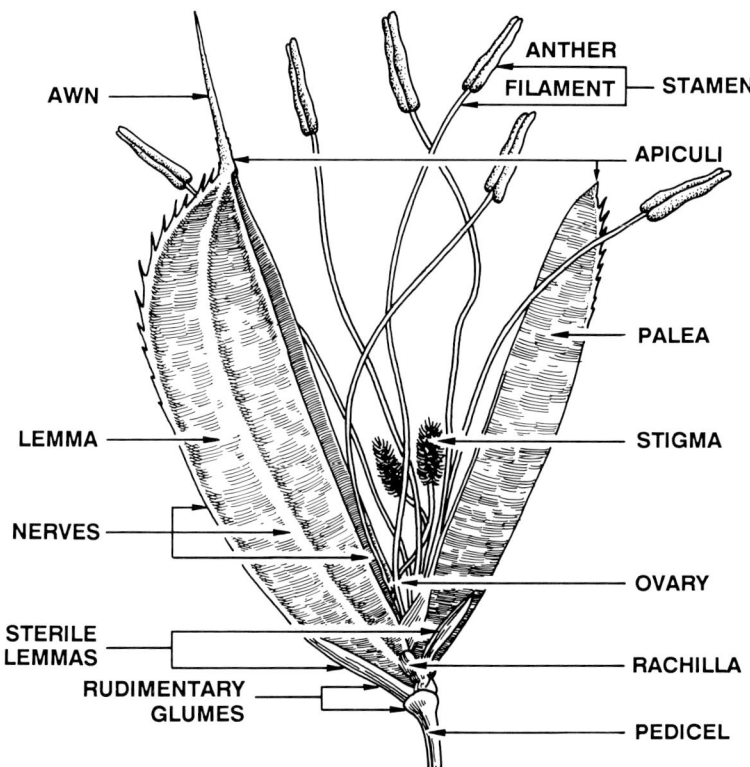

Fig. 8-19. Diagram of a mature rice spikelet. From Chang and Bardenas (1965) with permission of IRRI Publications Office.

Table 8-3 Selected Mutants of *Oryza* (2N = 24) Which Affect Flower Development

Gene Symbol	Gene Name	Description
cps	compact panicle sterile	Deformed and excessive development of all floral organs.
fes-1	female sterile-1	Normal pollen fertility but nonfunctional ovules: low seed fertility.
ms-1	male sterile-1	One of six g-mst genes listed.
ga-1	gametophyte gene-1	Responsible for differential fertilization caused by the lesser viability of pollen grains possessing ga causing F_2 segregation distortion.
g-1	long sterile lemmas-1	Sterile lemmas showing exceptional length, nearly as long as lemma and palea.
Hg	Hairy glume	Extremely long hairs on hulls, auricles, leaf margins and panicle branches.
mp-1	multiple pistils	2–7 pistils with 1–4 functional ovaries in each spikelet; high rate of double kernels.
Dn-1	Dense panicle-1	Compact panicle.

Source: From Kinoshita (1984).

metophyte genes (responsible for differential fertilization) and female steriles. In view of the fact that many of the dwarf mutants (fifty-seven listed) are associated with gibberellin metabolism (Suge 1984), many opportunities for observations on how gibberellins interact with flower development are available. No doubt the opportunities for molecular intervention will extend the genetic knowledge of development measurably.

Potential for Developmental Studies

In spite of the diminutive and non-showy appearance of the rice flower, a number of opportunities for developmental studies present themselves.

BASIC DESCRIPTION

As with most suggested model systems included in this book, considerable scope remains for basic description which attempts to document, both qualitatively and quantitatively, developmental aspects of genetic and environmentally based variability. In rice, opportunities abound for original work at all levels of organization, but because the literature on rice is scattered in journals and reports in a variety of languages and disciplines, thorough literature reviews are needed.

STUDIES OF MALE STERILES

Opportunities for both descriptive and experimental studies of g-mst and c-mst in rice are prevalent (Kinoshita 1984; Kaul 1988); many different expressions have been described. Despite the diminutive size of the rice flower, flower variants are available for study. For example, Trees and Rutger (1978) report on four different morphological male sterile strains which exhibited a number of different flower variables (Table 8-4).

Of particular interest is the report of the availability of a photoperiod-sensitive, temperature-sensitive g-mst (Xinggui and Jilin 1988). In Wuchang, China (30°27′N),

Table 8-4 Descriptions of Four Different Abnormal Flowers Associated with Different Strains of Male Sterile Strains of *Oryza sativa*

Strain	Description of mst Flowers
Calady	Flowers possessed small white anthers; otherwise normal.
Caloro	Flowers possessed reduced or missing palea, excess sterile glumes, elongated glumes and abnormally shaped stamens. Often fewer than 6 stamens per flower.
Earlyrose	mst flowers bore 5 to 7 stamens; filaments and anthers often shorter and more compressed. Gynoecia possessed extra stigmatic surfaces, elongated styles and anther-like structures protruding from the ovaries. Lodicules were enlarged.
CS-M3	Flowers often contained abnormally elongated ovary sacs at anthesis.

Source: From Trees and Rutger (1978).

this strain (Hubei Photoperiod-sensitive Genic Male Sterile Rice) is male-sterile if heading occurs before September 1. Male-fertile flowers form if heading occurs after this date. Clearly, the opportunities for descriptive and experimental studies on such a trait coupled with the practical applications that could arise from studies of such a trait are formidable. The potential opportunities for developing materials such as these for hybrid seed production are enormous.

Lolium (*Lolium temulentum* L.)

Lolium is selected as a model grass for flower development studies because some of its strains (e.g., Ceres) are sensitive LDP, requiring exposure to a single LD for the induction of flowering (Evans 1958). Further, an extensive literature on aspects of flower physiology has been developed (Evans 1969c; Evans and King 1985). Growth conditions are discussed in Appendix 1.

Origin of Inflorescence and Flowers

Morphology

The *Lolium* inflorescence is interpreted as a terminal spike bearing 8-30 sessile spikelets, depending on strain and growth conditions (Evans 1969c). Spikelets initiate and mature acropetally (Weber 1938) (Fig. 8-23).

Transition to Flowering

The precision and sensitivity of the inductive response of *Lolium* has permitted a large number of experiments on the metabolic events in meristems before and after the arrival of the inductive signal to the shoot apex. These have included basic morphological observations on the apical meristem in response to different daylengths (Fig. 8-20) and chemical (e.g., Rijven and Evans 1967a, 1967b) and histochemical (i.e., Knox and Evans 1966, 1968) changes in the meristem (Fig. 8-21 and Fig. 8-22).

More recently (McDaniel, King and Evans 1991), an *in vitro* culture method was used to test the nutritional requirements for successful shoot apical meristem culture of induced and uninduced shoot apical meristems and to establish the time at which the induced meristem is capable of initiating spikelets (florally induced). It appears from the data that the LD stimulus arrives at the apex about 22h after the start of the LD and is completed after a further 14h. At this time, all isolated apices are determined to flower. For determined meristems, GA_3 is required for floral development, as is the provision of a high (5%) sucrose concentration (Fig. 8-23).

In summary, *Lolium*, though used thus far primarily for studies on flower physiology, could serve as a model organism for the exploration of developmental

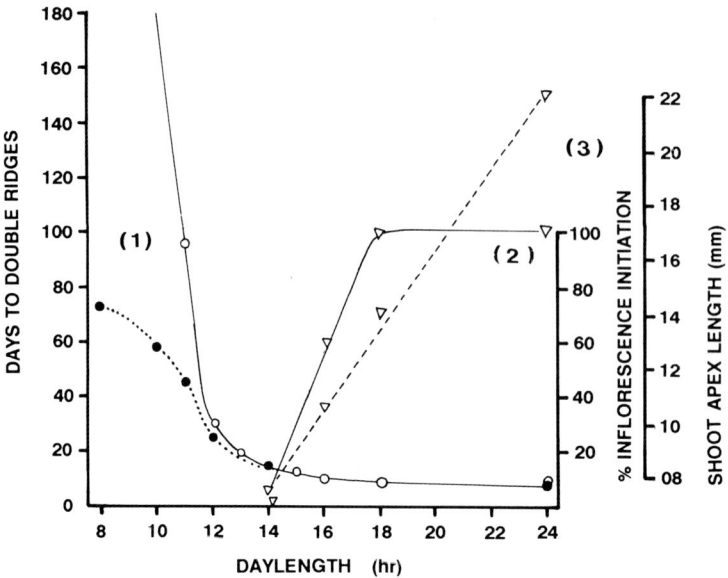

Fig. 8-20. Flowering response to daylength in *Lolium temulentum* as revealed by three different criteria: (1) Days to the appearance of the double ridges (left-hand scale). Symbols: ○ = plants grown in artificial light. ● = plants grown to intense summer daylight. The dotted curve indicates daylengths in which flowering occurred but inflorescences were abnormal and largely sterile. (2) Percent inflorescence initiation 3 weeks after exposure to 1 LD. (3) Shoot apex length 3 weeks after exposure to 1 LD. △ applies to both of the right-hand axes. Reproduced with permission from Evans and King (1985) in A. H. Halevy [ed.], *CRC Handbook of Flowering,* vol. 3, pp. 306–323. Copyright CRC Press Inc. Boca Raton, FL.

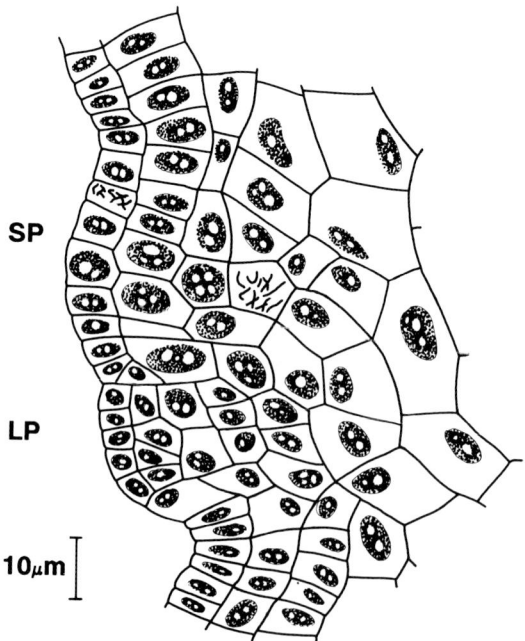

Fig. 8-21. Diagram of developing spikelet primordium from a DAY 5 apex. Abbreviations: LP = leaf primordium; SP = spikelet primordium. From Knox and Evans (1966).

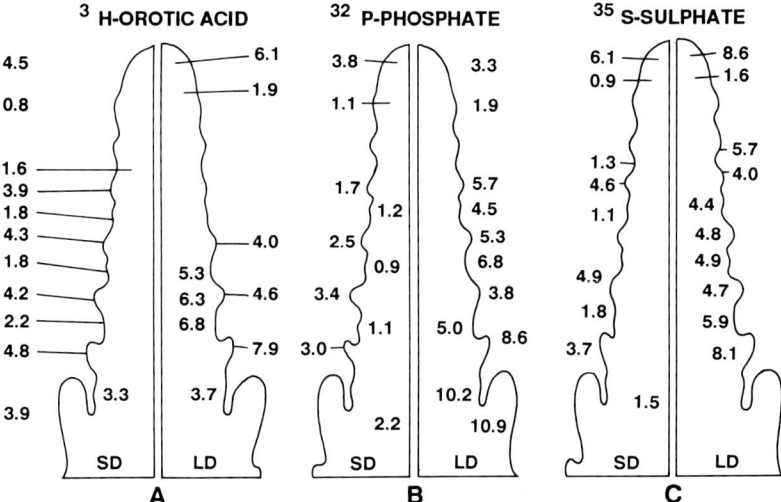

Fig. 8-22. Pattern of incorporation of (A) ^{3}H-orotic acid, (B) ^{32}P-phosphate, (C) ^{35}S-sulphate in vegetative (left-hand side) and LD-induced (right-hand side) apices of *Lolium temulentum*. The isotopes were applied and the apices excised and extracted on DAY 2. The numbers refer to the average number of grains per 10.9 u^2 in the regions indicated. Reproduced with permission from Evans and King (1985) in A. H. Halevy [ed.], *CRC Handbook of Flowering*, vol. 3, pp. 306–323. Copyright CRC Press Inc. Boca Raton, FL.

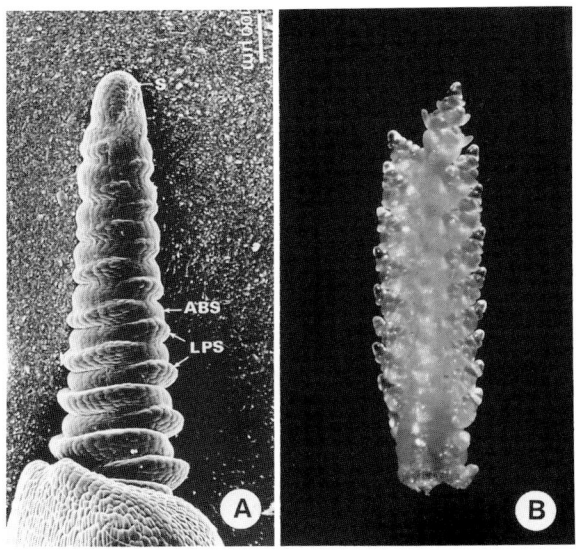

Fig. 8-23A–B. (A) SEM view of *Lolium temulentum* inflorescence meristem induced to flower by one long day, viewed after 5 days. Abbreviations: S = shoot apex; ABS = spikelet primordium; LPS = leaf primordium. Original photograph [with labels in French] courtesy of A. Havelange, P. B. Green and G. Bernier. (B) Developing inflorescence from *L. temulentum* strain Ceres plant. The plant was grown in a growth room for 7 weeks under short days and then induced with 3 long days. Twelve days after the beginning of the first long day, expanded leaves were removed to expose the apex, which was 2.5mm tall. Florets in the terminal spikelet had initiated the formation of anthers. Photo courtesy of Carl McDaniel.

factors that are involved in the normal elaboration of the inflorescence and the differentiation of flowers in the spikelets. Other observational and experimental studies can be anticipated. In addition, the potential to induce and select for mutant variability that would shed light on the cellular events that lead to inflorescence and flower differentiation should be followed.

9

Flower Organogenesis: Summary and Prospect

> In my opinion, the most important unsolved problem of biology is that presented by development and differentiation in higher plants and animals. How can cells containing the same genes come to be so different as a parenchyma cell and a wood fiber, a root hair and an ordinary epidermal cell, or a pollen tube and an egg cell?
>
> Stebbins (1964).

> The question of how a single cell, the fertilized egg, develops into a complex multicellular organism, is one of the most fundamental problems of developmental biology. Since the work of E. B. Wilson and T. H. Morgan it has become increasingly clear that the developmental program resides in the genome, and that in most cases the environment provides only general stimuli and relatively little specific information. Development is a highly ordered process which requires a precise temporal and spatial control of gene expression.
>
> Gehring (1984), with permission of Wiley-Liss, a division of John Wiley and Sons, Inc. Copyright © 1984

> The problem central to all of developmental biology is to understand how two daughter cells from mitotic cell division, or their respective mitotic derivatives, can come to have different phenotypes. All the cells in a mature plant are ultimately derived from a zygote.
>
> Christianson (1986).

It is increasingly apparent that unravelling the complexities involved in the developmental biology of flowers and inflorescences—the "site," "kind" and "time" questions that have been raised in earlier chapters—is both possible and desirable. In addition to providing a significant intellectual challenge in developmental biology, the results will have considerable basic impact on the understanding of aspects of plant development and should generate practical spin-offs to horticulture and agriculture. Further, and surprisingly, because of recent successes with the molecular biology of flower mutants and other technologies, the "central problem of development" alluded to in the three quotations above may be clarified more quickly with flower development than with other aspects of plant cell differentiation.

At present, however, answers to this central problem, as they apply to the production of flowers from induced meristems, are less well understood than for other plant organs and are primitive compared to those for animal differentiation.

Thus, it is not possible to state with conviction that developmental decisions of "site," "kind" and "time" are distinct or connected. Nor is it possible to decide with precision the relationship between processes in individual cells and those in aggregates of cells and tissues—interactive organismal processes (see Kaplan and Hagemann 1991). As discussed in Chapter 1, our ignorance arises both from the complexity of the problem and from scientific and philosophical choices made over the last century.

The limited pragmatic approach taken here (an extension of the position taken in Chapter 1) focuses first on the general properties that have been attributed to the process of flower development. These necessarily superficial and incomplete statements about flower development reflect our ignorance as much as they do our knowledge and understanding. In the absence of specific information, we are forced to use analogy and metaphor in our attempts to define the flowering process. These statements are, therefore, beginning positions or first approximations. In some cases they represent hypotheses from which tests can be performed and more precise articulations can be generated.

Our second objective is to identify those individual components, factors or developmental sequences of flower development that can be examined and manipulated at the present time. In some cases, considerable information exists relating to an individual component; in others, only minimal information is available.

The difficulty, of course, is to arrange the individual elements into an intelligible statement on flower formation. Notwithstanding, our final task in this overview is to explore the possibilities regarding where control and/or regulation of development might reside.

Throughout this summary chapter, our goal is to identify, as closely as possible, how the cellular elements of an induced floral meristem can specify what appear to be three, more or less, distinct activities in the induced floral meristem: (i) the siting of organ initiation (space); (ii) the establishment of different kinds of organs and tissues at these sites (kind); (iii) the maintenance of predictable sequences of organ types (time). References to these three concerns occur throughout the following discussion.

Questions About the General Developmental Properties of Flower Formation

Is Flower Formation a "Diffusion-Reaction" System?

An early attempt to connect the fragmented data about flower development into an integrated statement was Wardlaw's (1956–57) **diffusion-reaction** hypothesis. It was stimulated, in part, by the mathematical demonstration of Turing (1952) that an embryonic system with an initially homogenous distribution of metabolites could develop a patterned distribution spontaneously through the interaction of diffusing metabolites.

As the first such integrated hypothesis of flower morphogenesis, the diffusion-reaction concept helped stimulate experimental tests (i.e., Blake 1962; Tepfer et al. 1963; Hicks and Sussex 1971). Its basic assumptions, phrased to refer to a terminal meristem that formed a flower (as in a cyme), included: (i) the existence of "particular metabolic substances" which are assumed to direct the formation of different organ types; (ii) an initial homogenous distribution of morphogenetically active metabolites in the embryonic peripheral cells of the meristem; (iii) a subsequent patterned distribution of particular metabolites, their concentration in particular cellular sites preceding and determining (in conjunction with other factors [e.g. genetic] and relationships) any obvious cytological or morphological developments (Wardlaw 1956–57).

While deserving recognition as a first attempt and as a stimulus to explore the developmental biology of meristems, the deficiencies of this hypothesis should be noted: (i) the spontaneous component of the Turing model seems inapplicable to the floral meristem, for it seems doubtful that the floral meristem is ever homogenous and unpatterned in the way that is envisaged; (ii) the generality of the assumption about the diffusible organ-specific substances which stimulate the site-specific responses makes the theory difficult to test; (iii) no organ-specific diffusible molecules have been identified, though auxins, GAs, cytokinins, ethylene and polyamines no doubt can play a role (see below and Chapters 4, 5 and 6); (iv) it is technically difficult to manipulate and detect the localization of metabolites within the few cells involved; (v) undoubtedly, the movement of molecules in a cellular meristem would involve, in addition to diffusion, active transport and other non-random activities.

The notion, recently simulated on a polyacrilamide gel (Ouyang and Swinney 1991), that critical metabolites can become patterned prior to an alteration of cellular activities at those sites, and that these conditions might interact with the expression of gene products at these sites, is nevertheless a useful contribution. Tests of sites of one or a few cells are, however, technically difficult. Until recently it was not possible to explore these sites with technologies that possess the required sensitivity and specificity. In particular, the extension of *in situ* hybridization technology to identify and localize *de novo* synthesis of appropriate mRNA should assist in the objective. So, too, will the extension of sensitive immunohistochemistry technologies to identify site-specific proteins and other macromolecules. The combination of these technologies in conjunction with micro- and laser surgery, the micro-application of metabolites to selected cell clusters and the use of computer-assisted analysis will assist in these objectives.

Does Flower Formation Involve a "Relay" System?

Heslop-Harrison's (1964) more formal proposal of a sequential relay mechanism to explain sequential development within the flower meristem was based more firmly in the genetic theory of the time, being suggested by the regulator-gene/operon concept (Jacob and Monod 1961) of metabolic regulation in bacteria. This interpretation (Fig. 9-1) also stimulated experiments on the floral meristem, though its assumption of a rigid developmental sequence is contradicted by explicit data.

For example, the surgical studies on *Nicotiana* (Hicks and Sussex 1971) flower meristems document that petals can arise along the margins of cut surfaces in the

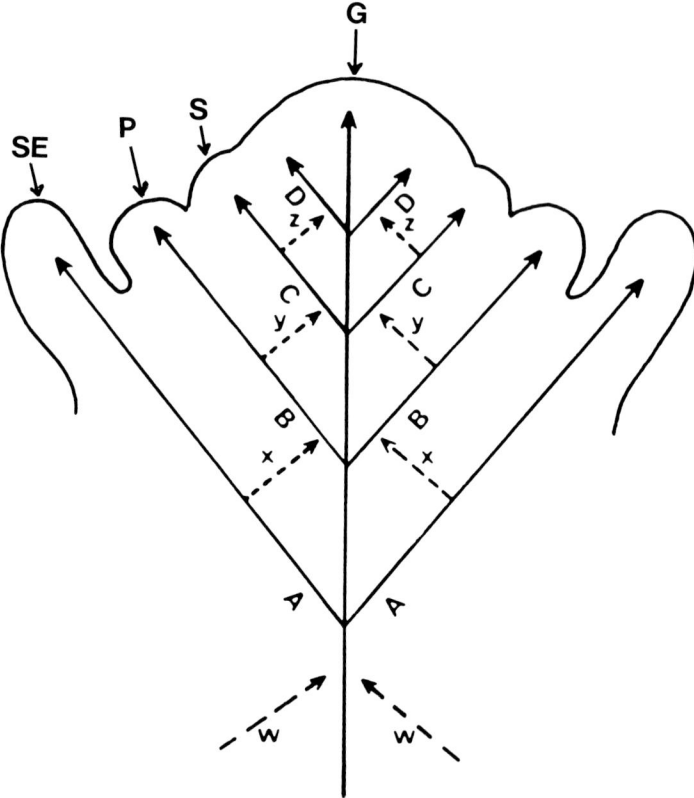

Fig. 9-1. A hypothesis for a developing flower meristem based upon the idea of control through a relay system of gene activation. The floral stimulus (stimuli) "w" initiates the floral transition. Cell lineages defined sequentially are destined to generate the different classes of lateral members: sepals (SE); petals (P); stamens (S); gynoecium (G). In the first lineage, gene complex A is activated and produces the inducer x. This diffuses to the next group of primordia, where it activates gene complex B, and so on. Redrawn from Heslop-Harrison (1964).

absence of sepal primordia in the immediate region. In *Nigella*, carpel primordia were formed on "single" flower meristems which had been cultured on a medium containing GA_3, without the prior initiation of stamens (Raman and Greyson 1978). Other examples of the lack of a rigidly determined sequence in the organ-forming capacity of the meristem are the reversion phenomena described in Chapters 6 and 9. Further, the accommodation of centrifugal patterns (described in Chapter 5) into the hypothesis, would require the insertion of some different assumptions in order to make the hypothesis fit their reality. Most serious for the relay hypothesis are the observations from genetic studies that give no hint of flower gene interaction in the manner postulated (Bowman, Smyth and Meyerowitz 1989; Coen and Meyerowitz 1991; Bowman et al. 1992).

While these criticisms are substantial, some can be countered by the realiza-

Table 9-1 Attributes of *Arabidopsis thalliana* Which Make It Suitable for Molecular Genetic Analysis of Flower Development

Attribute	Description
Length of life cycle	5–6 weeks[a,b]
Photoperiod requirements	Most strains are LDP[c]; commonly grown in continuous fluorescent light at 25°C[a]
Growth conditions	Routine greenhouse or growth cabinet conditions[a,b]
Availability of genetic stocks and lines	"Landsberg" ecotype containing the "erecta" mutant is the most commonly studied line. A number of flower mutants are available.[d,e,f,g]
Size of haploid genome	70,000 kb[h]
Number of chromosomes	N = 5[h]

References: [a]Meyerowitz (1987); [b]Smyth, Bowman and Meyerowitz (1990); [c]Napp-Zinn (1985); [d]Bowman, Smyth and Meyerowitz (1989); [e]Koornneef et al. (1983); [f]Estelle and Somerville (1986); [g]Haughn and Somerville (1988); [h]Bowman, Yanofsky and Meyerowitz (1988).

tion that the theory does not define just how much organ-specific information of the first type has to be available so that the next organ type can be initiated. It is possible that visible primordia are not required for the relay reaction to occur. Further, the possibility that a number of different regulatory patterns might function within the angiosperms must be considered.

Thus, both of these hypotheses, while containing elements that might contribute to future models, fail because of excessive generality (reaction system) or rigidity (relay system). The resolution of these difficulties in the form of a new generalized hypothesis is premature; more restricted, partial hypotheses are needed at present. Beyond the genetic models that are being generated (Haughn and Somerville 1988; Coen and Meyerowitz 1991), speculations about the mechanisms by which genetic information is expressed and communicated to different cells and regions of the meristem and its organs. This will require the development of a spectrum of sensitive and specific histochemical, immunological and molecular techniques whereby cell-to-cell molecular differences can be monitored and manipulated in organ-initiating meristems. Before embarking on such studies, investigators will have to decide on a short-list of research model species that are best suited to the resolution of these questions. The integration of modern cell analysis with *Arabidopsis* and other model species, in which the genetic component of flower development is available, deserves considerable effort. The attributes of *Arabidopsis* which make it a desirable model research plant are listed in Table 9-1.

Is Flower Formation a Multi-step or an All-or-None Process?

As discussed in Chapter 3, floral induction is a leaf response to a number of different environmental conditions and other factors. In some cases the induction can be mimicked by the application of chemicals. Evocation at the apex most often leads to inflorescence arrangements and mature flowers appropriate to the species and cultivar. Thus, in most cases, evocation and the development of flowers can be interpreted as a stable sequence of developmental processes that, once initiated, tends to go to completion (Kinet, Sachs and Bernier 1985; McDaniel 1992). For example, Lang considered the "flower initiation stage" as a "true turning-point" and that "the plant cannot initiate less than one flower primordium" (Lang 1965).

Considerable evidence, however, suggests that the sequence of events in some flowers and under some conditions, in keeping with many plant traits, is not irreversible or an all-or-none process as might be inferred. Analysis of examples of these atypical patterns of development and the conditions which produce them deserves special attention. Some of these are anecdotal, single-record examples of unknown origin (Worsdell 1916; Meyer 1966). Others arise under documented environmental, genetic or pathological conditions and are capable of experimental analysis. Battey and Lyndon (1990) summarize some of the different examples of flower and inflorescence reversal so far studied. While recognizing a range of possible structures which can be included under the term "partial and anomalous flowers," they concentrate on obvious cases of inflorescence and flower reversion.

The flower of *Impatiens balsamina*, which produces organs with vegetative traits in response to day-length manipulation (Krishnamoorthy and Nanda 1968), is an interesting case study (Battey and Lyndon 1984, 1986, 1988) of flower reversion (see also Chapter 6). When induced to flower under SD, followed by LD cycles of varying lengths, followed by a return to SD conditions, mosaic organs possessing different leaf-like and petal-like traits (Battey and Lyndon 1988) are produced. Evidence is presented that determination of primordia to leaf or petal development takes place after the primordium is 750μm long and is conditioned by the existing light conditions (Battey and Lyndon 1988). In these experiments on *Impatiens*, the phenomenon of late organ determination may be correlated with the reversion capacity.

It seems obvious that the collection and maintenance of genetic strains of plants such as *Impatiens*, which differ in their day-length sensitivity, is a desirable strategy for the exploration of the all-or-none hypothesis. It is probably wise to distinguish these day-length sensitivity variants from other abnormalities of flower development such as pathogen-induced proliferation or genetically based petalody. It may well turn out that some species or strains are more amenable in this reversion capacity than others, reflecting differences in the stability of developmental states and their expression.

Evidence of a somewhat less than all-or-none developmental regime also comes from surgical manipulation. Although a relatively traumatic act which could itself affect the outcome, surgical bisection of *Primula* (Cusick 1956), *Portulaca* (Soetiarto and Ball 1969) and *Nicotiana* (Hicks and Sussex 1971) not only succeeded in altering positions where some organs regenerate but in some cases resulted in failure of a complete organ-type to form on the half-meristems. Hicks and Sussex (1971), in comparing the three studies, suggest that the data, rather than providing evidence of an all-or-none process, can be interpreted as illustrating that the meristem progresses through sepal-, petal-, stamen-, and carpel-forming phases. In the early stages these can be interrupted. In later stages, the regenerative capacity of the meristem appears to decline.

How Does the Flower Compare to Other Developmental Systems?

Attempts to interpret, describe and define the events and processes that occur during the development of an animal embryo, or of embryonic tissue, have generated a number of concepts. These represent provisional statements about development and are rooted in the specific experiments and observations that suggested them. Our interest here is to learn whether these concepts are useful in studies on flower development. Some overlap is obvious between some of these concepts.

Competence/Determination

Various experiments and observations suggest that organs and tissues pass through different developmental stages as they grow and differentiate. The identification of these concepts arises from two questions about a developing cell, tissue or organ:

1. At what point in development does the cell tissue or organ become capable of differentiating along a certain developmental pathway? That is, when is it **competent?**
2. When is a cell, tissue or organ irrevocably committed (**determined**) to a particular developmental pathway? When is the commitment sufficiently stable to withstand the manipulations necessary to demonstrate the commitment (see Chapter 3)?

Because these concepts can be applied to myriad different stages from the prefloral meristem through to post-pollination events in the gynoecium, one must be explicit about the developmental state being assayed and the time of assessment. Further ambiguity is added where modifications of the internal or external conditions alter the answers one obtains. (Note: These are operational concepts, anchored in the particular experimental conditions of the particular situation. The results can change dramatically under different conditions.) These concepts, as they relate to induction and evocation of flowering, are discussed in Chapter 3.

Examples of competence during flower formation are not easily documented, nor has much effort been extended to uncover them. During evocation, meristems of plants that require both vernalization (chilling treatment) and an appropriate daylength requirement for the induction of flowering illustrate competence. Unvernalized meristems do not respond to inductive photoperiods; vernalized meristems are, however, competent (Chouard 1960). Uncovering examples of competence in tissues and organs during the maturation of the flower is more difficult. Perhaps one of the best tissues with which to study the timing and conditions associated with competence would be flower vasculature. It is well known from studies on vegetative tissues that IAA, in conjunction with other PGS, can be a sensitive inducer of vasculature (Sachs 1981). Another example which illustrates some of the features required is the phenomenon of haploid embryo or callus formation from microspores. Timing (i.e., competence) is critical for successful switching in some of these examples (Chapter 5).

Another process where analysis might benefit from the use of the concept of competence is megasporogenesis (Chapter 6). Typically, a single archespore differentiates out of a tissue of historically similar ovular cells. The exploration of the time and the details of that developmental decision should be of considerable interest. The related phenomenon of apomixis, and its potential exploitation in agriculture (Chapter 6), might be approached in a similar fashion.

The concept of determination, as related to induction and evocation, was discussed in Chapter 3. This concept as it relates to later developmental events in the tissues of flowers, has in general not been attempted, though many suitable candidates for experimentation suggest themselves. These include the differentiation of a number of tissues in anthers (e.g., endothecium, tapetum, sporogenous tissue and microspores) or, alternatively, tissues in the ovule associated with embryo sac formation. It is possible that *in vitro* culture experiments of anthers for normal microsporogenesis or, alternatively, sporophytic embryogenesis could be interpreted in terms of determination. Similarly, many different situations in tissue development in the maturation of the ovule (Chapter 6) could be studied and manipulated from the point of view of exposing the timing of determination and the factors that affect it.

Experimental documentation of determination in whole organs is more easily achieved. Surgical bisection of organs at different stages in development can, in some cases, reveal the point (stage, size, time, etc.) when determination is achieved. In theory, surgical bisection of an organ prior to organ determination should result in two separate organs. In vegetative tissues, for example, bisection of early leaf primordia generated two entire leaves (Sachs 1969). Bisection of older primordia produced, in general, two incomplete half-leaves. Other examples are discussed by Steeves and Sussex (1989).

The data from surgical bisections of whole-flower primordia can document determination phenomena, though the results are complex and the interpretations equivocal. Experiments have been carried out in *Primula* (Cusick 1956), *Portulacca* (Soetiarto and Ball 1969), *Nicotiana* (Hicks and Sussex 1971) and *Aquilegia* (Jensen 1971). The work on *Primula* and *Portulacca* was prepared *in vivo;* the work on *Nicotiana* and *Aquilegia* involved *in vitro* culture.

On the basis of their studies on immature flower primordia of *Nicotiana* and other species, Hicks and Sussex (1971) conclude that, in "presepal" and "sepal primordia" stages, extensive adjustment of anticipated organ positions takes place following bisection. Thus, at this stage, the positions of petal, stamen and gynoecia primordia are not fixed. Peripheral organs tended, however, to occur at expected sites. These experimental data were interpreted to indicate three organ-forming stages: sepal, petal/stamen and carpel. Ovary regeneration more closely resembled the leaf regeneration model (above), with whole bicarpellate gynoecia being initiated commonly from meristems with up to five primordial petals. Older meristems produced incomplete gynoecial components. These observations are also reviewed by Hicks (1980).

Waddington (1956, 1966b) introduced an alternative distinction in his discussions of determination which is useful in discussing some situations. He suggested that "determination" be used to describe the initial commitment or one of a number of paths within a "morphogenetic field," and that "canalization" be used to refer to the degree of commitment to that path. While Waddington's terminology is not frequently used in botanical presentations, it served as the basis for the discussion of the changes that were described in the tomato mutant (sl-2; Fig. 9-2) following different manipulations (Sawhney and Greyson 1979). Waddington's main contribution, in our view, was to begin the process of identifying sub-steps in the overall multi-step process of determination.

The ultimate goal in these studies will, of course, be to replace verbal generalities about development with objective biochemical and biophysical data about the many processes that transpire in the floral meristem as it matures. One aspect of this search will be to learn of any stage-specific molecules or conditions which might represent the chemical basis for any one of a multitude of competence and determination states that occur during morphogenesis and cell differentiation in flower formation.

Plasticity

An alternative articulation to the all-or-none concept and the determined state is the realization that flowers, like many plant structures and tissues, exhibit **"plasticity"** in certain situations (Bernier 1986). It is a common feature in the expression of many developmental processes of plants (Trewavas 1982; Jennings and Trewavas 1986). Examples of plasticity are found at all stages of flower development, from initiation to maturity, though decreasing plasticity is related to the progressive increase in the number of levels of organization involved (Bernier 1986).

Regulative/Mosaic

An additional dichotomy of properties of embryonic tissues are recognized by the terms **"regulative"** and **"mosaic."** These concepts, borrowed from animal embryology, attempt to assess the degree to which cells, tissues or organs mature and differentiate as a consequence of their cell lineage or, alternatively, in response to environmental, chemical or physical cues related to their location at the time of expression. Because of some major differences between animal and plant cellular organization, a debate is possible over the applicability of this analysis to plant tissues. One might expect a continuum of conditions from rigidly mosaic to highly regulative examples. In one sense these concepts have to do with the regulation of expression of different "kinds" of genetic expression; alternatively, they could represent modifications of the "time" at which determinative decisions are made in the tissues.

While experimental tests are feasible with surgery, grafting, localized application of PGS and *in vitro* culture, the most convincing tests are the genetic manipulations that generate sectorial chimeras. In these cases, tissues and organs initiate across sectors that differ with respect to a different gene or allele. In *Zea,* chimeric studies illustrate rather convincingly that the flower/inflorescence mutants ramosa-1 (ra), tunicate (Tu), tassel seed-6 (Ts6) and vestigial (Vg) are cell auton-

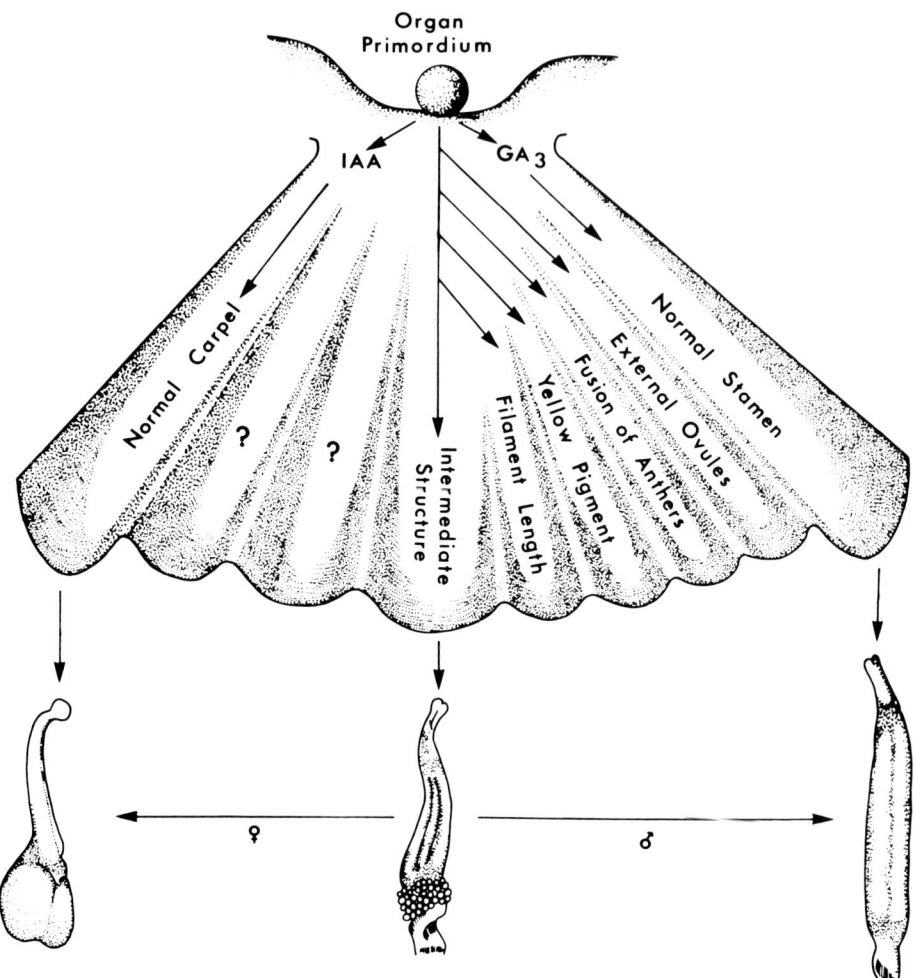

Fig. 9-2. Epigenetic landscape (sensu Waddington) of a stamen primordium of the stamenless-2 mutant of *Lycopersicon*. The ball represents the primordium at the time of initiation and would normally proceed along the central path and produce the mutant phenotype (the intermediate structure), which has a distal anther part with microspores, a short filament, and external ovules at the junction of the anther and filament. At early stages of development, however, the ball can be diverted to alternative pathways. Application of IAA to the plant or the inflorescence results in the formation of a near-normal carpel: a basal ovary region has internal ovules plus a style and stigma. Applications of GA_3 to the plant or the inflorescence results in the production of a near-normal stamen: elongate anther with fertile pollen, a short filament and no external ovules. GA_3 applied at later times in development results in stamens with more of the sl_2 phenotype. From Sawhney and Greyson (1979).

omous; that is, they express a phenotype that is appropriate to the genotype of their cells and are not apparently regulated by adjacent cells (Johri and Coe 1983).

Mosaicism is also apparent in many teratomas where chimeric organs are produced. Petals with distinct patches of stamen tissue are frequently reported. Various mosaic organs were produced in temperature-shift experiments of the ap3 mutants of *Arabidopsis* (Fig. 5-28). Rather than producing generalized changes throughout the whole organ, localized patches of tissue are scattered about the organ, though detailed cell-to-cell analysis has yet to be carried out (Bowman, Smyth and Meyerowitz 1989).

An example of mosaicism induced under experimental conditions is the production of stigmatoid structures on cultured stamens from *Nicotiana,* described by Hicks (1979). The induction of these patches, representing differentiation of traits normally restricted to the gynoecium, was most frequent on the youngest explants and was absent on older primordia.

Whether these two examples provide evidence that relates to the mosaic/regulative dichotomy, or whether they are a product of regeneration or plasticity in response to the experimental conditions, is of course debatable. It may turn out that the level of developmental plasticity in flower tissues, in response to manipulation, is enough to render the mosaic/regulative distinction moot in many systems.

Positional Information

The observations of the patterning of developmental systems on which the concept of **"positional information"** was proposed (Wolpert 1971) are both old and obvious in both animal and plant developmental situations (Wolpert and Stein 1984). They include segmentation patterns of *Drosophila* and other insects and invertebrates, regeneration patterns in *Hydra* and chick limb differentiation (Wolpert and Stein 1984). In plants, many examples of cell patterns and organ arrangements suggest themselves as candidates for inclusion in the concept (Barlow and Carr 1984). A notable example, appropriate to this discussion, is the whorled arrangement of different organ types in flowers.

The basic assumptions of the concept include: (a) developing and differentiating systems contain molecules (morphogens) which influence cell differentiation; (b) the spatial pattern of differentiation reflects the spatial distribution of these molecules; (c) the pattern of morphogen distribution reflects sites of synthesis and the direction of their transport; (d) transport is often polarized (Barlow and Carr 1984).

The extension of positional information theory and analysis to plant systems has, in general, been hesitant. Holder (1979) attempted to rephrase a number of plant developmental phenomena in the context of positional information terminology and implicated PGS as major candidates for the relevant morphogens. Lintilhac (1984), however, questioned the application of the concept to plants on the grounds that the original articulation was too rigidly tied to a genetic determinism. In Lintilhac's view, spatial patterning in plant meristems is more likely to be based in intercellular, mechanical stimuli within the meristem.

Notwithstanding, there are many situations in flower meristems where some notion of positional control [in the general sense of Barlow and Carr (1984)] seems to operate. For example, one must assume some method by which cells, cell-lines or small surface regions of the meristem become committed to express a sequence of organ-specific phenotypes. The common sequence of initiation on a typical whorled flower (calyx–corolla–androecium–gynoecium) serves as a model. In some manner, as different whorls are initiated, the primordia formed develop into different organ types, each with a unique phenotypic expression. Except for their small size, it would appear that floral meristems of *Arabidopsis,* expressing different genetic backgrounds (Coen and Meyerowitz 1991), could be appropriate material with which to explore the different regions of the meristem for positional information.

Cellular Components Involved in Flower Development

The convenient, and sometimes necessary, technique of compartmentalizing information into lists of categories and flow charts is common, and it is used frequently in this text. Weiss's diagram of "levels of organization" (Fig. 1-2) was used to emphasize our assumption that development proceeds at a number of interacting levels of organization simultaneously. Any penetrating understanding of the flower-forming process will therefore have to identify these levels and determine how, when and to what degree they interact.

The reality, of course, is that the processes we wish to understand transcend these arbitrary boundaries. Flower development is best assumed to be a multifactorial process, involving many more interactive components than we can conveniently identify. Many of these interactions may be subtle and difficult to resolve with available technology. Thus, any classification of levels of organization within the developing flower is provisional and of value only in so far as it assists in identifying the categories, in delineating their boundaries and in assessing the ways in which they interact. A taxonomy of levels, while necessary, will be arbitrary.

This reservation applies, for example, to the recognition of those components of regulation which are directly based upon differences in the genetic code, as compared to those differences which arise through differences in RNA transcription, translation of protein and different cytoplasm-based processes. These components are frequently identified, in many discussions, as a genetic/epigenetic duality. Unfortunately, too many complications in the use of this dichotomy militate against its use here. Most of the cellular activities (nuclear and cytoplasmic) that we might wish to implicate as causal components of development are epigenetic. Further, the segregation of developmental factors into genetic and non-genetic sectors reinforces a notion about exclusiveness which is not supported in current biological thought.

The genetic/epigenetic duality may be overly simplistic and irrelevant to the understanding of many developmental systems, but it reflects a long-standing tradition in flower morphogenesis and in a number of serious research thrusts. Until recently, many discussions of flower morphogenesis exhibited a common assumption that control of many flower-forming features had a cytoplasmic or physiological origin, and the potential for direct genetic input was ignored. Many discussions of the involvement of PGS in organogenesis and flower form, for example, create this impression.

Our compromise with these complications is to identify five components of cellular activity which have been seriously considered to make contributions to flower formation. These include: (i) biophysics of cell wall formation; (ii) cytoskeletal structures; (iii) PGS metabolism; (iv) mitochondria; and (v) the genetic component. It is appropriate to ask to what extent these components contribute to flower formation.

The Biophysics of Cell Wall Formation

In earlier chapters, attention was drawn to the early cellular patterns associated with the initiation of floral organ primordia. In most of the descriptions, initiation was associated with periclinal divisions in the surface cell layers (T1, T2 or T3) of the meristem. The assumed regularity of some of these division patterns, along with the knowledge that shape and size of plant organs are based upon a fixed position of plant cells relative to their site of formation, notwithstanding those cases where cell elongation alone, accounts for shape (Haber 1962). The final shape of most organs is strongly influenced by the pattern of cell divisions. Therefore, the location and pattern of divisions in organs as they grow have attracted many researchers of plant development (Sinnott 1960; Walbot 1989). The subtleties of these events have yet to be fully plumbed, but some useful insights have been obtained by Green and his colleagues (Green 1984*b*; 1988; Green and Poethig 1982; Selker, Steucek and Green 1992) through their studies on cellulose microfibril orientation in growing and differentiating cells.

Among the many specific questions we have about cells and organ growth is the following: Is the position of the cell plate during mitosis determined prior to or after the expression of unique site-specific, gene-related chemical differentiation? After many careful studies of the alignment of microfibrillar components in cells of vegetative tissues, Green (1980) chooses the first of these alternatives.

Green's studies have documented routinely that alterations in cell division patterns are associated with reorientation of the cellulosic microfibrils. It is postulated that these shifts in orientation are a response to physical forces acting throughout the growing meristem but focussing a strain at the site of organ initiation. It is assumed in this interpretation that the short chain of causality begins with biophysical forces as the prior event.

In an analysis of wall formation on the apical meristem of *Echeveria,* during transition to flowering and flower organ initiation, Green (1988, 1989) demonstrates that cell walls of flower-forming meristems conform to the observations on other systems. His data, too complex to detail here, make a strong case for the assumption that early stages of organ initiation, perhaps the determination of sites rather than organ types, are significantly influenced by biophysical factors.

Cytoskeletal Components

Another, similar hypothesis [termed "structural epigenesis" by Lintilhac (1984)] proposes that the positioning of cell plates during mitosis results from the ability of cells to sense changes in the distances across the cell in different directions, with the new cell plate organized on a plane perpendicular to this axis. The most likely cellular components thought to sense cellular tensions are assumed to be the stiff, non-contractile cytoplasmic elements such as the microtubules. The shearing of microtubules in cells at sites of focussed compression or tension could, it is argued, result in repair, in a cascade of chemical consequences and possibly in novel cell plate orientation.

Both Green's and Lintilhac's hypotheses underscore the possibility that regulation of some features of flower morphogenesis may reside in biophysical factors such as compression and tension, notwithstanding the significant genetic input to be discussed later. Whether this biophysical cue specifies the position of primordia (space) or, less likely, the kind and sequence (time) of organ differentiation remains open to debate.

That biophysical forces are considered to be important determinants of pattern is of course not new. Explanations of phyllotactic patterns based on assumed contact pressures of primordial packing have been proposed at different times (Adler 1974), though they have not been the most widely accepted nor served often as the basis for experimental tests. Field hypotheses, invoking diffusion of morphogenetically active molecules, have been most frequently postulated, tested and modelled (Mitchison 1977). Because these two biophysical hypotheses identify specific cellular components, cellulose microfibrils and microtubules, they are capable of experimental tests and should be considered seriously by those interested in organ initiation and differentiation.

PGS Involvement

As documented in earlier chapters, there are many suggestions that PGS are intimately involved in different aspects of flower differentiation. Much of the data are, unfortunately, correlative and inconclusive. The critical question, therefore, is this: Which examples of PGS involvement strongly support the assumption that PGS act in a decisive, regulatory role in flower development? Or further: If PGS regulation of flower development can be demonstrated, is its mode of action "directly" or "indirectly" related to differential gene action?

Of the many suggestions of PGS involvement in developing flowers, two types deserve attention. The first example is the determination of unisexual flowers in monoecious and dioecious species [e.g., *Begonia, Cucumis, Cannabis, Mercurialis, Spinacia, Zea* (Chapter 6)]. For each, many different observations are available to implicate PGS in the development of staminate or gynoecial flowers. In the most convincing examples, *Cucumis* and *Mercurialis,* evidence of appropriate changes in endogenous content is supported by data from exogenous application studies.

A second striking example of PGS involvement is the post-initiation growth of petals and stamens [*Cleome, Gaillardia, Glechoma, Lamium, Nigella, Pharbitis* (Chapters 4 and 5)]. In some cases the endogenous content has been correlated with growth and/or their sensitivity to applied substances has been noted. In both these situations, enough data have accumulated to consider individual examples as serious contenders for this search.

The evidence for a strong causal relationship for GA stimulation of the staminate flowers in *Cucumis* is supported from different kinds of data [reviewed by Pharis and King (1985)]: (i) analysis of endogenous content (Hayashi et al. 1971; Hemphill, Baker and Sell 1972); (ii) response to applied GAs (Fuchs, Atsmon and Halevy; 1977); (iii) response to inhibitors of GA metabolism (Chailakhyan and Khrianin 1987); (iv) *in vitro* response (Galun, Jung and Lang 1962).

If from the evidence one concludes that GAs are regulators of male flower development in *Cucumis,* at least two potential models are available. GAs could function by regulating activities in the cytoplasm which ultimately induce flower-specific gene expression and the development of the flower phenotype (**indirect** action), or alternatively GAs could more or less directly activate differential flower-specific gene synthesis (**direct** action).

Attempts to test either model rigorously will be difficult by current standards; however, the direct route may be more readily tested at present than the indirect route. Serious tests of the direct route would require: (i) the identification of flower-specific genes and gene products; (ii) the identification of specific activation of the genes and the gene products by the PGS; (iii) *in situ* localization of the flower-specific gene products in the early stages of flower initiation; (iv) the demonstration that the gene products are directly related to GA-level alteration or sensitivity. The ability to provide data of this quality must await future technical developments.

Critical tests of an "indirect" mode of action of PGS, while popular in the past, are in practice difficult to accomplish. Estimates of endogenous levels, whether by extraction or diffusion exudates and regardless of experimental quality, can correlate with development but cannot answer the question asked. In fact, the potential complexity of the physiological interactions of PGS within the cytoplasm precludes the execution of incisive experiments. We conclude, therefore, that serious future attempts to connect PGS metabolism with flower differentiation should be carried out on model systems that possess appropriate genetic backgrounds from which exploration will be directed at elucidating the "direct" genetic route rather than the "indirect" physiological approach.

In *Mercurialis,* the evidence of cytokinin involvement in the development of gynoecious flowers is particularly impressive. It is based upon a number of studies reviewed in Durand and Durand (1984, 1985, 1990). Evidence includes: (i) exogenous application of cytokinins stimulate formation of gynoecious flowers on genetically staminate plants; (ii) exogenous application of IAA induces staminate flowers on genetically female plants; (iii) endogenous levels of cytokinins, primarily zeatin, are elevated in female plants; (iv) correlations between cytokinin-binding to proteins and ribosomes and the sex of plant have been detected; (v) correlations between some esterase and peroxidase levels and plant form have been identified.

Once again, the data suggest a causal link but are inconclusive. As with *Cucumis,* critical tests of the "indirect" route, because of the large number of cytoplasmic possibilities that might be explored, are complex and difficult to execute. Tests of the "direct" genetic mode might be more feasible, as accessibility to molecular genetic approaches becomes available and the genetic basis of flowering is clarified in more species and with greater precision. For example, in some model species, genetic combinations of alternative PGS expression (Morris 1987; Horgan 1987) with genes determining organ kind should be tested. Parallel quantitative analyses of PGS status in localized sites and at different times will be a necessary corollary.

Despite the impressive documentation (Chailakhyan and Khrianin 1987; Durand and Durand 1990) which implicates PGS, either "indirectly" or "directly," as sensitive regulators of flower form, there are many reasons to be hesitant: (a) both GAs and cytokinins are known to be involved in a number of sometimes overlapping cellular actions; (b) GAs, while strongly associated with staminate flower development in *Cucumis,* are associated with gynoecial growth in *Zea* (Rood, Pharis and Major 1980) and inhibit stamen maturation in cultured *Zea* tassels (Bommineni and Greyson 1990); (c) the problem about whether PGS action is based upon "sensitivity" (i.e., the availability of receptors) or "concentration" has yet to be resolved (Trewavas 1987); (d) evidence of PGS involvement in many developmental events in some flowers is difficult to obtain.

In contrast to earlier, frequently monofactorial, studies, future research strategies on the involvement of PGS in the regulation of gynoecial and staminate development will undoubtedly require more complex assumptions. The adoption of multifactorial interpretations which not only assess levels of PGS but also examine sensitivity responses and interactions with other substances will be necessary. For many aspects of flower development, it does not appear that monofactorial logic, such as PESIGS Rules (Jacobs 1959), is adequate. Future studies of PGS and flower development will also involve more explicit examination of the direct activation of genetic components by the PGS (Baulcombe 1987). Future hypotheses and studies must also take account of the growing importance of PGS-binding molecules, their distribution and their molecular biology.

The second type of growth response during flower development in which PGS are strongly implicated as regulating or limiting factors occurs during petal and stamen growth. Of the examples described in Chapters 4 and 5, perhaps the most impressive is the documentation of the sensitivities of the petal and stamen of *Ipomoea* to GAs and ethylene (Raab and Koning 1988).

Despite the refinement of experimental design, the quality of the data and the fact that the cellular activities (mainly cell elongation) are well-documented correlations of the normal action of GAs and ethylene, it is still not possible to distinguish readily between an "indirect" versus a "direct" genetic mode of action.

Mitochondrial Component

Cytoplasmic male sterility (c-mst), discussed in Chapter 5, represents an obvious example of the involvement of the cytoplasm in flower maturation. There is strong

Table 9-2 Descriptions of the Floral Structures in Three Parental cms Cultivars of *Nicotiana tabacum* Used by Kofer, Glimelius and Bonnett (1990)

CMS cultivar	Corolla Length[a]	Corolla Form	Stamens	Pistil Length
Nta(big)S	Normal	Split	Filaments with sterile, flattened, fringed ends	Normal
Nta(sua)S	Normal	Fused	Absent or reduced to remnant.	Variable, frequently up to 1 cm shorter
Nta(und)S	Shortened by 1–1.5 cm	Fused	Petalodes, no filaments, occasionally tipped with stigmatoids.	Normal (protruding from corolla 1–1.5 cm)

Note: Each stable cultivar is of *N. tabacum* containing cytoplasm of *N. biglovii* (Nta(big)S); *N. sauveolens* (Nta(sua)S; and *N. undulata* (Nta(und)S. Each cultivar produced distinctive petal and stamen morphologies.

[a] In comparison to male-fertile *N. tabacum*.

correlative evidence from many flowers (particularly in cytoplasmic male sterility) that deviations in mitochondrial DNA (mtDNA) are frequently involved. In c-mst plants, temporal and spatial expression of the trait is for the most part restricted to the tapetum, the sporogenous cells or the microspores. All the examples point to a requirement of a permissive mtDNA component and, in view of the nuclear restorer (rf) genes of some species, a complementary nuclear condition (Kaul 1988).

Further evidence of involvement of mitochondria in the development of flowers is provided by the studies of some c-mst cultivars of *Nicotiana* (Rosenberg and Bonnett 1983; Kofer, Glimelius and Bonnett 1990, 1991). Following protoplast fusion between three c-mst lines of *N. tabacum*, nearly 200 cybrid calli were regenerated into plants. Most cybrid plants exhibited the parental male sterile phenotypes (Table 9-2). Some, however exhibited novel male-sterile phenotypes, or phenotypes combining traits of both parents. Others developed normal stamens and fertile pollen (Kofer, Glimelius and Bonnett 1990) and were fertile. From an analysis of the mtDNA, it was concluded (Kofer, Glimelius and Bonnett 1991) that specific mtDNA alterations correlated with specific morphological variations. The data suggested that several mitochondrial genes are involved at a number of post-initiation steps in the development of petals and stamens. It seems obvious that considerably more remains to be uncovered about the role of mitochondria and other cytoplasmic components in flower development and whether they are also involved in other species. Unfortunately, the difficulties and complexities in producing cybrids of other species will slow the exploration of this component and its general contribution to flower development.

The Genetic Components of Development

Until recently, there was a general reluctance by developmental botanists to explore the genetic component of development in plant tissues, and interest in the genetic basis of flower development was low (Marx 1983; Horsch 1990). One reason for this hesitation was the tendency on the part of morphologists, physiologists and other botanists to leave the problems of genetics to geneticists and to attempt to explore phenomena at a single level of interest. As discussed in Chapter 1, the compartmentalization of biology into separate disciplines has limited the amount of idea-exchange between plant morphology and plant physiology and has also affected the interaction between researchers and the exchange of ideas between these two botanical subject areas and genetics. No doubt, the fact that the administration of much biological investigation is based on support for individual investigators, as compared to groups, is also a factor.

It is also true that, regardless of how conversant a developmental botanist might have been with the genetic implications of the material (e.g., Sinnott 1958; Postlethwait and Nelson 1964; Stebbins and Yagill 1966), there were, until recently, many intellectual and technological limitations on attempts by researchers to connect genotype with phenotype. Recently, some of these deficiencies and barriers to understanding have begun to disappear (Drews and Goldberg 1989) and their removal has resulted in the encouragement of multi-disciplinary projects among many researchers. Answers to many perennial questions about flower development are, therefore, about to be uncovered through the exploration of the genetic component of development. Whether these studies will answer all our "site," "kind" and "time" questions, remains to be seen. The beginnings in this search are promising (Coen 1991; Coen and Meyerowitz 1991). Some of the opportunities for research and questions that are appropriate to the study of the genetic basis of flower development are discussed below.

The Identification, Collection and Maintenance of Genetically Based Variation

These necessary exercises have traditionally been relegated to geneticists and plant breeders and are frequently taken for granted by subsequent researchers. The dedication and effort given to these collections by some of the pioneer breeders and geneticists, frequently with little personal recognition, represents a monumental scientific resource which should be appreciated and protected by current workers who are fortunate to have access to these stocks of genetically based variation. Increasingly, developmentalists and morphologists will find it necessary to maintain and perform genetic experiments with their own material. They should as well support attempts to maintain germplasm banks and cooperative seed exchanges.

For genetic and physiological experiments, the maintenance of well-established inbred lines in which nuclear-based genetic variants are maintained is essential. Ideally, the identification and source of the cytoplasm should also be taken into account. Therefore, when possible, reciprocal pollinations should be attempted in experimental crosses. Documentation, record keeping and stock maintenance are necessary components of any research program and should be a routine budget item for research proposals. Published reports should fully identify the stocks and their sources.

Other activities traditionally carried out by geneticists include: (a) the observations on mutants in different backgrounds; (b) preparation and observations on combinations of mutants (dihybrid and trihybrid crosses) (Bowman, Smyth and

Meyerowitz 1989; 1991), which can reveal different types of epistasis (Avery and Wasserman 1992); (c) performing linkage tests; (d) the preparation of linkage maps; and (e) cytogenetical observations and analyses. The results of these activities are made available in informal reports and seed collections exchanged through cooperative working groups (see Appendix 2).

Observations of Mutant Expression in Different Backgrounds

This essentially descriptive exercise establishes a basis for further genetic and experimental studies. For morphological mutants, contemporary standards of description should apply, though considerable variation in format and detail exists in practice. Items to consider include:

1. Initial qualitative description of the mutant and a comparison with the normal phenotype. Macrophotography, SEM and epi-illumination microscopy, camera lucida tracings and hand drawings can provide appropriate documentation. Special attention should be given to positions of organs, relative sizes of individual parts, presence of chimeric sectors and color patterns on different organs. Examples of each of these types of observations are presented in earlier chapters.
2. Quantitative comparisons of organ growth and development and the characterization of the mature status of normal and mutant phenotypes. In some instances, allometry (Stebbins and Yagill 1966) is appropriate. Size differences can be related to differences in cell number and cell size. The dynamics of mitosis should be observed directly in order to establish its timing, frequency and distribution. Time-lapse video experiments of marked organs and growth can provide insights that might not otherwise be available. The integration of these technologies with computer-assisted monitoring and data accumulation will be of considerable value to future studies. An example of an early technical model of these methods is provided in Berg and Peacock (1991).
3. In morphological descriptions, sampling procedures should be as systematic as in other scientific exercises. This requires that care be taken in the conditions under which the plants are grown, in the selection of the items to be described and in the examination of replicate samples. Particular attention should be paid to the position on the inflorescence from which samples are taken. When possible, quantitative data should be collected and handled in a statistically appropriate fashion. Researchers should expect, recognize, document and report variation. The phenotypic frequency and the range of expression can provide information from which estimates of the **penetrance** and **expressivity** (Weaver and Hedrick 1989) can be determined. Recent studies of flower mutants which include some of these research attributes are: in *Arabidopsis*, "pistillata" (Hill and Lord 1989), "agamous," "apetala" and "pistillata" (Bowman, Smyth and Meyerowitz 1989; Komaki et al. 1988) (Table 5-1); and in *Lycopersicon*, "solanifolia" (Chandra Sekhar and Sawhney 1987) (Table 5-2).

Expression of Mutants in Different Environmental and Experimental Situations

While many phenotypic features of normal flowers are stable within broad ranges of temperature, day length and other growth conditions, many flower mutants exhibit plasticity in their phenotypic expression. This possibility requires that considerable care be given to growing mutant material under a range of prescribed conditions and reporting these conditions in detail in all reports.

For example, in *Lycopersicon,* stamenless-2 (Table 5-2) near-normal, fertile flowers develop at 18/15°C (day/night), whereas plants grown at 23/18°C (day/night) develop the mutant male-sterile phenotype (Sawhney 1983; Sawhney and Polowick 1986). In *Arabidopsis,* the mutants ap2-1 and ap-3 (Bowman, Smyth and Meyerowitz 1989) and ap2-2, ap2-8 and ap2-9 (Bowman, Smyth and Meyerowitz 1991) exhibit a variety of phenotypes when grown at 16°C, 25°C or 29°C. Other flower mutants (e.g., ag and pi-1) are not temperature sensitive.

As mentioned in Chapter 5, testing for and documenting environmental variability of male sterile mutants is of particular interest because, in a few cases and under some conditions, reversion to fertility has been demonstrated (Kaul 1988). For example, the barley mutant ms_9 expressed g-mst in Finland but was partially fertile in the United States (Ahokas and Hockett 1977), presumably in response to a combination of environmental and geographic variables. Individual environmental items to consider would be: (a) temperature regime (extremes and mean values); (b) photoperiod; (c) light levels; (d) spectral distribution; (e) humidity levels; and (f) soil conditions.

Generation of Mutants

The experimental induction and selection of morphological and physiological mutants, with chemicals (e.g., ethylmethanesulfonate (EMS)) or radiation, from seeds employs well-standardized procedures (Redei 1975; Meinke and Sussex 1979; Neuffer 1982b; Estelle and Somerville 1986). A number of flower mutants have been produced in this fashion. Flower variants have also been isolated from suspension cultures of *Nicotiana* following UV irradiation and methylglyoxalbis(guanylhydrazone) (MGBG) resistance selection. Cell lines derived from these procedures had altered polyamine metabolism and produced plants possessing a range of line-specific flower variability (Malmberg and McIndoo 1984). Some of these, when successfully bred, have been shown to be transmissible mutants (Malmberg et al. 1985). A range of variants and mutants affecting flower development have been documented.

In future, lines that have been produced via genetic techniques such as alterations by mobile genetic elements, genetic transformation and transposon tagging, coupled with tissue culture technologies, should be of interest to those with basic research objectives as well as to those looking for novel horticultural and agronomic specimens. In the case of infertile examples, methods for vegetative propagation are, of course, required.

Characterization of "Homeotic" Mutants

Homeosis, referring to the development of organs and tissues in unusual locations, has come to describe a number of different situations since it was first introduced (Bateson 1894). The use of "homeotic" is also variable and perhaps controversial (Sattler 1988b). In its broadest sense, homeosis describes the expressions of anatomical, physiological or morphological features in unusual settings (Sattler 1988). In flowers, examples of petaloid stamens or sepaloid petals would fit this interpretation. An example from contemporary animal developmental studies characterizes genetically based replacement of organs (e.g., the antennae on the head of the fruit fly *(Drosophila)* that are transformed into an additional pair of second legs (Gehring and Hiromi 1986). For some homeotic genes, a unique, conserved DNA sequence (a homeobox) has been demonstrated.

Current usage in flower research, (e.g., Schwarz-Sommer et al. 1990; Bowman, Smyth and Meyerowitz 1991; Coen 1991) parallels that of animal biology. In these cases, genetically based organ or tissue transference occurs with, in some cases, the demonstration of conserved DNA sequences within the gene—for example, the *deficiens* locus of *Antirrhinum* (Sommer et al. 1990), exhibiting a number of alleles, in which the five petals become sepaloid and leaf-like. In the *globifera* allelic form, in place of the four stamens and single staminodium, a penta-locular gynoecium is produced which bears some stigmatic areas. The normal bilocular gynoecium is missing (Sommer et al. 1990). In this case, sepals arise in whorls normally occupied by petals, carpellary tissues arise in whorls normally occupied by stamens and the normal gynoecium is missing. The *deficiens* locus encodes the DEF A protein, which showed a high level of homology to the conserved DNA binding and dimerization domains of two known transcription factors in animals and yeast (Schwarz-Sommer et al. 1990). *In situ* localization of *deficiens* cDNA documents enhanced hybridization signals of the gene in petals and stamens, the organs normally affected when the gene is mutated (Schwarz-Sommer et al. 1990). Other genes of *Antirrhinum* that are assumed to be homeotic are indicated in Table 4-2.

In *Arabidopsis,* the mutant *agamous (ag)* presents some analogous features. The basic phenotypic modification of this mutant includes the replacement of the six stamens by six petals and the replacement of the gynoecial whorl by a whole flower. "Thus *ag* flowers consist of ten petals (four normal and six in the place where stamens are normally found) inside the four sepals and inside the petals are again four sepals and ten petals" (Yanofsky et al. 1990). Data from cloning and sequencing studies suggest that the *AG* locus encodes a protein that has a high degree of sequence similarity with a class of transcription factors from humans, yeast and the *deficiens* gene of *Antirrhinum*. Other homeotic genes of *Arabidopsis* are listed in Table 5-1. The attributes of Arabidopsis as a model research organism are listed in Table 9-1.

The above observations of morphological and molecular expression and interaction of these and other mutants of *Arabidopsis* and *Antirrhinum* have sparked attempts to explain what these mutants reveal about normal flower morphogenesis

(Haughn and Somerville 1988; Schwarz-Sommer et al. 1990; Bowman, Smyth and Meyerowitz 1991; Meyerowitz et al. 1991; Coen 1991; Coen and Meyerowitz 1991). These attempts vary in scope and documentation, but share a number of elements. The most extensive reports (Coen 1991; Coen and Meyerowitz 1991), based on breeding studies and molecular analysis in *Arabidopsis* and *Antirrhinum*, make the following points:

1. They distinguish between homeotic genes which act early in development and modify the normal inflorescence (flo in Tables 4-2 and lfy and ap1 in Table 5-1) and those which express later in development and affect the kinds of organs formed in the different whorls of flowers.
2. They assume that the floral meristem of *Arabidopsis* and *Antirrhinum* normally generate organs within four concentric whorls. These whorls are occupied normally by sepals, petals, stamens and gynoecium.
3. Floral mutants are assumed to have three fields of expression which are superimposed upon these four whorls (Fig. 9-3 and Table 9-3). The basis of this assumption rests in the observation that most of these mutant variants involve two adjacent whorls.
4. The site of action of mutants is restricted to individual fields (Table 9-3).
5. Homeotic mutants are occasionally unstable and show temperature sensitivity.
6. Additional components which further specify "site," "kind" and "time" related expressions are yet to be uncovered.

Concerning this last point, in addition to further genetic sequences which modify and regulate the genes so far identified, one should expect to integrate data of a physiological and biophysical nature which impinge on the genetic expression. For example, it is not clear that cellular decisions about the "site" of organs are explained in this model; "kind" at a site appears to be the most obvious item controlled. The "time" component is more difficult to characterize. Attempts to distinguish homeosis from heterochrony are generally unsatisfactory (Coen 1991).

Fig. 9-3 (facing page). A schematic representation of a proposed model depicting how two classes of floral homeotic genes (ap2 and ag) could specify the identity of the four whorls of floral organs on a flower meristem of *Arabidopsis*. A section through one-half of a floral primordium is represented as a set of boxes, with the regions representing each whorl shown at the top of each column. Each box represents a single field. These are also shown at the top of the first column. The genotype under consideration is listed at the left, with the predicted distribution of gene products present in each genotype indicated by uppercase letters within the boxes. The predicted phenotype of the organs in each whorl is shown under the diagrams. Abbreviations: Se = sepal; P = petal; St = stamen; C = carpel; ST/P = petaloid stamen; L = leaves and carpeloid leaves. The * is a reminder that, in each genotype containing ag, there are several whorls of organs interior to the fourth whorl. The whorls interior to whorl 9 under the ag-containing genotypes are not shown. A schematic drawing of a longitudinal section of genotype of each flower is depicted at the right. Each organ type is shaded differently: sepals are black; petals are unshaded; stamens are stippled; carpels are hatched lower left to upper right; petaloid stamens are hatched upper left to lower right; leaves and carpeloid leaves are cross-hatched. From Bowman, Smyth and Meyerowitz (1991).

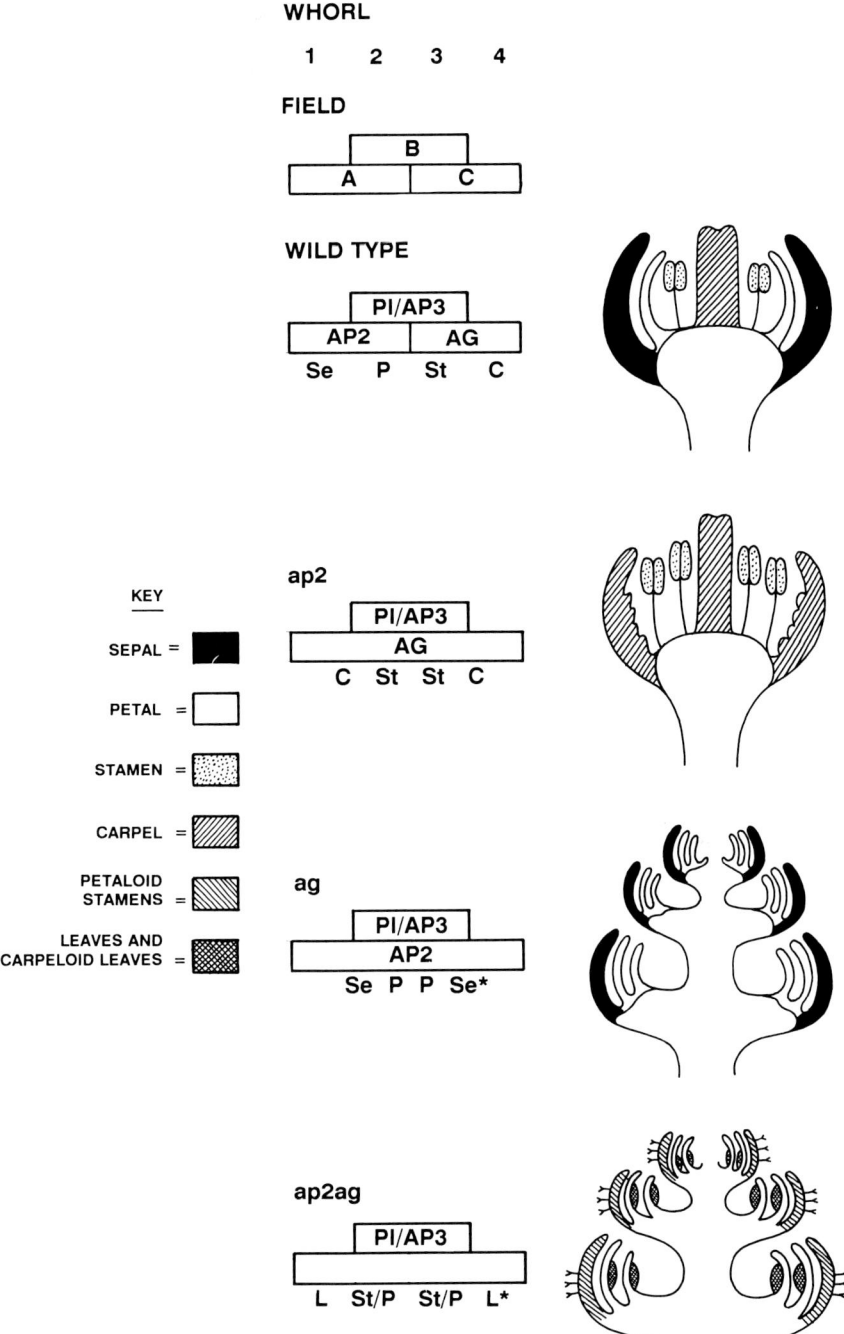

Table 9-3 Phenotype of Some Organ Identity Mutants in *Antirrhinum* and *Arabidopsis*

Genotype[a]	Phenotype				Region Affected
	Whorl 1	Whorl 2	Whorl 3	Whorl 4	
Wild type	Sepal	Petal	Stamen	Carpel	
ovu, ap2	Carpel	Stamen	Stamen	Carpel	A
def, glo, sep, pl, ap3	Sepal	Sepal	Carpel	Carpel	B
plena, ag	Sepal	Petal	Petal	Variable[b]	C

[a]The *Arabidopsis* mutations are apetala2 (ap2), pistillata (pi), apetala3 (ap3) and agamous (ag). The Antirrhinum mutations are ovulata (ovu), deficiens (def), sepaloidea (sep), globosa (glo) and plena (ple). Several mutations at a locus called pleniflora have recently been described in Antirrhinum. Implementation tests have shown that these are allelic to the classic plena mutation (R. Carpenter and E.S.C., unpublished results).

[b]In Antirrhinum, whorl 4 can be petaloid, sepaloid, carpelloid or a mixture of these; in Arabidopsis this whorl contains sepals. In both species additional petaloid and sepaloid whorls are produced interior to whorl 4.

Source: From Coen and Meyerowitz (1991). Reprinted by permission from *Nature* vol. 353, pp. 31–37; copyright © 1991 MacMillan Magazines Limited.

The molecular data so far accumulated do not explain the positioning of whorls nor the apportioning of different mutant expressions to them. For example, while hybridization and immunological studies document general localization of molecular expression to the flower and zones within the meristem, distribution patterns with the sharp cell-to-cell boundaries that might be expected if concentration of gene product was the only determining variable have yet to be presented. Perhaps, in addition to the levels of gene products of the flower genes so far explored, interaction with other distribution patterns is required.

Enzyme Differentiation Associated with Flower Differentiation

The ability to detect differential protein synthesis associated with development is a potent and sensitive tool for investigating the interface between the genetic and epigenetic components of flower development. Early reports [Marushige and Marushige (1962) and Barber and Steward (1968)] illustrated a variety of organ- and stage-related patterns of soluble, non-dissociated proteins. In addition, tests for esterases, malic acid dehydrogenase and alkaline phosphatases identified some organ-specific and some stage-specific differences (Barber and Steward 1968).

Advances in the resolution and sensitivity of the technology, as it relates to flower differentiation, include: (a) one- and two-dimensional gels of SDS-dissociated soluble proteins; (b) stage-specific differential polypeptide synthesis (Bommineni et al. 1990); (c) the demonstration of organ-specific isozymes such as stamen-specific peroxidases in *Mercurialis* (Kahlem 1976) and esterase, peroxidase and dehydrogenases in *Petunia* (Sawhney and Nave 1986) and *Lycopersicon* (Bhadula and Sawhney 1987); (d) isolation of organ-specific monoclonal antibodies of *Nicotiana* (Holaway, Evans and Malmberg 1989); organ-specific arabinogalactan-proteins of *Lycopersicon* (van Holst and Clarke 1986); (e) flower-specific proteins and antibodies from *Lycopersicon* (Lifschitz 1988); and (f) monoclonal antibodies to flower-specific antigens, possessing mainly carbohydrate epitopes (Evans, Holaway and Malmberg 1988). Extensions of these technologies to other species and isolates and the development of refinements and novel technologies can be anticipated.

Molecular Genetics of Flower Development

Perhaps the most currently provocative and productive topic in flower development is the burgeoning exploration of its genetic component (Goldberg 1988; Gasser 1991; Coen 1991). Because of the increasing availability of a large number of techniques and opportunities for its investigation, coupled with a resurgence of interest in the flower, one can anticipate the accumulation of a large data base and insight which will represent a very major contribution to flower development in particular and, as well, to an understanding of overall plant developmental biology.

While an examination of this genetic component is not essential to the objectives of this text, and indeed any extended discussion would be premature, a short description of the explorations into the molecular genetic components of flower development serves to complete the mandate of the text and point to areas of future research.

The objectives of a molecular genetic analysis of flower development will include:

1. Preparation of complementary DNA (cDNA) and genomic libraries of flower-specific or organ-specific genes (Gasser et al. 1989; Smith et al. 1990; Koltunow et al. 1990).
2. The molecular characterization of the DNA sequence of individual flower-specific genes.
3. *In situ* localization of expressed mRNA (Koltunow et al. 1990)
4. Uncovering and evaluating the relative contributions of the different steps in gene expression (e.g., differential processing and transport from the nucleus, mRNA stability, translational efficiency and protein stability) (Gasser 1991) to the overall expression of the phenotypic trait.
5. Identification and evaluation of the role of "transcription factors" in the regulation of gene expression.
6. The characterization of the involvement of mitochondria.

To varying degrees, these and other objectives are currently being achieved in a number of species but mainly in *Arabidopsis, Antirrhinum, Nicotiana* and *Zea*. It is enough for this discussion to express our enthusiastic encouragement for these diverse and complex analyses. We agree with Chasan (1991) that these studies "reveal that the levels of control of floral initiation and morphogenesis are likely to be very complicated indeed. It will be some time before the journey to an understanding of the regulation of flowering is complete."

A Three-dimensional Spacing (Siting) Mechanism Operating in the Outer Cell Layers of the Floral Meristem

In spite of the preceeding discussion on the role of genetic and cytoplasmic factors in flower development, it seems important to give special emphasis to the problem of understanding the spacing of primordia on flower meristems. In spite of all the insights thus far uncovered, this problem has not been described satisfactorily nor examined in sufficient detail. As discussed in Chapter 4, research on the factors involved in the positioning of organs on the plant has been primarily on the helical generation of leaves on the vegetative meristem. It has involved careful description, computer simulation, surgical and chemical manipulation and other probes. Because of their large size, the helical patterns of some flowers and inflorescences (e.g., *Helianthus;* see Chapter 7) should be considered as legitimate research subjects.

For helically arranged organs (e.g., *Ranunculus*) an extrapolation of the assumptions about mechanisms in vegetative meristems seems justified. However, many flowers exhibit other, non-helical patterns. Perhaps the most common is the whorled arrangement where groups of organs arise, more or less simultaneously, in concentric rings on the evoked meristem. Alternative complications include unidirectional, centrifugal and zygomorphic patterns. Each of these different conditions necessitate new sets of assumptions and speculations about the mechanisms that underlie initiation. Unfortunately, little is known about how chemical and biochemical factors, the necessary preconditions to the cellular patterning, become patterned. This is an area of research which requires further development of technical innovation.

At the cellular level, primordium initiation of floral organs, as with leaves, involves an alteration of the orientation of the cell plate and in some cases an alteration in the length of the cell cycle. These events take place in the peripheral 2–3 layers of the meristem. As mentioned, considerable variation has been detected in the sequence of events in the initiation of different kinds of organs and between the same organs from different species. At this stage, it appears that the differences described relate more to different strategies than to basic causal relationships.

Attempts to disentangle the causal relationships involved in organ initiation have produced only modest returns. Insights about the factors regulating cell cycle (Van't Hof 1985; Jacobs 1992), orientation of spindles and the methods by which cells communicate is, however, progressing. Additionally, technologies with which to assess ion flow, evaluate cell division kinetics and observe the meristem continue to improve and develop the necessary resolution for understanding the different processes.

Finally, Where Does Control Reside?

Our goal throughout this book has been to generate objective responses to the many questions that biologists pose about the cellular and physiological features that accompany and regulate the initiation and development of flowers. In the process, we have evaluated data, taken sides on differing interpretations and pointed out where future effort might be expended. The assumption underlying this approach has been that it will only be from a deliberate, dedicated dissection of the cellular, biochemical, biophysical and genetic components of the process that we will learn enough to focus sharply on future research strategies. This is the traditional approach of the biochemist, cell biologist and molecular biologist. Whether this reductionist technique is adequate to explain the total process of flower development in all angiosperms is impossible to predict at present. Certainly, biologists philosophically inclined to organismic and holistic interpretations of biological organization will have doubts about the wisdom of this assumption.

Scientific exploration usually involves an active interplay between the activities of theory generation, theory testing, data collection and intuition. For flower development, it appears premature to attempt an elaborate and comprehensive model for flower development or to speculate too deeply on where control of the process might reside. Researchers, at this time, are forced to choose the level of operation where they feel that the greatest research opportunities exist and where partial models can be proposed and tested.

For many, the general assumption that the process of flower formation is "primarily genetically determined" (Gasser 1991) represents enough of a theoretical assumption on which to base research. The time is propitious for a variety of genetically oriented explorations into flower development. There is scope as well for research at all other levels which participate in the total process. Many model studies at other levels have been discussed in earlier chapters.

It is not possible to assign a priority to the level at which control of the process might ultimately be identified. Indeed, some argue that the concept of "control" by genes, for example, is indefensible (Goodwin 1985). In this view, the multidimensional nature of the interplay between environment, cell physiology, biophysical components and genetic information in the total process of flower and inflorescence formation precludes simplistic interpretations. This is not just a problem for flower biologists, but applies as well to students of morphogenesis in general (Bard 1990; Nijhout 1990).

In addition to the complexity of the flower-forming process, flower variability should caution us from taking too seriously our hypotheses and hierarchical assumptions. Apart from the assumed conserved homology of meiosis expressed in microsporogenesis and megasporogenesis, there is enough variability in organ initiation and growth patterns to make us wary of too rigid a view of what is correct.

In conclusion, we should acknowledge that an unsatisfactory conclusion about the identity of the elements, their number involved in morphogenesis and the relationships between them is not unique to flower biology. The following quotations, taken from Wolpert's introduction to a symposium on the cellular basis of morphogenesis, are equally pertinent to the topic of this text and serve as a suitable cautionary final statement:

> From the beginning of our subject there has been a tension between the "left wing" and the "right wing." On the right I see those who thought that pattern formation, the organization of spatial differentiation, was something internal to the cell with little influence of cell interactions. Those people believed in cytoplasmic localization and autonomous cell lineages generating diversity between cells, following in the tradition of A. Weismann and E. B. Wilson. What I regard as the left wing were the people who thought that things were much more global, more interactive, in the tradition of Hans Driesch. I think that distinction is still

present today—on the one hand are those who pursue the lineage cytoplasmic line and on the other hand are those who prefer gradients and more global interactions. In the middle are those who think in terms of local interactions, such as induction.   (Wolpert 1989).

Finally, I would hope that the "left" and the "right" and the "middle" will ultimately come together. I would be very sad if it turned out that there were lots of different mechanisms of morphogenesis that bear no relation to each other whatsoever. I hope that general mechanisms for going from genes to flesh and blood are going to emerge. But I may just have to face that fact that I am going to be disappointed.   (Wolpert 1989)

In view of Wolpert's caution and because of the complexity of the problem, it would be easy to despair of acquiring precise understandings about the myriad aspects of flower morphogenesis. Such a conclusion is unwarranted, and some perspective is appropriate.

As reviewed in Chapter 1, attempts to treat flower development with the same intellectual rigor as animal embryologists apply to embryos of frogs, fruit flies and roundworms are rare and recent. But experimental animal embryology has nearly a century of solid discovery behind it by many renowned biologists, whereas developmental botany has attracted fewer researchers, with fewer resources, and has done so relatively recently.

If plant developmentalists use the current and future insights in cell biology, plant physiology, biophysics, genetics and molecular biology, progress towards the level of understanding desired should be possible. In the process, opportunities for many exciting and intellectually stimulating discoveries await those who make the effort to inquire and explore. By that time, perhaps the "left," "right" and "middle" will have found some accommodation.

Glossary

Many terms are defined in the body of the manuscript and can be accessed through the index. This short glossary includes terms and definitions from flower biology.

Aestivation The opening and unfolding of the floral organs during flower opening.

Anticlinal In apical meristems, divisions in which the cell plates are perpendicular to the surface.

Chasmogamous An open, normal-functioning, frequently out-crossing flower.

Chimeras Organs or tissues with patches or segments of cells which differ genetically from surrounding cells.

Cleistogamous Flowers, often small and unopened, which are self-fertilizing.

Glume One of two scale-like bracts at the base of the spikelet of a grass inflorescence.

Inflorescence An aggregation of flowers (see box, p. 174).

Lemma The lower of two chaffy bracts enclosing the grass flower.

Lodicule A small organ of the grass flower, proximal to stamens, which swells during anthesis, forcing the glumes apart.

Megasporangium The portion of the ovule in which megasporogenesis occurs. In some treatments it is equated with the ovule; more correctly it is equivalent to the nucellus.

Microsporangium An organ of the anther in which microsporogenesis occurs.

Morphometry Obtaining quantitative data about the amount and distribution of structures and organelles in cells from microscope slide preparations or from photomicrographs. The most widely used method is based on stereology (Berlyn and Miksche 1976).

Parastichy A line connecting primordia on a floral or inflorescence meristem where a number of primordia have been initiated.

Periclinal In apical meristems, divisions in which the cell plates are oriented parallel to the surface.

Petalody Developmental transformation of flower organs to resemble petals.

Phyllotaxy The arrangement of leaves (helical (spiral), decussate, whorled, distichous, bijugate, etc.). Also used to describe the pattern of arrangement of floral organs on the flower.

Plant Growth Substance(s) (PGS) A variety of naturally occurring molecules and sythetic analogues which have profound stimulatory or inhibitory effects on growth. They also, in many cases, affect differentiation. Substances normally included in this class of chemicals are auxins, gibberellins, cytokinins, ethylene, abscisic acid and polyamines.

Plastochron The time interval between the initiation of successive leaves (floral organs and primordia) on a meristem.

Sepalody Developmental transformation of flower organs to resemble sepals.

Stereology A statistically based sampling method of obtaining information about the amount and distribution of structures within sectioned objects. See Briarty (1985) for a review of its use in botanical studies.

Teratology The study of abnormalities and monstrosities.

Teratoma An abnormal growth or an abnormal form of an organ.

Tunica Layers Outer cell layers of the shoot apical meristem which result from a limitation of division to anticlinal planes. (Note: T1 and T2 refer to tunica layers in the meristem; LI, LII and LIII refer to germ layers as demonstrated through periclinal chimeras.)

Appendix 1: Culture Protocols for Rearing Research Plants

The objective of any set of culture procedures for the production of research plants for developmental studies is a population of plants which possess a minimum of variability in their time-to-flower and in their culture- and environment-based morphological variation. Procedures to minimize between-experiment variability throughout the year are also necessary. The reproducibility and value of developmental botanical studies can be limited by the degree to which these objectives are attained.

Data reports should identify the culture conditions with precision and accuracy. Attention to light conditions (intensity, daylength, spectral distribution), temperature conditions (mean, maximum, minimum, day/night), humidity and the fertilizer regime can be particularly critical in some studies. Careful identification of the seed stock is also essential.

For many species, standard greenhouse and incubator/growth chamber conditions and procedures will suffice. Examples of these procedures are illustrated in the following descriptions for raising *Arabidopsis* and *Lycopersicon* plants. For other plants, special conditions are sometimes useful. Some examples are given below.

Antirrhinum majus

Detailed conditions for the culture of snapdragons are described by Rogers (1980), and the requirements for flowering are discussed by Cockshull (1985a). The flower of *A. majus* is described by Hickey and King (1988), Maginnes and Langhans (1961) and Singh and Jain (1979).

Arabidopsis thaliana

The commonly used ecotypes, Columbia and Landsberg, have a generation time of approximately 6 weeks at 25°C under continuous light. A typical protocol is taken from Bowman et al. (1992):

> Seeds were planted on a peat moss/potting soil/sand (3:3:1, v:v:v:) mixture. The plants were grown in incubators under constant cool-white fluorescent light at 25°C (unless otherwise stated) and 70% relative humidity.

Lolium temulentum

According to Evans (1969c), *Lolium* is easily grown with no germination problems. Plants grow in a number of soils but do best when grown individually in small pots of perlite or vermiculite. Growth is optimal at about 25/20°C (day/night). Knox and Evans (1968) detail a procedure for the Ceres strain.

Lycopersicon esculentum

Seeds were germinated in peat pots ("Jiffy sevens"), and seedlings with three to four leaves were transferred to 15cm plastic pots containing a mixture of loam/sand/peat (1:1:1). Plants were grown in a growth chamber illuminated with Gro-Lux wide-spectrum fluorescent tubes (16h/day at an intensity of 180 $\mu E/m^{-2}/s^{-1}$). Temperature was maintained at 23 ± 1°C (day) and 18 ± 1°C (night). Plants were supplied weekly with a commercial fertilizer (20:20:20). This procedure is drawn from Chandra Sekhar and Sawhney (1984).

Oryza sativa

The following procedure for greenhouse rice culture was suggested for the cv. "Shiokari," common to Hokkaido, by Dr. T. Kinoshita, University of Hokkaido, Faculty of Agriculture, Sapporo, Japan:

1. Disinfect seed with a 6–24h soak in "Benlate" solution (1:500 or 1:1000 dilution). Allow seeds to dry for at least 24h.
2. Soak seeds in water for 2–6 days at 4–6°C in the dark.
3. Germinate seeds on moistened filter paper in petri dish at 30°C and in the dark.
4. Broadcast germinated seed to a nursery bed of sandy loam soil (pH 5.0–5.5) containing N; P_2O_5 and K_2O fertilizer (2g each/25g soil). After irrigating the seed bed, cover seeds with soil (˜1–2cm) and allow seeds to grow for 1–2 days at 30°C in the light. Irrigate daily.
5. Transplant seedlings (4–6 leaves) to pots (plastic or Wagner (1/5000a)). Fill pots with a paddy-like mixture of soil and water containing fertilizer (N, P_2O_5 and K_2O at 0.3–0.6g each per pot. Grow the seedlings under greenhouse conditions at 15–25°C. Plantings of 1–2 plants/pot are common, while dense plantings reduce the tillering. Irrigate daily. Pots with drainage holes should be placed in a tray of water. In northern latitudes, in winter, use supplemental illumination to maintain a 14h day length.
6. Flowers and inflorescences will be available within 50–100 days, depending upon cultivar and the stage desired.

Triticum aestivum

Spring wheat is considered to be a quantitative LDP (Pinthus 1985). In studies of cv. Marquis, Friend, Fisher and Helson (1963) demonstrated that, under continuous illumination, floral initiation was earlier with each increase in light intensity from 200 to 2500 ft-c and with each increase in temperature between 10°C and 30°C.

One method to produce a population of reasonably uniform wheat plants for research was described by Friend (1960). It included the selection of uniform fruits and uniformly germinated seedlings. Initially, eight seedlings per pot were planted, and these were thinned to four plants per pot after 1–2 weeks. Seedlings were sown in quartz sand and irrigated from below with a modified Hoagland's solution. Obviously, many variants of these procedures are possible for wheat, and they can be adapted for other species.

Zea mays

For many studies on the flowers of maize, the early-maturing sweet corn varieties are useful. Even earlier-maturing isolates such as cv. Gaspé Flint can be of value in certain projects. Genetically based variation has typically been maintained in field corn inbreds. Typically these are later maturing than the sweet corn cultivars.

While many different growth protocols will produce maize plants suitable for research (Neuffer 1982a), the following procedures are known to produce, under greenhouse conditions, suitably uniform populations of plants for experimental studies.

Fungicide-treated seeds are first sown in 5cm peat pots, in a standard greenhouse soil mixture of loam/sand/peat (8/5/4). After 7–10 days, when plants have germinated, the peat pots are carefully planted, one per pot, in 20cm plastic pots containing the standard greenhouse soil mixture including a slow release fertilizer. A standard, weekly fertilizer regime should also be followed. Field-grown plants are more variable than greenhouse-grown material and require more effort to obtain sterility for *in vitro* culture studies.

The day/night, greenhouse temperature regime is maintained at $24/18 \pm 2°C$ (day/night). An 18/6 day/night light regime of a mixture of natural, incandescent and sodium vapor illumination was found to be satisfactory. In view of the fact that maize is a quantitative SDP (Hanway and Ritchie 1985), experiments with longer dark periods might be tested when producing donor material for developmental experiments on flower development. For some experiments, borderline flower-induction conditions may be beneficial. As well, because temperature values and level of illumination can affect the condition of the plants, it is important to control and record these conditions as closely as possible when conducting experiments and comparing their data. Routine insect and pathogen control is assumed.

Appendix 2: Sources of Information about Genetic Stocks of Model Species for Flower Studies

Detailed studies of the morphological and physiological basis of delopmental phenomena eventually require model organisms which have been especially selected, bred and maintained as genetically defined lines and stocks. The maintenance of these genetic stocks and the collation and dissemination of information about these model species are frequently the product of voluntary efforts among interested investigators. A variety of informal working groups, cooperatives and organizations have developed to stimulate and coordinate the information about these individual organisms and preserve and distribute germ plasm. The existence and encouragement of these groups is indispensable for the continued exploration of the developmental biology of flowers.

A selected list of examples of these cooperative organizations, stock centers and publications which are relevant to flower studies is provided below. Similar organizations exist to assist with research on other genera, and information about them may be obtained from active researchers.

Antirrhinum majus

Mutant stocks of *Antirrhinum* are collected and maintained by the John Innes Institute and are available for research and teaching by writing to Dr. Rosemary Carpenter, Department of Genetics, John Innes Institute, John Innes Centre, Norwich Research Park, Colney, Norwich NR4 7UH, U.K.

Arabidopsis thaliana

A summary of services available to researchers who wish to use *Arabidopsis* in research is the National Science Foundation Publication entitled "The Multinational Coordinated *Arabidopsis thaliana* Genome Research Project" (NSF Publication No. 92-112). The following three centers provide information and genetics stocks.

Arabidopsis Biological Resource Center (ABRC) maintains and distributes a large collection of flower mutant stocks and DNA libraries of *Arabidopsis*. Information about the services and the current *Seed and DNA Stock List* can be obtained from the *Arabidopsis* Biological Resource Center at Ohio State University, 1735 Neil Avenue, Columbus, OH 43210, FAX (614) 292 0603.

Arabidopsis Information Management System (AIMS) is available for stock information and ordering from ABRC. AIMS is accessible worldwide through

Telnet login and can also be queried through electronic mail. Informaton on how to access AIMS can be obtained by sending an electonic mail message to the following address: inquire-aims @ genesys.cps.msu.edu. If you send a blank message to this address, introductory information about AIMS, including the AIMS tutorial, will be sent.

The Nottingham *Arabidopsis* Stock Centre (NASC) collects, maintains, catalogs and distributes mutant strains and ecotype collections of *Arabidopsis*. NASC is part of an *Arabidopsis* Resource Centre network with a European DNA Stock Centre at the Max-Delbruck-Laboratorium, Koln, Germany, and the ARBC at Ohio State University, Columbus, OH. 43210. Information is available from Dr. Mary Anderson, Director, NASC, Department of Life Science, University of Nottingham, University Park, Nottingham NG7 2RD, U.K.

Brassica (Rapid cycling varieties)

Brassica species are increasingly of interest as model systems in plant physiology and biotechnical applications. The development of rapid cycling lines (Rcb) represents a major contribution to flower physiology and no doubt will eventually involve studies of flower morphology and development as well (Williams and Hill 1986). Information and stocks of these Rcbs and other crucifers *(Capsella, Sinapis, Lepidium, Thlaspi, Sisymbrium* and *Roroippa)* are available from Crucifer Genetics Cooperative, Department of Plant Pathology, University of Wisconsin–Madison, 495 Russell Labs, 1630 Linden Drive, Madison, WI 53706.

A few of the Wisconsin Rapid Cycling *Brassica* stocks are commercially available as Wisconsin Fast Plants ™ from Carolina Biological Supply, 2700 York Rd., Burlington, NC 27215. Stocks are available in kit form for use in classrooms and teaching laboratories.

Lycopersicon esculentum

Information regarding mutant stocks and wild species accessions, as well as seed thereof, can be obtained from the Charles M. Rick Tomato Genetics Resource Center, Department of Vegetable Crops, University of California, Davis, CA 95616. Lists of available stocks are published annually in the *Report of the Tomato Genetics Cooperative,* available from the Tomato Genetic Cooperative, Departments of Plant Breeding, Biometry and Agronomy, 1017 Bradfield Hall, Cornell University, Ithaca NY 14853-1901.

Nicotiana tabacum

A tobacco and *Nicotiana* collection is maintained at the USDA Crops Laboratory in Oxford, North Carolina. The current collection consists of 2000-plus different accessions. These include representatives of the recognized *Nicotiana* species. Specific information about stocks and seed samples can be obtained from Dr. Verne A. Sisson, Research Geneticist, *Nicotiana* Germ Plasm Collection, USDA-

ARS-SAA, Tobacco Research Laboratory, P.O. Box 1168, Oxford, NC 27565-1168.

Pisum sativum

Pisum Genetics is issued annually by the Pisum Genetics Association under the editorship of Dr. Ian C. Murfet, Department of Plant Science, University of Tasmania, Hobart, Tasmania 7001, Australia. Information regarding membership, fees etc. can be obtained from Dr. Murfet, phone 61 02 20 2605 or FAX 61 02 202698.

Information about the availability of seed stocks can be obtained from (1) John Innes *Pisum* Collection, c/o Mr. Michael Ambrose, John Innes Centre, Colney Lane, Norwich NR4 7UH, U.K., FAX 0603 56844; or (2) The Marx Genetic Stock Collection, c/o Dr. James R. McFerson, USDA-ARS, Plant Genetics Resources Unit, Cornell University, Geneva, NY 14456-0462, FAX (315) 787 2397.

Zea mays

The Maize Genetics Cooperation Newsletter is an annual compilation of information by maize researchers. Information about subscriptions and current maize gene lists may be obtained from Dr. E. H. Coe, Editor, *Maize Genetics Cooperation Newsletter*, Curtis Hall, University of Missouri, Columbia, MO 65211, phone (314) 882 2768, FAX (314) 874 4063. E mail internet: ed @ teosinte.agron.missouri.edu.

Mutations affecting flower morphology and formation of gametes, in addition to numerous other types of heritable variants, are maintained by the Maize Genetics Cooperation Stock Center. For information regarding samples of genetic stocks that are available upon request, contact Dr. Martin Sachs, Director, Maize Genetics Stock Center, s-123 Turner Hall, Agronomy Department, University of Illinois, 1102 S. Goodwin Avenue, Urbana, IL 61801, phone (217) 333 6631, FAX (217) 333 6064. E mail: maize @ uxl.cso.uiuc.edu.

Informal descriptions of ongoing research projects and other information about research on flowers is available in two annual issues of *Flowering Newsletter*. Write to International Working Group on Flowering, c/o Georges Bernier, Editor, Department of Botany, University of Liège, Sart Tilman B22, B-4000 Liège, Belgium.

References

Abbe, E. C., and B. O. Phinney. 1940. The effect of the gene d_1 on the developmental pattern and cellular constitution of the stem in maize. Amer. J. Bot. 27:1s (Abstr.).

———. 1942. The action of the gene dwarf$_1$ in the ontogeny of the stem in maize. Genetics 27:129 (Abstr.).

Adler, I. 1974. A model of contact pressure in phyllotaxis. J. Theo. Biology 45:1–79.

Aghion-Prat, D. 1965. Neoformation de fleurs *in vitro* chez *Nicotiana tabacum* L. Physiol. Végét. 3:229–303.

Ahokas, H. 1976. Evidence of a pollen esterase capable of hydrolysing sporopollenin. Experientia 32:175–77.

Ahokas, H., and E. A. Hockett. 1977. Male sterility mutants of barley. IV. Different fertility levels of msg9ci (cv. Vantage) to an ecoclinal response. Barley Genet. Newslett. 7:10–11.

Alberch, P. 1980. Ontogenesis and morphological diversification. Amer. Zool. 20:653–667.

———. 1985. Problems with the interpretation of developmental sequences. Syst. Zool. 34:46–58.

Albertson, M. C., and R. L. Phillips. 1978. Ethidium bromide ineffectiveness and production of all male-sterile progeny by combining tillering and genic male sterility. Maize Genet. Coop. Newslett. 54:110–112.

———. 1981. Developmental cytology of 13 genetic male sterile loci in maize. Can. J. Genet. Cytol. 23:195–208.

Allsopp, A. 1967. Heteroblastic development in vascular plants. Adv. Morphog. 6:127–171.

Ammirato, P. V. 1983. Embryogenesis. In D. A. Evans, W. R. Sharp, P. V. Ammirato, and Y. Yamada [eds.], Handbook of plant cell culture, vol. 1, Techniques for Propagation and Breeding (New York: Macmillan), pp. 82–123.

Arber, A. 1934. The Gramineae: A study of cereal, bamboo and grass. Cambridge: Cambridge University Press.

———. 1946. Goethe's botany. Chron. Botan. 10(2):63–126.

———. 1970. The natural philosophy of plant form. Darien, CT: Hafner.

Armstrong, T. A.; T.-S. Soong; and M. A. W. Hinchee. 1987. Culture of detached spikes and the early development of the fourth floret caryopsis in wheat. J. Plant Physiol. 131:305–314.

Arnison, P. G.; P. Donaldson; L. C. C. Ho; and W. A. Keller. 1990. The influence of various physical parameters on anther culture in broccoli *(Brassica oleracea* var. *italica)*. Plant Cell, Tissue and Organ Culture 20:147–155.

Arnott, H. J., and S. C. Tucker. 1963. Analysis of petal venation in *Ranunculus*. I. Anastomoses in *R. repens* v. *pleniflorus*. Amer J. Bot. 50:821–830.

———. 1964. Analysis of petal venation in *Ranunculus*. II. Number and position of dichotomies in *R. repens* var. *pleniflorus*. Bot. Gaz. 125:13–26.

Ashby, E. 1948*a*. Studies on the morphogenesis of leaves. 1. An essay on leaf shape. New Phytol. 47:153–176.

———. 1948b. Studies in the morphogenesis of leaves. II. The area, cell size and cell number of leaves of *Ipomoea* in relation to their position on the shoot. New Phytol. 47:177–195.

Avery, L., and S. Wasserman. 1992. Ordering gene function: The interpretation of epistasis in regulatory hierarchies. Trends in Genet. 8:312–316.

Bachmann, K. 1983. Evolutionary genetics and genetic control of morphogensis in flowering plants. Evol. Biol. 16:157–208.

Bachmann, K., and K. L. Chambers. 1978. Pappus part number in annual species of *Microseris* (Compositae, Cichoriaceae). Pl. Syst. Evol. 129:119–134.

Bachmann, K.; K. L. Chambers; and H. J. Price. 1981. Genetic determination of pappus part number in the annual hybrid *Microseris* B 87 (Asteraceae-Lactuceae). Pl. Syst. Evol. 138:235–246.

Bachmann, K.; K. L. Chambers; H. J. Price; and A. Konig. 1982. Four additive genes determining pappus part numbers in *Microseris* annual hybrid C34 (Asteraceae/Lactuceae). Pl. Syst. Evol. 141:123–141.

Bajaj, Y. P. S. 1983. *In vitro* production of haploids. In D. A. Evans, W. R. Sharp, P. V. Ammirato, and Y. Yamada [eds.], Handbook of plant cell culture, vol. 1, Techniques for propagation and breeding (New York: Macmillan), pp. 228–287.

Baker, H. G. 1948. Corolla size in gynodioecious and gynomonoecious species of flowering plants. Proc. Leeds Phil. Lit. Soc. 5:136–39.

Baker, H. G., and I. Baker. 1983. A brief historical review of the chemistry of floral nectar. In B. Bently and T. Elias [eds.], The biology of nectaries (New York: Columbia University Press), pp. 126–152.

Ballard, L. T. A. 1969. *Anagallis arvensis* L. In L. T. Evans [ed.], The induction of flowering—Some case histories (Ithaca, NY: Cornell University Press), pp. 376–392.

Barber, J. T., and F. C. Steward. 1968. The proteins of *Tulipa* and their relation to morphogenesis. Dev. Biol. 17:326–349.

Bard, J. 1990. Morphogenesis: The cellular and molecular processes of devlopmental anatomy. Cambridge: Cambridge University Press.

Barlow, P. W., and D. J. Carr [eds]. 1984. Positional controls in plant development. Cambridge: Cambridge University Press.

Barnard, C. 1955. Histogenesis of the inflorescence and flower of *Triticum aestivum* L. Aust. J. Bot. 3:1–20.

———. 1957. Floral histogenesis in the monocotyledons. 1. The Gramineae. Aust. J. Bot. 5:1–20.

———. 1958. Floral histogenesis in the monocotyledons. 111. The Juncaceae. Aust. J. Bot. 6:285–298.

Bashaw, E. C. 1980. Apomixis and its application in crop improvement. In W. R. Fehr and H. H. Hadley [eds.], Hybridization of crop plants (Madison, WI: American Society of Agronomy), pp. 45–63.

Bateson, W. 1894. Materials for the study of variation. London: MacMillan.

Battey, N. H., and R. F. Lyndon. 1984. Changes in apical growth and phyllotaxis on flowering and reversion in *Impatiens balsamina* L. Ann. Bot. 54:553–567.

———. 1986. Apical growth and modifications of the development of primordia during re-flowering of reverted plants of *Impatiens balsamina* L. Ann. Bot. 58:333–341.

———. 1988. Determination and differentiation of leaf and petal primordia in *Impatiens balsamina*. Ann. Bot. 61:9–16.

———. 1990. Reversion of flowering. Bot. Rev. 56:162–189.

Baulcombe, D. C. 1987. Do plant hormones regulate gene expression during development? In G. V. Hoad, J. R. Lenton, M. B. Jackson and R. K. Aitkin [eds.], Hormone action in plant development—A critical appraisal (London: Butterworths), pp. 63–70.
Beadle, G. W. 1980. The ancestry of corn. Sci. Am. 242:112–119.
Bell, A. D. 1991. Plant form: An illustrated guide to flowering plant morphology. Oxford: Oxford University Press.
Bell, G. 1985. On the function of flowers. Proc. R. Soc. Lond. B. 224:223–265.
Bell, P. R. 1975. Physical interactions of nucleus and cytoplasm in plant cells. Endeavour 34:19–22.
Bennett, M. D. 1977. The time and duration of meiosis. Phil. Trans. Roy. Soc. Ser. B (Biol. Sci.) 277:201–226.
Bennett, M. D.; M. K. Rao; J. B. Smith; and M. W. Bayliss. 1973. Cell development in the anther, the ovule and the young seed of *Triticum aestivum* L. var. Chinese Spring. Phil. Trans. Roy. Soc. Lond. 266B:39–81.
Benson, L. 1957. Plant Classification. Boston: D. C. Heath.
Berg, A. R., and K. Peacock. 1991. Growth patterns in nutating and nonnutating sunflower hypocotyls. Amer. J. Bot. 79:77–85.
Berghoef, J., and J. Bruinsma. 1979a. Flower development in *Begonia franconis* Liebm. I. Effects of growth-regulating substances and environmental conditions on the composition of the inflorescence. Zeit. fur Phlanzenphys. 93:303–315.
———. 1979b. Flower development in *Begonia franconis* Liebm. II. Effects of nutrition and growth-regulating substances on the growth of flower buds *in vitro*. Zeit. fur Phlanzenphys. 93:345–357.
———. 1979c. Flower development in *Begonia franconis* Liebm. III. Effects of growth-regulating substances on organ initiation in flower buds *in vitro*. Zeit. fur Phlanzenphys. 93:377–386.
Berlyn, G. P., and J. P Miksche. 1976. Botanical microtechnique and cytochemistry. Ames, IA: Iowa State University Press.
Bernier, G. 1986. The flowering process as an example of plastic development. In D. H. Jennings and A. J. Trewavas [eds.], Plasticity in plants. Symp. Soc. Exp. Biol. No. 40. (Cambridge: Society for Experimental Biology), pp. 257–286.
———. 1988. The control of floral evocation and morphogenesis. In W. R. Briggs, R. L. Jones and V. Walbot [eds.], Annual Review Plant Physiology and Plant Molecular Biology, vol. 39 (Palo Alto, CA: Annual Reviews), pp. 175–219.
Bernier, G.; J.–M. Kinet; and R. Bronchart. 1967. Cellular events at the meristem during floral induction in *Sinapis alba* L. Physiol. Végét. 5:311–324.
Bernier, G.; J.–M. Kinet; and R. M. Sachs. 1981a. The physiology of flowering, vol. 1. Boca Raton, FL: CRC Press.
———. 1981b. The physiology of flowering, vol. 2. Boca Raton, FL: CRC Press.
Beyer, E. M, Jr., and O. Sundin. 1978. $^{14}C_2H_4$ metabolism in morning glory flowers. Pl. Physiol. 61:896–899.
Bhadula, S. K., and V. K. Sawhney. 1987. Esterase activity and isozymes during the ontogeny of stamens of male fertile *Lycopersicon esculentum* Mill., a male sterile stamenless-2 mutant and the low temperature–reverted mutant. Plant Science 52:187–194.
Bhandari, N. N. 1984. The microsporangium. In B. M. Johri [ed.], Embryology of angiosperms (Berlin: Springer-Verlag), pp. 53–121.
Bhar, D. S. 1970. *In vitro* studies of floral shoot apices of *Pharbitis nil*. Can. J. Bot. 48:1355–1358.

Blackmore, S., and R. B. Knox [eds.]. 1991. Microspores—Evolution and ontogeny. London: Academic Press.
Blake, J. 1962. Normal and abnormal developments of the stem apex in carnation. Ann. Bot. 26:95–104.
———. 1966. Flower apices cultured *in vitro*. Nature 211:990–991.
Blakeslee, A. F.; S. Satina; and A. G. Avery. 1940. Utilization of induced periclinal chimeras in determining the constitution of organs and their origins from the three germ layers in *Datura*. Science 91:423.
Boke, N. H. 1948. Development of the perianth of *Vinca rosea* L. Amer. J. Bot. 35:413–423.
———. 1949. Development of the stamens and carpels in *Vinca rosea* L. Amer. J. Bot. 36:535–547.
Bommineni, V. R.; B. G. Atkinson; R. I. Greyson; and D. B. Walden. 1990. Polypeptides synthesized during the maturation of flower organs from tassel and ear inflorescences of *Zea mays* L. Maydica 35:165–201.
Bommineni, V. R., and R. I. Greyson. 1987. In vitro culture of ear shoots of *Zea mays* and the effect of kinetin on sex expression. Amer. J. Bot. 74:883–890.
———. 1990. Regulation of flower development in cultured ears of maize (*Zea mays* L.). Sex Plant Reprod. 3:109–115.
Bonner, J. T. 1974. On development: The biology of form. Cambridge, MA: Harvard University Press.
Bonnett, O. T. 1936. The development of the wheat spike. J. Agric. Res. 53:445–451.
———. 1940. Development of the staminate and pistillate inflorescences of sweet corn. J. Agric. Res. 60:25–37.
———. 1948. Ear and tassel development in maize. Ann. Mo. Bot. Gard. 35:269–288.
———. 1953. Developmental morphology of the vegetative and floral shoots of maize. Univ. Ill. Agric. Exp. Sta. Bull. No. 568.
———. 1966. Inflorescences of maize, wheat, rye, and oats: Their initiation and development. Univ. Ill. Agric. Exp. Sta. Bull. No. 721.
Boothe, J. G., and D. B. Walden. 1990. Gene expression in embryos and seedlings of maize. Maydica 35:187–194.
Bouman, F. 1984. The ovule. In B. M. Johri [ed.], Embryology of Angiosperms (Berlin: Springer-Verlag), pp. 123–157.
Bowman, J. L.; H. Sakai; T. Jack; D. Weigel; U. Mayer; and E. M. Meyerowitz. 1992. SUPERMAN, a regulator of floral homeotic genes in *Arabidopsis*. Development 114:599–615.
Bowman, J. L.; D. R. Smyth; and E. M. Meyerowitz. 1989. Genes directing flower development in *Arabidopsis*. Plant Cell 1:37–52.
———. 1991. Genetic interactions among floral homeotic genes of *Arabidopsis*. Development 112:1–20.
Bowman, J. L.; M. F. Yanofsky; and E. M. Meyerowitz. 1988. *Arabidopsis thaliana*: A review. Oxford Survey of Plant Mol. and Cell Biol. 5:57–87.
Brenner, S. 1975. Closing remarks: The genetic outlook. In Cell patterning (Amsterdam: Associated Science Publishers), pp. 343–345. CIBA Symposium 29.
Brewbaker, J. L. 1967. The distribution and phylogenetic significance of binucleate and trinucleate pollen grains in the angiosperms. Amer. J. Bot. 54:1069–1083.
Brewbaker, J. L., and B. Y. Kwack. 1963. The essential role of calcium ion in pollen germination and pollen tube growth. Amer. J. Bot. 50:859–865.
Briarty, L. G. 1985. Quantitative morphological analysis of botanical micrographs. In A. W. Robards [ed.], Botanical microscopy 1985 (New York: Oxford University Press), pp. 91–127.

Briggle, L. W., and B. C. Curtis. 1987. Wheat worldwide. In E. G. Heyne [ed.], Wheat and wheat improvement, 2d ed. (Madison, WI: American Society of Agronomy), pp. 1–32.

Briggs, B. G., and L. A. S. Johnson. 1979. Evolution in the Myrtaceae—Evidence from inflorescence structure. Proc. Linn. Soc. N. S. Wales 102:157–256.

Brooks, R. M. 1940. Comparative histogenesis of vegetative and floral apices in *Amygdalus communis*, with special reference to the carpel. Hilgardia 13:249–306.

Brossard-Chriqui, D., and B. K. Tripathi. 1984. Comparaison des aptitudes morphogénétiques des étamaines fertiles ou stériles d'*Actinidia chinensis* cultivées *in vitro*. Can. J. Bot. 62:1940–1946.

Browder, L. W. 1980. Developmental biology. Philadelphia: Saunders College.

Brulfert, J. 1965. Etude expérimentale du développement végétatif et floral chez *Anagallis arvensis* L. ssp. *phoenicea* Scop. Formation de fleurs proliferes chez cette meme espece. Rev. Gén. Bot. 72:641–694.

Brulfert, J.; D. Fontaine; and C. Imhoff. 1985. *Anagallis arvensis*. In A. H. Halevy [ed.], CRC handbook of flowering, vol. 1 (Boca Raton, FL: CRC Press), pp. 434–449.

Buvat, R. 1952. Structure, evolution et functionnement en meristeme apical de quelques dicotyledons. Ann. Sci. Nat. Bot. Ser. 11 13:199–300.

Carlquist, S. 1969. Toward acceptable evolutionary interpretations of floral anatomy. Phytomorphology 19:332–362.

Carlson, W. R. 1986. The B-chromosome in maize. Crit. Rev. Plant Sci. 3:201–226.

———. 1988. The cytogenetics of corn. In G. F. Sprague and J. W. Dudley [eds.], Corn and corn improvement, 3 ed. (Madison, WI: American Society of Agronomy), pp. 259–343.

Carpenter, R.; A. Hudson; T. Robbins; J. Almeida; C. Martin; and E. Coen. 1987. Genetic and molecular analysis of transposable elements in *Antirrhinum majus*. In O. Nelson [ed.], Plant transposable elements. (New York: Plenum Press).

Cass, D. D., and G. C. Fabi. 1990. Early ovule development in *Papaver rhoeus*. Can. J. Bot. 68:258–265.

CBE Style Manual, 5th ed. 1983. Bethesda, MD: Council of Biology Editors.

Chailakhyan, M. Kh., and V. N. Khrianin. 1978. Effect of growth regulators and the role of roots in the sex expression in spinach plants. Planta 142:207–210.

———. 1987. Sexuality in plants and its hormonal regulation. New York: Springer-Verlag.

Chandra Sekhar, K. N., and V. K. Sawhney. 1984. A scanning electron microscope study of the development and surface features of floral organs of tomato *(Lycopersicon esculentum)*. Can. J. Bot. 62:2403–2413.

———. 1987. Ontogenetic study of the fusion of floral organs in the normal and "solanifolia" mutant of tomato *(Lycopersicon esculentum)*. Can. J. Bot. 65:215–221.

———. 1990. Regulation of the fusion of floral organs by temperature and gibberellic acid in the normal and solanifolia mutant of tomato *(Lycopersicon esculentum)*. Can. J. Bot. 68:713–718.

Chang, T.-T., and E. A. Bardenas. 1965. The morphology and varietal characteristics of the rice plant. IRRI Technical Bulletin No. 4. The International Rice Research Institute. Manilla, Philippines.

Chao, M. D. 1947. Growth of the dandelion scape. Plant Physiol. 22:393–406.

Chapman, G. P. 1986. Mitochondrial delivery via the male gametophyte and the prospects for recombination. In S. H. Mantell, G. P. Chapman and P. F. S. Street [eds.], The chondriome—Chloroplast and mitochondrial genomes. (New York: John Wiley & Sons), pp. 61–68.

———. 1987. The tapetum. In K. L. Giles and J. Prakash [eds.], Int. Rev. Cyt. 107:111–125.

Chapman, G. P., and W. E. Peat. 1992. An introduction to grasses. Wallingford, UK: C·A·B International.
Charpentier, A.; L. Brouillet; and D. Barabe. 1989. Organogenese de la fleur pistillee du *Begonia dregei* et de l'*Hillebrandia sandwicensis* (Begoniaceae). Can. J. Bot. 67:3625–3639.
Chasan, R. 1991. An odyssey of flowering. Plant Cell 3:745–746.
Cheng, P. C.; R. I. Greyson; and D. B. Walden. 1979. Comparison of anther development in genic male-sterile (ms10) and in male fertile corn *(Zea mays)* from light microscopy and scanning electron microscopy. Can. J. Bot. 57:578–596.
———. 1983. Organ initiation and the development of unisexual flowers in the tassel and ear of *Zea mays*. Amer. J. Bot. 70:450–462.
Chouard, P. 1960. Vernalization and its relation to dormancy. In L. Machlis and W. R. Briggs [eds.], Annu. Rev. Plant Physiol. 11:191–238.
Christianson, M. L. 1986. Fate map of the organizing shoot apex in *Gossypium*. Amer. J. Bot. 73:947–958.
Clifford, H. T. 1987. Spikelet and floral morphology. In T. R. Soderstrom; K. W. Hilu; C. S. Campbell; and M. E. Barkworth [eds.], Grass systematics and evolution. (Washington: Smithsonian Institution Press), pp.21–30.
Cockshull, K. E. 1985a. *Antirrhinum majus*. In A. H. Halevy [ed.], CRC handbook of flowering, vol. 1 (Boca Raton, FL: CRC Press), pp. 476–481.
———. 1985b. *Chrysanthemum morifolium*. In A. H. Halevy [ed.], CRC handbook of flowering, vol. 2 (Boca Raton, FL:CRC Press), pp.238–257.
Coe, E. H., Jr.; M. G. Neuffer; and D. A. Hoisington. 1988. The genetics of corn. In G. F. Sprague and J. W. Dudley [eds.], Corn and corn improvement, 3d ed. (Madison, WI: American Society of Agronomy), pp. 81–258.
Coen, E. S. 1991. The role of homeotic genes in flower development and evolution. Annu. Rev. Plant Physiol. Plant Mol. Biol. 42:241–279.
Coen, E. S., and E. M. Meyerowitz. 1991. The war of the whorls: Genetic interactions controlling flower development. Nature 353:31–37.
Coen, E. S.; J. M. Romero; S. Doyle; R. Elliott; G. Murphy; and R. Carpenter. 1990. Floricaula: A homeotic gene required for flower development in *Antirrhinum majus*. Cell 63:1311–1322.
Considine, J. A.; R. B. Knox; and O. H. Frankel. 1982. Stereological analysis of floral development and quantitative histochemistry of nucleic acids in fertile and base-sterile varieties of wheat. Ann. Bot. 50:647–663.
Cornish, E. C.; J. M. Pettit; and A. E. Clarke. 1988. Self-incompatability genes in flowering plants. In D. P. S. Verma and R. B. Goldberg [eds.], Temporal and spatial regulation of plant genes. (Vienna: Springer-Verlag), pp. 117–130.
Cornu, A. 1984. Genetics. In K. C. Sink. [ed.], Petunia. (Berlin: Springer-Verlag), pp. 34–48.
Cornu, A.; E. Farcy; D. Maizonnnier; M. Haring; W. Veerman; and A. G. M. Gerats. 1990. *Petunia hybrida* (2N = 14). In S. J. O'Brien [ed.], Genetic maps, 5th ed. (New York: Cold Spring Harbor), pp.6.113–6.123.
Corson, G. E., Jr. 1969. Cell division studies in the shoot apex of *Datura stramonium* during transition to flowering. Amer. J. Bot. 56:1127–1134.
Cousson, A., and K. Tran Thanh Van. 1981. *In vitro* control of *de novo* flower differentiation from tobacco thin cell layers cultured on a liquid medium. Physiologia Plantarum 51:77–84.
Cresti, M.; P. Gori; and E. Pacini [eds.]. 1988. Sexual reproduction in higher plants. Berlin: Springer-Verlag.

Crone, W., and E. M. Lord. 1991. A kinematic analysis of gynoecial growth in *Lilium longiflorum*: Surface growth patterns in all floral organs are triphasic. Dev. Biol. 143:408–417.

Cronquist, A. 1981. An integrated system of classification of flowering plants. New York: Columbia University Press.

———. 1988. The evolution and classification of flowering plants. 2d ed. Bronx, NY: New York Botanical Garden.

Crouch, M. L. 1990. Debating the responsibilities of plant scientists in the decade of the environment. Plant Cell 2:275–277.

Crouch, M. L., and I. M. Sussex. 1981. Development and storage-protein synthesis in *Brassica napus* L. Planta 153:64–74.

Crozier, A. 1981. Aspects of the metabolism and physiology of gibberellins. In H. W. Woolhouse [ed.], Advances in botanical research, vol. 9 (London: Academic Press), pp. 33–149.

Cusick, F. 1956. Studies of floral morphogenesis. 1. Median bisections of flower primordia of *Primula bulleyana* Forrest. Roy. Soc. (Edinburgh) Trans. 63:153–166.

———. 1966. On phylogenetic and ontogenetic fusions. In E. G. Cutter [ed.]. Trends in plant morphogenesis (London: Longmans), pp. 170–183.

Cusset, G. 1982. The conceptual bases of plant morphology. In R. Sattler [ed.], Axioms and principles of plant costruction. (The Hague: Martinus Nijhoff/Dr. W. Junk Publishers), pp. 8–86.

———. 1986. La morphogenèse du limbe des Dicotylédones. Can. J. Bot. 64:2807–2839.

Cutler, H. C., and M. C. Cutler. 1948. Studies on the structure of the maize plant. Mo. Bot. Gard. Ann. 35:301–316.

Cutter, E. G. 1957*a*. Studies of morphogenesis in the Nymphaeaceae. I. Introduction: Some aspects of the morphology of *Nuphar lutea* (L.) SM. and *Nymphaea alba* L. Phytomorphology 7:45–56.

———. 1957*b*. Studies of morphogenesis in the Nymphaeaceae. II. Floral development in *Nuphar* and *Nymphaea:* Bracts and calyx. Phytomorphology 7:57–73.

———. 1967. Morphogenesis and developmental potentialities of unequal buds. Phytomorphology 17:437–445.

———. 1971. Plant anatomy: Experiment and interpretation. Part 2. London: Edward Arnold.

Dale, J. E., and F. L. Milthorpe [eds.]. 1983. The growth and functioning of leaves. London: Cambridge University Press.

Dangl, J. L.; K. D. Hauffe; S. Lipphardt; K. Hahlbrock; and D. Scheel. 1987. Parsley protoplasts retain differential responsiveness to u.v. light and fungal elicitor. EMBO J. 9:2551–2556.

Daniel, E., and R. Sattler. 1978. Development of the perianth tubes of *Solanum dulcamara:* Implications for comparative morphology. Phytomorphology 28:151–171.

Darwin, C. 1862. On the two forms, or dimorphic condition, in the species of *Primula*, and on their remarkable sexual relations. Proc. Linn. Soc. (Botany) 6:105–139.

———. 1868. The variation of animals and plants under domestication. London: John Murray.

———. 1877. The different forms of flowers on plants of the same species. London: John Murray.

———. 1897. The different forms of flowers on plants of the same species. New York: D. Appleton.

Dauphin-Guerin, B.; G. Teller; and B. Durand. 1980. Different endogenous cytokinins between male and female *Mercurialis annua* L. Planta 148:124–129.

Davis, E. L.; P. Rennie; and T. A. Steeves. 1979. Further analytical and experimental studies on the apex of *Helianthus annuus:* Variable activity in the central zone. Can. J. Bot. 57: 971–980.

Davis, G. L. 1966. Systematic embryology of the angiosperms. New York: John Wiley & Sons.

Davis, P. H, and J. Cullen. 1989. The identification of flowering plants, 3d ed. Cambridge: Cambridge University Press.

Dawe, R. K., and M. Freeling. 1990. Clonal analysis of the cell lineages in the male flower of maize. Dev. Biol. 142:233–245.

Deitzer, G. F. 1989. Interaction between phytochrome and the circadian clock mechanism to control the photoperiodic induction of flowering. In E. Lord and G. Bernier [eds.], Plant reproduction: From floral induction to pollination. (Rockville, MD: American Society of Plant Physiologists), pp. 1–9.

Delozier, G.; K. Eckard; M. Greene; and E. Lord. 1987. A computer graphics program for the three dimensional reconstruction of plant organs from serial section. Amer. J. Bot. 74:136–140.

De Mason, D. A.; K. W. Stolte; and B. Tisserat. 1982. Floral development in *Phoenix dactilifera*. Can. J. Bot. 60:1439–1446.

Deming, J.; L. Zebing; and W. Jingmeng. 1988. Use of photoperiod-sensitive genic male sterility in rice breeding. In Hybrid rice, pp. 267–268. Proceedings of the International Symposium on Hybrid Rice. International Rice Research Institute. P. O. Box 933. 1099 Manila, Philippines.

Dengler, N. G. 1972. Ontogeny of the vegetative and floral apex of *Calycanthus occidentalis*. Can. J. Bot. 50:1349–1356.

Diboll, A. G., and D. A. Larson. 1966. An electron microscope study of the mature megagametophyte in *Zea mays*. Amer. J. Bot. 53:391–402.

Dickens, C. W. S., and J. van Staden. 1988. The induction and evocation of flowering *in vitro*. S. Afr. J. Bot. 54:325–344.

Dickinson, H. G. 1987. The physiology and biochemistry of meiosis in the anther. Int. Rev. Cytol. 107:79–109.

Dickinson, H. G., and J. Heslop-Harrison. 1977. Ribosomes, membranes and organelles during meiosis in angiosperms. Phil. Trans. R. Soc. Lond. Ser. B 277:327–342.

Dickerson, T. A. 1978. Epiphylly in angiosperms. Bot. Rev. 44:181–232.

Dickinson, T. A., and R. Sattler. 1974. Development of the epiphyllous inflorescence of *Phyllonoma integerrima* (Turez.) Loes.: Implications for comparative morphology. Bot. J. Linn. Soc. 69:1–13.

Donovan, G. R., and J. W. Lee. 1977. The growth of detached wheat heads in liquid culture. Plant Sci. Lett. 9:107–113.

———. 1978. Effect of the nitrogen source on grain development in detached wheat heads in liquid culture. Aust. J. Plant Physiol. 5:81–87.

Douglas, G. E. 1944. The inferior ovary. Bot. Rev. 10:125–186.

———. 1957. The inferior ovary. II. Bot. Rev. 23:1–46.

Doyle, J. A., and M. J. Donoghue. 1986. Seed plant phylogeny and the origin of angiosperms: An experimental cladistic approach. Bot. Rev. 52:321–431.

Drews, G. N., and R. A. Goldberg. 1989. Genetic control of flower development. Trends in Genet. 5:256–261.

Dubuc-Lebreux, M.-A. and R. Sattler. 1984. Quantitative distribution of mitotic activity during early corolla development of *Solanum dulcamara* L. Bot. Gaz. 145:22–25.

———. 1985. Quantitative distribution of mitotic activity during early corolla development of *Nicotiana tabacum*. Phytomorphology 35:17–23.

Dumas, C. 1978. Stigmates sécréteurs et lipides neutres sécrétes. Bull. Soc. Bot. Fr. 125:61–68.

Dumas, C.; R. B. Knox; C. A. McConchie; and S. D. Russell. 1984. Emerging physiological concepts in fertilization. What's New in Plant Physiol. 15:17–20.

Duncan, W. G., and J. D. Hesketh. 1968. Net photosynthetic rates, relative leaf growth rates, and leaf numbers of 22 races of maize grown at eight temperatures. Crop Science 8:670–674.

Dunwell, J. M. 1985a. Anther and ovary culture. In S. W. J. Bright and M. G. K. Jones [eds.], Cereal tissue and cell culture (Dordrecht: Martinus Nijhoff/Dr. W. Junk Publishers), pp. 1–44.

———. 1985b. Haploid cell cultures. In R. A. Dixon [ed.], Plant cell culture—A practical approach (Oxford: IRL Press), pp. 21–36.

Durand, R., and B. Durand. 1984. Sexual differentiation in higher plants. Physiol. Plant. 60:267–274.

———. 1985. *Mercurialis*. In A. H. Halevy [ed.], CRC handbook of flowering, vol. 3 (Boca Raton, Fl: CRC Press), pp. 376–387.

———. 1990. Sexual determination and sexual differentiation. CRC Crit. Rev. in Plant Sci. 9:295–316.

Durkee, L. T. 1983. The ultrastructure of floral and extrafloral nectaries. In B. Bentley and T. Elias [eds.], The biology of nectaries. (New York: Columbia University Press), pp. 1–29.

Durkee, L. T.; D. J. Gaal; and W. H. Reisner. 1981. The floral and extra-floral nactaries of *Passiflora*. 1. The floral nectary. Amer. J. Bot. 68:453–462.

Eames, A. J. 1931. The vascular anatomy of the flower with refutation of the theory of carpel polymorphism. Amer. J. Bot. 18:147–188.

———. 1961. Morphology of the angiosperms. New York: McGraw-Hill.

Eames, A. J., and C. L. Wilson. 1928. Carpel morphology in the cruciferae. Amer. J. Bot. 15:251–270.

Echlin, P. 1971. The role of the tapetum during microsporogenesis of angiosperms. In J. Heslop-Harrison [ed.], Pollen: Development and physiology (London: Butterworth), pp. 41–61.

Edwards, J. W., and G. M. Coruzzi. 1990. Cell-specific gene expression in plants. Annu. Rev. Genet. 24:275–303.

Ellstrand, N. C. 1983. Floral formula inconstancy within and among plants and populations of *Ipomopsis aggregata* (Polemoniaceae). Bot. Gaz. 144:119–123.

Emerson, R. A. 1920. Heritable characters of maize. II. Pistillate-flowered maize plants. Jour. Hered. 11:65–76.

Endress, P. K. 1990. Patterns of floral construction in ontogeny and phylogeny. J. Linn Soc. 39:153–175.

Erbar, C. 1986. Investigations on the development of the spiral flower in *Stewartia pseudocamellia* (Theaceae). (In German.) Bot. Jahrb. Syst. 106:391–407.

Erickson, R. O. 1948. Cytological and growth correlations in the flower bud and anther of *Lilium longiflorum*. Am. J. Bot. 35:729–739.

Esau, K. 1977. Anatomy of seed plants, 2d ed. New York: John Wiley & Sons.

Estelle, M. A., and C. R. Somerville. 1986. The mutants of *Arabidopsis*. Trends in Genet. 2:89–93.

Evans, L. T. 1958. *Lolium temulentum* L., a long day plant requiring only one inductive photocycle. Nature 182:197–198.

———. 1960. The influence of environmental conditions on inflorescence development in some long-day grasses. New Phytol. 13:429–440.

———. 1964. Inflorescence initiation in *Lolium temulentum* L. VI. Effects of some inhibitors of nucleic acid, protein and steroid biosynthesis. Aust. J. Biol. Sci. 17:24–35.

———. 1966. Abscisin II. Inhibitory effect on flower induction in a long day plant. Science 151:107–108.

———. [ed.]. 1969a. The induction of flowering. Melbourne: MacMillan.

———. 1969b. The nature of flower induction. In L. T. Evans [ed.], The induction of flowering. (Melbourne: MacMillan), pp. 457–480.

———. 1969c. *Lolium temulentum* L. In L. T. Evans [ed.], The induction of flowering: Some case histories (Ithaca, NY: Cornell University Press), pp. 328–349.

———. [ed.]. 1969d. The induction of flowering: Some case histories. Ithaca, NY: Cornell University Press.

Evans, L. T., and R. W. King. 1985. *Lolium temulentum*. In A. H. Halevy [ed.], CRC handbook of flowering, vol. 3 (Boca Raton, FL: CRC Press), pp. 306–323.

Evans, L. T.; R. W. King; A. Chu; L. N. Manden; and R. P. Pharis. 1990. Gibberellin structure and florigenic activity in *Lolium temulentum*, a long-day plant. Planta 182:97–106.

Evans, L. T., and I. F. Wardlaw. 1966. Independent translocation of ^{14}C-labelled assimilates and of the floral stimulus. Planta 68:310–326.

Evans, P. T.; B. L. Holaway; and R. L. Malmberg. 1988. Biochemical differentiation in the tobacco flower probed with monoclonal antibodies. Planta 175:259–269.

Evans, P. T., and R. L. Malmberg. 1989. Alternative pathways of tobacco placental development: Time of commitment and analysis of a mutant. Dev. Biol. 136:273–283.

Eyde, R. H. 1975. The foliar theory of the flower. Amer. Scientist 63:430–437.

Fahn, A. 1979a. Secretory tissues in plants. London: Academic Press.

———. 1979b. Ultrastructure of nectaries in relation to nectar secretions. Amer. J. Bot. 66:977–985.

———. 1982. Plant anatomy, 3d ed. Oxford: Pergamon Press.

———. 1990. Plant Anatomy, 4th ed. Oxford: Pergamon Press.

Farr, C. H. 1915. The origin of the inflorescence of *Xanthium*. Bot. Gaz. 59:136–148.

Ferrand, A., and G. Cusset. 1983. La morphogenese du calice de *Fittonia vershaffeltii*: Données biometriques. Can. J. Bot. 61:2694–2701.

Fick, G. N. 1976. Genetics of floral color and morphology of sunflowers. J. Hered. 67:227–230.

———. 1978. Breeding and genetics. In J. F. Carter [ed.], Sunflower science and technology. (Madison, WI: American Society of Agronomy), pp. 279–338. Number 19 in series Agronomy.

Foarde, D. E. 1971. The initial protrusion of a leaf primordium can form without concurrent periclinal divisions. Can J. Bot. 49:1601–1603.

Folsum, M. W., and D. D. Cass. 1989. Embryo sac development in soybean: Ultrastructure of megasporogenesis and early megagametogenesis. Can. J. Bot. 67:2841–2849.

———. 1990. Embryo sac development in soybean: Cellularization and egg apparatus expansion. Can. J. Bot. 68:2135–2147.

Foster, A. S., and E. M. Gifford, Jr. 1974. Comparative morphology of the vascular plants, 2d ed. San Francisco: W. H. Freeman.

Francis, D., and R. F. Lyndon. 1978. The cell cycle in the shoot apex of *Silene* during the first day of floral induction. Protoplasma 96:81–88.

Franck, D. H. 1976. The morphological interpretation of epiasciaidate leaves—An historical perspective. Bot. Rev. 42:345–388.

Frankel, O. H. 1976. Floral initiation in wheat. Proc. Roy. Soc. Lond. 192B:273–298.

Frankel, O. H.; R. B. Knox; and J. A. Considine. 1981. The development of the wheat flower: Genetics and physiology. In L. T. Evans and W. J. Peacock [eds.], Wheat science—Today and tomorrow (Cambridge: Cambridge University Press), pp.167–190.

Frankel, O. H.; B. Shineberg; and A. Munday. 1969. The genetic basis of an invariant character in wheat. Hered. 24:571–591.

Freeling, M. 1992. A conceptual framework for maize leaf development. Dev. Biol. 153:44–58.

Friedman, W. E. 1990. Double fertilization in *Ephedra*, a non-flowering seed plant: Its bearing on the origin of the angiosperms. Science 247:951–954.

Friend, D. J. C. 1960. The control of chlorophyll accumulation in leaves of Marquis wheat by temperature and light intensity. 1. The rate of chlorophyll accumulation and maximal absolute chlorophyll contents. Physiol. Plant. 13:776–785.

Friend, D. J. C.; J. E. Fisher; and V. A. Helson. 1963. The effect of light intensity and temperature on floral initiation and inflorescence development of Marquis wheat. Can. J. Bot. 41:1663–1674.

Frijters, D. 1978a. Principles of simulation of inflorescence development. Ann. Bot. 42:549–560.

———. 1978b. Mechanisms of developmental integration of *Aster novae-angliae* L. and *Hieracium murorum*. Ann. Bot. 42:561–575.

Frost, H. B. 1915. The inheritance of doubleness in *Matthiola* and *Petunia*. 1. The hypotheses. Am. Nat. 49:623–636.

Fuchs, E.; D. Atsmon; and A. H. Halevy. 1977. Adventitious staminate flower formation in gibberellin treated gynoecious cucumber plants. Plant Cell Physiol. 18:1193–1201.

Galinat, W. C. 1985. Domestication and diffusion of maize. In R. I. Ford [ed.], Prehistoric food production in North America. Anthropological Papers No. 75. (Ann Arbor, MI: Museum of Anthropology, University of Michigan), pp. 245–278.

Galun, E.; Y. Jung; and A. Lang. 1962. Culture and sex modification of male cucumber buds *in vitro*. Nature 194:596–598.

Ganders, F. R. 1979. The biology of heterostyly. N. Z. J. Bot. 17:607–635.

Garner, W. W., and H. A. Allard. 1920. Effect of the relative length of day and night and other factors of the environment on growth and reproduction in plants. J. Agric. Res. 18:553–606.

Garrison, R., and R. H. Wetmore. 1961. Studies in shoot-tip abortion: *Syringa vulgaris*. Amer. J. Bot. 48:789–795.

Gasser, C. S. 1991. Molecular studies on the differentiation of floral organs. Annu. Rev. Plant Physiol. Plant Mol. Biol. 42:621–649.

Gasser, C. S.; K. A. Budelier; A. G. Smith; D. M. Shah; and R. T. Fraley. 1989. Isolation of tissue-specific cDNAs from tomato pistils. Plant Cell 1:15–24.

Gaude, T., and C. Dumas. 1987. Molecular and cellular events of self-incompatability. Int. Rev. Cytol. 107:333–366.

Gauthier, R. 1950. The nature of the inferior ovary in the genus *Begonia*. Contrib. Inst. Bot. Univ. Montreal 66:1–93.

Gebhardt, J. S., and C. N. McDaniel. 1987. Induction and floral determination in the terminal bud of *Nicotiana tabacum* L. cv. Maryland Mammoth, a short-day plant. Planta 172:526–530.

———. 1991. Flowering response of dayneutral and short-day cultivars of *Nicotiana ta-*

bacum L.: Interactions among roots, genotype, leaf ontogenic position and growth conditions. Planta 185:513–517.
Gehring, W. J. 1984. Homeotic genes and the control of cell determination. In E. H. Davidson and R. A. Firtel [eds.], Molecular biology of development (New York: A. R. Liss), pp. 3–22.
Gehring, W. J., and Y. Hiromi. 1986. Homeotic genes and the homeobox. Annu. Rev. Genet. 20:147–173.
George, E. F., and P. D. Sherrington. 1984. Plant propagation by tissue culture. Eversely, UK: Exegetics Ltd.
Gifford, E. M., Jr., and G. E. Corson. 1971. The shoot apex in seed plants. Bot. Rev. 37:143–229.
Gifford, E. M., and A. S. Foster. 1989. Morphology and evolution of vascular plants, 3d ed. New York: W. H. Freeman.
Goebel, K. 1900. Organography of plants. Part 1. General organography. Oxford: Clarendon Press.
Goethe, J. W. von. 1790. Versuch die metamorphose der phlanzen zu erklaren. Gotha. pp. 5, 40–45, 73, 173, 191, 195.
Goldberg, R. B. 1988. Plants: Novel developmental processes. Science 240:1460–1467.
Good, R. 1956. Features of evolution in flowering plants. London: Longmans.
Goodman, M. M., and W. L. Brown. 1988. In G. F. Sprague and J. W. Dudley [eds.], Corn and corn improvement, 3d ed. (Madison, WI: American Society of Agonomy), pp. 33–79. No. 18 of Agronomy.
Goodwin, B. C. 1985. What are the causes of morphogenesis? Bioessays 3:32–36.
Gould, F. W., and R. B. Shaw. 1968. Grass systematics. College Station, TX: Texas A&M University Press.
Gould, K. S., and E. M. Lord. 1988. Growth of anthers in *Lilium longiflorum:* A kinematic analysis. Planta 173:161–171.
———. 1989. A kinematic analysis of tepal growth in *Lilium longiflorum*. Planta 177:66–73.
Gould, S. J. 1966. Allometry and size in ontogeny and phylogeny. Biol. Rev. 41:587–640.
———. 1977. Ontogeny and phylogeny. Cambridge, MA: Belknap Press of Harvard University Press.
Gray, J. E.; B. A. McClure; I. Bönig; M. A. Anderson; and A. E. Clarke. 1991. Action of the style product of the self-compatibility gene of *Nicotiana alata* (S-RNase) on *in vitro*-grown pollen tubes. Plant Cell 3:271–283.
Green, P. B. 1980. Organogenesis—A biophysical view. Annu. Rev. Plant Physiol. 31:51–82.
———. 1984a. Analysis of axis extension. In P. W. Barlow and D. J. Carr [eds.], Positional controls in plant development (London: Cambridge University Press), pp. 53–82.
———. 1984b. Shifts in plant cell axiality: Division direction influences cellulose orientation in the succulaent *Graptopetalum*. Dev. Biol. 103:18–27.
———. 1988. A theory for inflorescence development and flower formation based on morphological and biophysical analysis in *Echeveria*. Planta 175:153–169.
———. 1989. Shoot morphogenesis, vegetative through floral, from a biophysical perspective. In E. Lord and G. Bernier [eds.], Plant reproduction from floral induction to pollination (Rockville, MD: American Society of Plant Physiologists), pp. 58–75.
Green, P. B., and R. S. Poethig. 1982. Biophysics of the extension and initiation of plant

organs. In S. Subtelny and P. B. Green [eds.], Developmental order: Its origin and regulation. (New York: Alan R. Liss), pp. 485–509.

Greyson, R. I. 1973. Alternative goals for flower morphology. In Y. S. Murty, B. M. Johri, H. Y. Mohan Ram, and T. M. Varghese [eds.], Advances in plant morphology, Professor V. Puri commemoration volume. (Meerut, India: Sarita Prakashan), pp. 28–32.

Greyson, R. I.; P. C. Cheng; and D. B. Walden. 1980. LM, TEM and SEM observations of anther development in the genic male-sterile (ms-9) mutant of corn *(Zea mays)*. Can. J. Genet. Cytol. 22:153–166.

Greyson, R. I., and K. Raman. 1975. Differential sensitivity of "double" and "single" flowers of *Nigella damascena* (Ranunculaceae) to emasculation and to GA_3. Amer. J. Bot. 62:531–536.

Greyson, R. I., and V. K. Sawhney. 1972. Initiation and early growth of flower organs of *Nigella* and *Lycopersicon:* Insights from allometry. Bot. Gaz. 133:184–190.

Greyson, R. I., and S. S. Tepfer. 1966. An analysis of stamen filament growth of *Nigella hispanica*. Amer. J. Bot. 53:485–490.

———. 1967. Emasculation effects on the stamen filament of *Nigella hispanica* and their partial reversal by gibberellic acid. Amer. J. Bot. 54:971–976.

Greyson, R. I.; D. B. Walden; and W. J. Smith. 1982. Leaf and stem heteroblasty. Bot. Gaz. 143:73–78.

Guern, J. 1987. Regulation from within: The hormone dilemma. Ann. Bot. 60:75–102.

Guha, S., and S. C. Maheshwari. 1964. *In vitro* production of embryos from anthers in *Datura*. Nature 204:497.

Haber, A. H. 1962. Nonessentiality of concurrent cell divisions for degree of polarization of leaf growth. I. Studies with radiation-induced mitotic inhibition. Amer. J. Bot. 49:583–589.

Habermann, H. M., and D. B. Sekulow. 1972. Development and aging in *Helianthus annuus* L.: Effects of the biological milieu of the apical meristem on patterns of development. Growth 36:339–349.

Hackett, W. P. 1985. Juvenility, maturation and rejuvenation in woody plants. Horticultural Reviews 7:109–155.

Hackett, W. P., and R. M. Sachs. 1985. *Bougainvillea*. In A. H. Halevy [ed.], CRC handbook of flowering, vol. 2 (Boca Raton, FL: CRC Press), pp. 38–47.

Hagemann, W., and D. R. Kaplan. 1989. The concept of marginal meristem in angiosperm leaf development. Amer. J. Bot. 76(Supp.): 36.

Hahlbrock, K., and D. Scheel. 1989. Physiology and molecular biology of phenylpropanoid metabolism. Annu. Rev. Plant Physiol. Plant Mol. Biol. 40:347–369.

Halevy, A. H. [ed.]. 1985. CRC handbook of flowering, vols. 1–6. Boca Raton, FL: CRC Press.

Halperin, W. 1978. Organogenesis at the shoot apex. Annu. Rev. Plant Physiol. 29:239–262.

Hansen, D. J.; S. K. Bellman; and R. M. Sacher. 1976. Gibberellic acid-controlled sex expression of corn tassels. Crop Sci. 16:371–374.

Hanway, J. J., and S. W. Ritchie. 1985. *Zea mays*. In A. H. Halevy [ed.], CRC handbook of flowering, vol. 4. (Boca Raton, FL: CRC Press), pp. 525–541.

Harada, H.; M. Kyo; and J. Imamura. 1988. The induction of embryogenesis in *Nicotiana* immature pollen in culture. In CIBA Foundation Symposium 137. Applications of plant cell and tissue culture. (Chichester: John Wiley & Sons), pp. 59–74.

Harberd, N. P., and M. Freeling. 1989. Genetics of dominant gibberellin-insensitive dwarfism in maize. Genetics 121:827–838.

Haring, V.; J. E. Gray; B. A. McClure; M. A. Anderson; and A. E. Clarke. 1990. Self-incompatability: A self-recognition system in plants. Science 250:937–941.

Harris, E. M.; S. C. Tucker; and L. E. Urbatsch. 1991. Floral initiation and early development in *Erigeron philadelphicus* (Asteraceae). Amer. J. Bot. 78:108–121.

Harte, C. 1980. *Antirrhinum majus* L. In R. C. King [ed.], Handbook of genetics, vol. 2, Plants, plant viruses and protists (New York: Plenum Press), pp. 315–331.

Haughn, G. W., and C. R. Somerville. 1988. Genetic control of morphogenesis in *Arabidopsis*. Dev. Gen. 9:73–89.

Haupt, W. 1969. *Pisum sativum* L. In L. T. Evans [ed.]. The induction of flowering: Some case histories (Ithaca, NY: Cornell University Press), pp. 393–408.

Havelange, A., and G. Bernier. 1974. Descriptive and quantitative study of ultrastructural changes in the apical meristem of mustard in transition to flowering. I. The cell and nucleus. J. Cell Sci. 15:633–644.

Hayashi, F.; D. R. Boerner; C. E. Peterson; and H. M. Sell. 1971. The relative content of gibberellins in seedlings of gynoecious and monoecious cucumber (*Cucumis sativus*). Phytochemistry 10:57–62.

Heberle-Bors, E. 1985. *In vitro* haploid formation from pollen: A critical review. Theor. Appl. Genet. 71:361–374.

Heide, O. 1969a. Environmental control of sex expression in *Begonia*. Z. Phlanzenphysiol. 61:279–285.

———. 1969b. Evidence of an inhibitory role of endogenous auxin and gibberellin in flower initiation of the short-day plant *Begonia* x *cheimantha* Everett. Physiol. Plant. 22: 1001–1012.

Heide, O. M., and W. Runger. 1985. *Begonia*. In A. H. Halevy [ed.], CRC handbook of flowering, vol. 2 (Boca Raton, FL: CRC Press), pp. 4–14.

Heide, O., and F. Skoog. 1967. Cytokinin activity in *Begonia* and *Bryophyllum*. Physiol. Plant. 43:266–270.

Heller, W., and G. F. Forkmann. 1988. Biosynthesis. In J. B. Harborne [ed.], The flavonoids: Advances in research since 1980 (London: Chapman & Hall), pp. 399–425.

Hemphill, D. D.; L. R. Baker; and H. M. Sell. 1972. Different sex phenotypes of *Cucumis sativus* and *C. melo* and their endogenous gibberellin activity. Euphytica 21:285–291.

Hernandez, L. F., and J. H. Palmer. 1988. Regeneration of the sunflower capitulum after cylindrical wounding of the receptacle. Amer. J. Bot. 75:1253–1261.

Herr, J. M., Jr. 1971. A new clearing-squash technique for the study of ovule development in angiosperms. Amer. J. Bot. 58:785–790.

———. 1973. The use of Nomarski-interference microscopy for the study of structural features in cleared ovules. Acta Bot. India 1:35–40.

Hesketh, J. D.; S. S. Chase; and D. K. Nanda. 1969. Environmental and genetic modification of leaf number in maize, sorghum and Hungarian millet. Crop Science 9:460–463.

Heslop-Harrison, J. 1952. A reconsideration of plant teratology. Phyton (Ann. R. Bot.) 4:19–34.

———. 1957. The experimental modification of sex expression in flowering plants. Biol. Rev. 32:38–90.

———. 1958. The unisexual flower—A reply to criticism. Phytomorphology 8:177–184.

———. 1959. Growth substances and flower morphogenesis. J. Linn. Soc. Lond. (B) 56:269–281.

———. 1961. The experimental control of sexuality and inflorescence strucure in *Zea mays* L. Proc. Linn. Soc. London 172:108–123.

———. 1964. Sex expression in flowering plants. In J. P. Miksche et al. [eds.], Meristems and differentiation. Brookhaven Symposia in Biology, No 16. Brookhaven National Laboratory. Upton. New York.

———. 1971. The pollen wall: Structure and development. In J. Heslop-Harrison [ed.], Pollen: Development and physiology. (London: Butterworths), pp. 75–98.

———. 1972. Sexuality of angiosperms. In F. C. Steward [ed.], Plant physiology: A treatise, vol. 6c. (New York: Academic Press), pp. 133–289.

———. 1980. Compartmentation in anther development and pollen wall morphogenesis. In L. Nover, F. Lynen, and K. Mothes [eds.], Cell compartmentation and metabolic channeling. (Jena: Fischer, and Amsterdam: Elsevier/North/Holland Biomedical Press), pp. 471–493.

———. 1987. Pollen germination and pollen-tube growth. Int. Rev. Cyt. 107: 1–78.

Heslop-Harrison, J. S., Y. Heslop-Harrison and B. J. Reger. 1987. Anther filament extension in *Lilium:* Potassium ion movement and some anatomical features. Ann. Bot. 59:505–515.

Heslop-Harrison, Y. 1953. *Nuphar intermedia* Ledeb., a presumed relict hybrid, in Britain. Watsonia 3:7–25.

Heslop-Harrison, Y.; J. Heslop-Harrison; and K. R. Shivanna. 1981. Heterostyly in *Primula*. 1. Fine structural and cytochemical features of the stigma and style in *Primula vulgaris* Huds. Protoplasma 107:171–187.

Heslop-Harrison, Y., and K. R. Shivanna. 1977. The receptive surface of the angiosperm stigma. Ann. Bot. 41: 1233–1258.

Heslop-Harrison, Y., and I. Wood. 1959. Temperature induced meristic and other variations in *Cannabis sativa*. J. Linn. Soc. Lond. 56:290–293.

Hess, K., and D. J. Morre. 1978. Fine structure analysis of the elongation zone of Easter lily *(Lilium longiflorum)* staminal filaments. Bot. Gaz. 139:312–321.

Heuther, C. A., Jr. 1968. Exposure of natural genetic variability underlying the pentamerous corolla constancy in *Linanthus androsaceus* spp. *androsaceus*. Genetics 60:123–146.

———. 1969. Constancy of the pentamerous corolla phenotype in natural populations of *Linanthus*. Evolution 23:572–588.

Hickey, L. J. 1973. Classification of the architecture of dicotyledonous leaves. Amer. J. Bot. 60:17–33.

Hickey, M., and C. King. 1988. 100 families of flowering plants. Cambridge: Cambridge University Press.

Hicks, G. S. 1973. Initiation of floral organs in *Nicotiana tabacum*. Can. J. Bot. 51:1611–1617.

———. 1979. Feminized outgrowths on the stamen primordia of tobacco *in vitro*. Pl. Sc. Lett. 17:81–89.

———. 1980. Pattern of organ development in plant tissue culture and the problem of organ determination. Bot. Rev. 46:1–23.

Hicks, G. S., and I. M. Sussex. 1971. Organ regeneration in sterile culture after median bisection of the flower primordia of *Nicotiana tabacum*. Bot. Gaz. 132:350–363.

Hill, J. B.; L. O. Overholts; H. W. Popp; and A. R. Grove. 1960. Botany—A textbook for colleges, 3d ed. New York: McGraw-Hill.

Hill, J. P., and E. M. Lord. 1989. Floral development in *Arabidopsis thaliana*: A comparison of the wild type and the homeotic pistillata mutant. Can. J. Bot. 67:2922–2936.

Hinata, N. T. K. 1982. Comparative studies on S-glycoproteins purified from different S-

genotypes in self-incompatible *Brassica* species. 1. Purification and chemical properties. Genetics 100:641–647.
Hockett, E. A.; P. S. Baenziger; and G. L. Steffens. 1978. A proposal for increased research on chemical induction of fertility in genetic male sterile barley. Euphytica 27:109–111.
Hogenboom, N. G. 1975. Incompatability and incongruity: Two different mechanisms for the non-functioning of intimate partner relationships. Proc. Roy. Soc. London (B) 188:361–375.
Holaway, B. L.; P. T. Evans; and R. L. Malmberg. 1989. Monoclonal antibodies and floral development. In R. Goldberg [ed.], The molecular basis of plant development (New York: Alan R. Liss), pp. 181–190.
Holder, N. 1979. Positional information and pattern formation in plant morphogenesis and a mechanism for the involvement of plant hormones. J. Theo. Biol. 77:195–212.
Hole, C. C., and R. C. Hardwick. 1976. Development and control of the number of flowers per node in *Pisum sativum* L. Ann. Bot. 40:707–722.
Holliday, R. 1984. The biological significance of meiosis. In C. W. Evans and H. G. Dickinson [eds.], Controlling events in meiosis: Soc. Exp. Biology Symp. 38:381–394.
Horgan, R. 1987. Cytokinin genes. In G. V. Hoad, J. R. Lenton, M. B. Jackson and R. K. Aitkin [eds.], Hormone action in plant development—A critical appraisal. London: Butterworths), pp. 119–130.
Horner, H. T., Jr. 1977. A comparative light and electron microscopic study of microsporogenesis in male-fertile and cytoplasmic male sterile sunflower *(Helianthus annuus)*. Amer. J. Bot. 64:745–759.
Horner, M., and E. Street. 1978. Pollen dimorphism—Origin and significance in pollen plant formation. Ann. Bot. 42:763–777.
Horsch, R. B. 1990. Bouquets of genes. Plant Cell 2:951–952.
Houghtaling, H. B. 1935. A developmental analysis of size and shape of tomato fruits. Bull. Torrey Bot. Club 62:243–252.
Hufford, L. D. 1988. Roles of early ontogenetic modifications in the evolution of floral form of *Eucnide* (Loasaceae). Bot. Jahrb. Syst. 109:289–333.
Huxley, J. 1932. Problems of relative growth. London: Methuen.
Iltis, H. H. 1983. From teosinte to maize: The catastrophic sexual transmutation. Science 222:886–894.
Irish, E. E., and T. M. Nelson. 1988. Development of maize plants from cultured shoot apices. Planta 175:9–12.
———. 1989. Sex determination in monoecious and dioecious plants. Plant Cell 1:737–744.
———. 1991. Identification of multiple stages in the conversion of maize meristems from vegetative to floral development. Development 112:891–898.
———. 1993. Development of tassel seed 2 inflorescences in maize. Amer. J. Bot. 80:292–299.
Iwanami, Y.; T. Sasakuma; and Y. Yamada. 1988. Pollen: Illustrations and scanning electron micrographs. Tokyo: Kodansha.
Jacob, F., and J. Monod. 1961. On the regulation of gene activity. Cold Spring Harbor Symp. Quant. Biol. 26:193–211.
Jacobs, T. 1992. Control of the cell cycle. Dev. Biol. 153:1–15.
Jacobs, W. P. 1952. The role of auxin in differentiation of xylem around a wound. Amer. J. Bot. 39:301–309.

———. 1959. What substance normally controls a given biological process? I. Formulation of some rules. Dev. Biol. 1:527–533.
Jacobs, W. P., and I. B. Morrow. 1957. A quantitative study of xylem development in the vegetative shoot apex of *Coleus*. Amer. J. Bot. 44:823–842.
Janssen, J. M., and A. Lindenmayer. 1987. Models for the control of branch position and flowering sequences of capitula in *Mycelis muralis* (L.) Dumont Compositae. New Phytol. 105:191–220.
Jean, R. V. 1984. Mathematical approach to pattern and form in plant growth. New York: John Wiley & Sons. New York.
Jeffcoat, B., and K. E. Cockshull. 1972. Changes in the levels of endogenous growth regulators during development of the flowers of *Chrysanthemum morifolium*. J. Exp. Bot. 23:722–732.
Jegla, D. E., and I. M. Sussex. 1989. Cell lineage patterns in the shoot meristem of the sunflower in the dry seed. Dev. Biol. 131:215–225.
Jennings, D. H. and A. J. Trewavas. (eds.) 1986. Plasticity in plants. Symp. Soc. Exp. Biol. No. 40. Society for Experimental Biology. Cambridge.
Jensen, L. C. W. 1971. Experimental bisection of *Aquilegia* floral buds cultured *in vitro*. I. The effect on growth, primordia initiation and apical regeneration. Can. J. Bot. 49:487–493.
Jensen, W. A. 1972. The embryo sac and fertilization in angiosperms. H. L. Lyon Arboretum Lecture No. 3. H. L. Lyon Arboretum, University of Hawaii, 3860 Manoa Rd., Honolulu, HI 96822.
Johri, B. M. 1963. Female gametophyte. In P. Maheshwari [ed.], Recent advances in the embryology of angiosperms (Delhi: Int'l. Soc. Plant Morphologists, Univ. Delhi), pp. 69–103.
———. [ed.] 1982. Experimental embryology of vascular plants. Berlin: Springer-Verlag.
———. [ed.] 1984. Embryology of angiosperms. Berlin: Springer-Verlag.
Johri, M. M., and E. H. Coe, Jr. 1983. Clonal analysis of corn plant development. I. The development of the tassel and the ear shoot. Dev. Biol. 97:154–172.
Johri, V. 1951. The role of floral anatomy in the solution of morphological problems. Bot. Rev. 17:471–553.
Jones, R. L., and I. D. J. Phillips. 1966. Organs of gibberellin synthesis in light-grown sunflower plants. Plant Physiol. 41:1381–1386.
Jurgens, G. 1992. Genes to greens: Embryonic pattern formation in plants. Science 256:487–488.
Kahlem, G. 1976. The isolation and localization by histoimmunology of isoperoxidases specific for male flowers of the dioecious species *Mercurialis annua* L. Dev. Biol. 50:58–67.
Kaihara, S., and A. Takimoto. 1983. Effect of plant growth regulators on flower opening of *Pharbitis nil*. Plant and Cell Physiol. 24:309–316.
Kaldewey, H. 1957. Wachstumsverlauf, Wuchsstoffbildung und Nutationsbewegungen von *Fritillaria meleagris* L. im Laufe der Vegetationsperiode. Planta (Berl.) 49:300–344.
Kamalay, J. C., and R. B. Goldberg. 1984. Organ-specific nuclear RNAs in tobacco. Proc. Nat. Acad. Sci. USA 81:2801–2805.
———. 1980. Regulation of structural gene expression in tobacco. Cell 19:935–946.
Kandasamy, M. K.; D. J. Paolillo; C. D. Faraday; J. B. Nasrallah; and M. E. Nasrallah. 1989. The S-locus specific glycoproteins of *Brassica* accumulate in the cell wall of developing stigma papillae. Dev. Biol. 134:462–472.

Kapil, R. N., and A. K. Bhatnagar. 1981. Ultrastructure and biology of female gametophyte in flowering plants. Int. Rev. Cyt. 70:291–341.

Kaplan, D. R. 1967. Floral morphology, organogenesis and interpretation of the inferior ovary in *Downingia bacigalupilii*. Amer. J. Bot. 54:1274–1290.

———. 1968. Structure and development of the perianth in *Downingia bacigalupilii*. Amer. J. Bot. 55:406–420.

———. 1971. On the value of comparative development in phylogenetic studies—A rejoinder. Phytomorphology 21:134–140.

———. 1973. Comparative developmental analysis of the heteroblastic leaf series of axillary shoots of *Acorus calamus* L. (Araceae). La Cellule 69:253–290.

Kaplan, D. R., and W. Hagemann. 1991. The relationship of cell and organism in vascular plants. Bioscience 41:693–703.

Karpoff, A. J. 1974. Control of *in vitro* sepal development of excised floral buds of *Aquilegia* (Ranunculaceae). Amer. J. Bot. 61:778–786.

Kaufman, P. B. 1959a. Development of the shoot of *Oryza sativa* L. I. The shoot apex. Phytomorphology 9:228–242.

———. 1959b. Development of the shoot of *Oryza sativa* L. II. Leaf histogenesis. Phytomorphology 9:277–311.

———. 1959c. Development of the shoot of *Oryza sativa* L. III. Early stages in histogenesis of the stem and ontogeny of the adventitious root. Phytomorphology 9:382–404.

Kaul, M. L. H. 1988. Male sterility in higher plants. Berlin: Springer-Verlag.

Keijzer, C. J.; I. H. S. Hoek; and M. T. M. Willemse. 1987. The development of the staminal filament of *Gasteria verrucosa*. Acta Bot. Neerl. 36:2271–2282.

Keller, W. A., and K. C. Armstrong. 1979. Stimulation of embryogenesis and haploid production in *Brassica campestris* anther cultures by elevated temperature treatments. Theor. Appl. Genet. 55:63–67.

Khait, A. 1986. Hormonal mechanisms for size measurement in living organisms in the context of maturing juvenile plants. J. Theor. Biol. 118:471–484.

Khush, G. S., and T. Kinoshita. 1991. Rice karyotype, marker genes and linkage groups. In G. S. Khush and G. H. Toenniessen [eds.], Rice biotechnology (Wallingford, UK: C·A·B International), pp. 83–108.

Khush, G. S., and G. H. Toenniessen [eds.], 1991. Rice biotechnology. Wallingford, UK: C·A·B International.

Kiesselbach, T. A. 1949. The structure and reproduction of corn. Univ. Neb. Coll. Agric. Res. Bull. 161.

Kim, C. H., and J. N. Rutger. 1988. Heterosis in rice. In Hybrid rice: Proceedings of the International Symposium on Hybrid Rice. International Rice Research Institute. P.O. Box 933, 1099 Manila, Philippines.

Kinet, J.-M.; M. Bodson; A. M. Alvinia; and G. Bernier. 1971. The inhibition of flowering in *Sinapis alba* after the arrival of the floral stimulus at the meristem. Z. Pflanzenphysiol. 667:49–63.

Kinet, J.-M., R. M. Sachs; and G. Bernier. 1985. The physiology of flowering, vol. 3 The development of flowers. Boca Raton, FL: CRC Press.

King, P. J. 1988. Plant hormone mutants. Trends in Genet. 4:157–162.

King, R. W., and L. T. Evans. 1969. Timing of evocation and development of flowers in *Pharbitis nil*. Aust. J. Biol. Sci. 22:559–572.

King, R. W.; L. T. Evans; and I. F. Wardlaw. 1968. Translocation of floral stimulus in *Pharbitis nil* in relation to that of assimilates. Zeit. fur Pflanzenphysiol. 59:377–388.

Kinoshita, T. 1984. Gene analysis and linkage map. In S. Tsunoda and N. Takahashi [eds.], Biology of rice (Tokyo: Japan Scientific Societies Press, and Amsterdam: Elsevier), pp. 187–273.
Kirby, E. J. M. 1974. Ear development in spring wheat. J. Agric. Sci. Camb. 82:437–447.
Kirk, S. C.; I. B. Morrow; and W. P. Jacobs. 1967. Developmental morphology of the *Xanthium* shoot apex after photo-induction. Phytomorphology 17:410–419.
Kiss, H. G., and R. E. Koning. 1990. Emasculation effects on filament growth in *Ipomoea nil* (Convolulaceae). Amer. J. Bot. 77:41–45.
Klee, H., and M. Estelle. 1991. Molecular genetic approaches to plant hormone biology. Annu. Rev. Plant Physiol. Plant Mol. Biol. 42:529–551.
Knowles, P. F. 1978. Morphology and anatomy. In J. F. Carter [ed.], Sunflower science and technology (Madison, WI: American Society of Agronomy), pp. 55–87. Number 19 in series Agronomy.
Knox, R. B. 1967. Apomixis: Seasonal population differences in a grass. Science 157:325–326.
———. 1984*a*. The pollen grain. In B. M. Johri [ed.], Embryology of angiosperms (Berlin: Springer-Verlag), pp. 197–271.
———. 1984*b*. Pollen-pistil interactions. In H. F. Linskens and J. Heslop-Harrison [eds.], Cellular interactions; new series vol. 17 of Encyclopedia of plant physiology (Berlin: Springer-Verlag), pp. 508–608.
Knox, R. B., and A. E. Clarke. 1976. Cell recognition in plants. In C. F. Graham and P. F. Wareing [eds.], Developmental control in animals and plants, 2d ed. (Oxford: Blackwell Scientific Publications), pp. 191–217.
Knox, R. B., and L. T. Evans. 1966. Inflorescence initiation in *Lolium temulentum* L. VIII. Histochemical changes at the shoot apex during induction. Aust. J. Biol. Sci. 19:232–243.
———. 1968. Inflorescence initiation in *Lolium temulentum* L. XII. An autoradiographic study of evocation in the shoot apex. Aust. J. Biol. Sci. 21:1083–1094.
Knuth, P. 1906–9. Handbook of flower pollination, vols. 1, 2 and 3. Trans. J. R. Ainsworth Davis. Oxford: Oxford University Press.
Koes, R. E.; C. E. Spelt; and J. N. M. Mol. 1989. The chalcone synthase multigene family of *Petunia hybrida* (V30): Differential, light regulated expression during flower development and UV light induction. Plant Mol. Biol. 12:213–225.
Koes, R. E.; C. E. Spelt; J. N. M. Mol; and A. G. M. Gerats. 1987. The chalcone synthase multigene family of *Petunia hybrida* (V30): Sequence homology, chromosomal localization and evolutionary aspects. Plant Mol. Biol. 10:159–169.
Koevenig, J. L. 1973. Floral development and stamen filament elongation in *Cleome hassleriana*. Amer. J. Bot. 60:122–129.
Kofer, W.; K. Glimelius; and H. T. Bonnett. 1990. Modifications of floral development in tobacco indused by fusion of protoplasts of different male sterile cultivars. Theor. Appl. Genet. 79:97–102.
———. 1991. Modifications of mitochondrial DNA cause changes in floral development in homeotic-like mutants of tobacco. Plant Cell 3:759–769.
Koltunow, A. M.; J. Trettner; K. H. Cox; M. Wallroth; and R. B. Goldberg. 1990. Different temporal and spacial gene expression patterns occur during anther development. Plant Cell 2:1201–1224.
Komaki, M. K.; K. Okada; E. Nishino; and Y. Shimura. 1988. Isolation and characterization of novel mutants of *Arabidopsis thaliana* defective in flower development. Development 104:195–203.

Konar, R. N., and S. Kitchlue. 1982. Flower culture. In B. M. Johri [ed.], Experimental embryology of vascular plants. (Heidelberg: Springer-Verlag), pp. 53–78.

Koning, R. E. 1983. The roles of auxin, ethylene, and acid growth in filament elongation in *Gaillardia grandiflora* (Asteraceae). Amer. J. Bot. 70:602–610.

———. 1984. The roles of plant hormones in the growth of the corolla of *Gaillardia grandiflora* (Asteraceae) ray flowers. Amer. J. Bot. 71:1–8.

———. 1986. The role of ethylene in corolla unfolding in *Ipomoea nil* (Convolvulaceae). Amer. J. Bot. 73:152–155.

Koning, R. E., and M. M. Raab. 1987. Parameters of filament elongation in *Ipomoea nil* (Convolvulaceae). Amer. J. Bot. 74:510–516.

Koornneef, M. 1990. Linkage map of *Arabidopsis thaliana*. In S. J. O'Brien [ed.]. Genetic maps, 5th ed. (New York: Cold Spring Harbor), pp. 6.95–6.99.

Koornneef, M.; J. van Eden; C. J. Hanhart; P. Stam; F. J. Braaksma; W. J. Feenstra. 1983. Linkage map of *Arabidopsis thaliana*. J. Hered. 74:265–272.

Krishnamoorthy, H. N., and K. K. Nanda. 1968. Flower bud reversion in *Impatiens balsamina* under non-inductive photoperiods. Planta 80:43–51.

Krishnamoorthy, H. N., and A. R. Talukdar. 1976. Chemical control of sex expression in *Zea mays* L. Zeit. fur Pflanzenphysiolog. 79:91–94.

Kubicki, B. 1969. Investigation on sex determination in cucumber *(Cucumis sativus)*. Genet. Pol. 10:3–143.

Kudirka, D. T.; G. W. Schaeffer; and P. S. Baenziger. 1986. Wheat: Genetic variability through anther culture. In Y. P. S. Bajaj [ed.], Biotechnology in agriculture and forestry, vol. 2: Crops I. (Berlin: Springer-Verlag), pp. 39–54.

Kudirka, D. T. and J. Van't Hof. 1980. G2 arrest and differentiation in the petal of *Tradescantia* Clone 4430. Exp. Cell Res. 130:443–450.

Kunst, L.; J. E. Klenz; J. Martinez-Zapater; and G. W. Haughn. 1989. ap2 gene determines the identity of perianth organs in flowers of *Arabidopsis thaliana*. Plant Cell 1:1195–1208.

Kurata, N., and T. Omura. 1984. Chromosome analysis. In S. Tsunoda and N. Takahashi [eds.], Biology of rice. (Tokyo: Japan Scientific Societies Press, and Amsterdam: Elsevier), pp. 305–320.

Lacroix, C., and R. Sattler. 1988. Phyllotaxis theories and tepal-stamen superposition in *Basella rubra*. Amer. J. Bot. 75:906–917.

Lagrimini, L. M.; S. Bradford; and S. Rothstein. 1990. Peroxidase-induced wilting in trangenic tobacco plants. Plant Cell 2:7–18.

Lakshmanan, K. K., and K. B. Ambegaokar. 1984. Polyembryony. In B. M. Johri [ed.], Embryology of angiosperms. (Berlin: Springer-Verlag), pp. 445–474.

Lam, H. J. 1950. Stachyospory and phyllospory as factors in the natural system of the Cormophyta. Svensk. Bot. Tidskr. 44:512–534.

Lance, A. 1957. Sur l'edification du rameau florifere du *Xanthium pennsylvanicum* Wallr. (Ambrosiacées). C. R. Hebd. Seance Acad. Sci. 244:927–930.

Lang, A. 1961. Auxins in flowering. In W. Ruhland [ed.], Encyclopedia of plant physiology, vol. 14 (Berlin: Springer-Verlag), pp. 909–950.

———. 1965. Physiology of flower formation. In W. Ruhland [ed.], Encyclopedia of plant physiology, vol. 15/1. (Berlin: Springer-Verlag), pp. 1380–1536.

Lang, A. 1965. Physiology of flower formation. In W. Ruhland [ed.], Encyclopedia of plant physiology, vol. 15/1 (Berlin: Springer-Verlag), pp. 1380–1536.

———. 1987. Prospectives in flowering research. In L. McIntosh and J. Key [eds.], Plant gene systems and their biology (New York: A. R. Liss), pp. 3–24.

Larkin, J. C.; R. Felsheim; and A. Das. 1990. Floral determination in the terminal bud of the short-day plant *Pharbitis nil*. Dev. Biol. 137:434–443.
Larson, P. R. 1983. Primary vascularization and the siting of primordia. In J. E. Dale and F. L. Milthorpe [eds.], The growth and functioning of leaves (Cambridge: Cambridge University Press), pp. 25–51.
La Rue, C. D. 1942. The rooting of flowers in sterile culture. Bull. Torrey Bot. Club 69:332–341.
Laser, K. D., and N. R. Lersten. 1972. Anatomy and cytology of microsporogenesis in cytoplasmic male sterile angiosperms. Bot. Rev. 38:425–454.
Lash, J., and J. R. Whittaker. 1974. Concepts of development. Stamford, CT: Sinauer Associates.
Laughnan, J. R., and S. Gabay-Laughnan. 1983. Cytoplasmic male sterility in maize. Annu. Rev. Genet. 17:27–48.
Lay-Yee, M.; R. M. Sachs; and M. S. Reid. 1987. Changes in cotyledon mRNA during floral induction of *Pharbitis nil* cv. Violet. Planta 170:104–109.
Leaver, C. J., and M. W. Gray. 1982. Mitochondrial genome organization and expression in higher plants. Annu. Rev. Plant. Physiol. 33:373–402.
Leavings, C. S., III, and G. G. Brown. 1989. Molecular biology of plant mitochondria. Cell 56:171–179.
Lecocq, M. 1977. Le gynecee du *Begonia tuberhybrida* et ses variations. Can. J. Bot. 55:525–541.
Lecocq, M., and C. Dumas. 1975. Histophysiologie des stigmates normaux et des formations stimatoides chex *Begonia tuberhybrida*. I. Observations preliminaires. Can. J. Bot. 53:1252–1258.
Ledbetter, M. C., and K. R. Porter. 1970. Introduction to the fine structure of plant cells. New York: Springer-Verlag.
Lee, S.-L. J.; E. D. Earle; and V. E. Gracen. 1980. The cytology of pollen abortion in S cytoplasmic male-sterile corn anthers. Amer. J. Bot. 67:237–245.
Lee, S.-L. J.; V. E. Gracen; and E. D. Earle. 1979. The cytology of pollen abortion in C cytoplasmic male-sterile corn anthers. Amer. J. Bot. 66:141–148.
Leins, P. 1972. Das Karpell im ober- und unterstandigen Gynoeceum. Ber. Deutsch. Bot. Ges. 85:291–294.
———. 1975. Die Beziechungen zwischen multistaminaten und einfachen Androeceen. Bot. Jahrb. Syst. 96:231–237.
Leins, P., and C. Erbar. 1985. The early gynoecial development in Apiaceae. [In German.] Bot. Jahrb. Syst. 106:53–60.
Leng, E. R. 1951. Time-relationships in tassel development of inbred and hybrid corn. Jour. Amer. Soc. Agron. 9:445–449.
Leonard, M.; J.-M. Kinet; M. Bodson; A. Havelange; A. Jacqmard; and G. Bernier. 1981. Flowering in *Xanthium strumarium*: Initiation and development of female inflorescence and sex expression. Plant Physiol. 67:1245–1249.
Lersten, N. R. 1987. Morphology and anatomy of the wheat plant. In E. G. Heyne [ed.], Wheat and wheat improvement, 2d ed. (Madison, WI: American Society of Agronomy), pp. 33–75.
Libbenga, K. R., and A. M. Mennes. 1987. Hormone binding and its role in hormone action. In P. J. Davies [ed.], Plant hormones and their role in plant growth and development. (Dordrecht: Martinus Nijhoff), pp. 194–221.
Lifschitz, E. 1988. Molecular markers for the floral program. Flowering Newsletter No. 6:16–20.

Lincoln, C.; J. H. Britton; and M. Estelle. 1990. Growth and development of the axr1 mutants of *Arabidopsis*. Plant Cell 2:1071–1080.

Lindenmayer, A. 1984. Positional and temporal control mechanisms in inflorescence development. In P. W. Barlow and D. J. Carr [eds.], Positional controls in plant development. (Cambridge: Cambridge University Press), pp. 461–486.

Linsmaier, E. M., and F. Skoog. 1965. Organic growth factor requirement for tobacco tissue culture. Physiol. Plant. 18:100–127.

Lintilhac, P. M. 1984. Positional controls in meristem development: A caveat and an alternative. In P. W. Barlow and D. J. Carr [eds.], Positional controls in plant development. (Cambridge: Cambridge University Press), pp. 83–105.

Lord, E. M. 1981. Cleistogamy: A tool for the study of floral morphogenesis, function and evolution. Bot. Rev. 47:421–449.

———. 1982. Floral morphogenesis in *Lamium amplexicaule* L. (Labiatae) with a model for the evolution of the cleistogamous flower. Bot. Gaz. 143:63–72.

Lord, E. M., and A. M. Mayers. 1982. Effects of gibberellic acid on floral development *in vivo* and *in vitro* in the cleistogamous species, *Lamium amplexicaule* L. Ann. Bot. 50:301–307.

Lovett Doust, J., and L. Lovett Doust. 1988. Plant reproductive ecology—Patterns and strategies. New York: Oxford University Press.

Lu, W.; K. Enomoto; Y. Fukunaga; and C. Kao. 1988. Regeneration of tepals, stamens and ovules in explants from perianth of *Hyacinthus orientalis* L.: Importance of explant age and exogenous hormones. Planta 175:478–484.

Lyndon, R. F. 1970. Rates of cell division in the shoot apical meristem of *Pisum*. Ann. Bot. 34:1–17.

———. 1978. Flower development in *Silene:* Morphology and sequence of initiation of primordia. Ann. Bot. 42:1343–1348.

———. 1985. *Silene*. In A. H. Halevy [ed.], CRC handbook of flowering, vol. 4 (Boca Raton, FL: CRC Press), pp. 313–319.

———. 1990. Plant development; The cellular basis. London: Unwin Hyman.

Mabberley, D. J. 1989. The plant book. Cambridge: Cambridge University Press.

Maginnes, E. A., and R. W. Langhans. 1961. The effect of photoperiod and temperature on initiation and flowering induction of snapdragon (*Antirrhinum majus*—variety Jackpot). Proc. Am. Soc. Hort. Sci. 77:600–607.

Maheshwari, P. 1950. An introduction to the embryology of the angiosperms. New York: McGraw-Hill.

Maksymowych, R. 1973. Analysis of leaf development. Cambridge: Cambridge University Press.

Maksymowych, R. 1990. Analysis of growth and development of *Xanthium* (Developmental and Cell Biology Series No. 21). Cambridge: Cambridge University Press.

Malmberg, R. L., and J. McIndoo. 1984. Ultraviolet mutagenesis and genetic analysis of resistance to methylglyoxal-bis (guanylhydrazone) in tobacco. Mol. Gen. Genet. 196:28–34.

Malmberg, R. L.; J. McIndoo; A. C. Hiatt; and B. A. Lowe. 1985. Genetics of polyamine synthesis in tobacco: Developmental switches in the flower. Cold Spring Harbor Symp. 50:475–482.

Mangelsdorf, P. C. 1974. Corn: Its origins, evolution, and improvement. Cambridge, MA: Harvard University Press.

Mangelsdorf, P. C., and W. C. Galinat. 1964. The tunicate locus in maize dissected and reconstituted. Proc. Nat. Acad. Sci. USA 51:147–150.

Marc, J., and J. H. Palmer. 1981. Photoperiodic sensitivity of inflorescence initiation and development in sunflower. Field Crop Res. 4:155–164.

———. 1982. Changes in the mitotic activity and cell size in the apical meristem of *Helianthus annuus* during the transition to flowering. Amer. J. Bot. 69:768–775.

Mariani, C.; M. de Beuckeleer; J. Truettner; J. Leemans; and R. B. Goldberg. 1990. Induction of male sterility in plants by a chimaeric ribonuclease gene. Nature 347:737–741.

Martin, C.; R. Carpenter; E. S. Coen; and T. Gerats. 1987. The control of floral pigmentation in *Antirrhinum majus*. In H. Thomas and D. Grierson [eds.], Developmemtal mutants in higher plants (Cambridge: Cambridge University Press), pp. 19–52.

Marushige, K., and Y. Marushige. 1962. An electrophoretic study of tissue extracts from leaf and flower of *Pharbitis nil* Chois. Plant & Cell Physiol. 3:319–322.

Marx, G. A. 1983. Developmental mutants in some annual seed plants. Annu. Rev. Plant Physiol. 34:389–417.

Mascarenhas, J. P. 1975. The biochemistry of angiosperm pollen development. Bot. Rev. 41:259–314.

———. 1989. The male gametophyte of flowering plants. Plant Cell 1:657–664.

———. 1990. Gene activity during pollen development. Annu. Rev. Plant Physiol. Plant Mol. Biol. 41:317–338.

Mason, H. L. 1957. The concept of the flower and the theory of homology. Madrono 14:81–95.

Masters, M. R. 1869. Vegetable teratology. London: Ray Society, Robert Hardwicke.

Matsushima, H.; T. Itoyama; Y. Mashiko; and T. Mizukoshi. 1974. Critical time of floral differentiation in *Pharbitis nil* shoot apex. In Plant growth substances 1973 (Tokyo: Hirokawa), pp. 967–973.

Matzke, E. B. 1932. Flower variations and symmetry patterns in *Stellaria media,* and their underlying significance. Amer. J. Bot. 19:477–507.

———. 1938. Inflorescence patterns and sexual expression in *Begonia semperflorens*. Amer. J. Bot. 25:465–478.

Mauseth, J. D. 1980. A morphimetric study of the ultrastructure of *Echinocereus engelmannii* (Cactaceae). I. Shoot apical meristems at germination. Amer. J. Bot. 67:173–181.

Mayer, U.; R. A. T. Ruiz; T. Berleth, and S. Mesera. 1991. Mutations affecting body organization in *Arabidopsis* embryo. Nature 353:402–407.

McClintock, B. 1950. The origin and behavior of mutable loci in maize. Proc. Nat. Acad. Sci. USA 36:344–355.

McClure, B. A.; J. E. Gray; M. A. Anderson; and A. E. Clarke. 1990. Self-incompatibility in *Nicotiana alata* involves degradation of pollen rRNA. Nature 347:757–760.

McDaniel, C. N. 1980. Influence of leaves and roots on meristem development in *Nicotiana tabacum* L. cv. Wisconsin 38. Planta 148:462–467.

———. 1984*a*. Shoot meristem development. In D. J. Carr and P. W. Barlow [eds.], Positional controls of plant development (Cambridge: Cambridge University Press), pp. 319–347.

———. 1984*b*. Competence, determination and induction in plant development. In G. M. Malacinski and S. V. Bryant [eds.], Pattern formation—A primer in developmental biology (New York: Macmillan), pp. 393–412.

———. 1989. Floral initiation as a developmental process. In E. Lord and G. Bernier [eds.], Plant reproduction: From floral induction to pollination (Rockville, MD: American Society of Plant Physiologists), pp. 51–57.

———. 1992. Determination to flower in *Nicotiana*. In R. A. Pederson [ed.], Current topics in developmental biology, vol 27 (New York: Academic Press), pp. 1–37.
McDaniel, C. N.; R. W. King; and L. T. Evans. 1991. Floral determination and *in vitro* floral differentiation in isolated shoot apices of *Lolium temulentum* L. Planta 185:9–16.
McDaniel, C. N., and R. S. Poethig. 1988. Cell lineage patterns in the shoot apical meristem of the germinating maize embryo. Planta 175:13–22.
McDaniel, C. N.; K. A. Sangrey; and D. E. Jegla. 1989. Cryptic floral determination: Stem explants from vegetative tobacco plants have the capacity to regenerate floral shoots. Dev. Biol. 134:437–478.
McDaniel, C. N.; S. R. Singer; J. S. Gebhardt; and K. A. Dennin. 1987. Floral determination: A critical process in meristem ontogeny. In J. G. Atherton [ed.], Manipulation of flowering (London: Butterworths), pp. 109–120.
McDaniel, C. N.; S. R. Singer and S. M. E. Smith. 1992. Developmental states associated with the floral transition. Dev. Biol. 153:59–69.
McDonald, A. D. 1980. Organogenesis of the female reproductive structure of *Myrica pennsylvanica*. Can. J. Bot. 58:2001–2006.
McGinnis, W.; R. L. Garber; J. Wirz; A. Kuroiwa; and W. J. Gehring. 1984. A homologous protein-coding sequence in *Drosophila* homeotic genes and its conservation in other metazoans. Cell 37:409–414.
McHughen, A. 1977. Development of tobacco petal *in vitro*. Ann. Bot. 41:1073–1076.
———. 1982. Inducing organ generation *in vitro:* Sepal/petal structures from flower buds. Can. J. Bot. 60:845–849.
McIntosh, R. A., and J. E. Cusick. 1987. Linkage map of hexaploid wheat. In E. G. Heyne [ed.], Wheat and wheat improvement, 2d ed. (Madison, WI: American Society of Agronomy), chapter 5H, pp. 289–297.
McIntyre, G. I. 1977. The role of nutrition in apical dominance. Symp. Soc. Exp. Biol. 31:251–273.
Meeks-Wagner, D. R.; E. S. Dennis; K. Tran Thanh Van; and W. J. Peacock. 1989. Tobacco genes expressed during in vitro floral initiation and their expression during normal development. Plant Cell 1:25–35.
Meeuse, A. D. J. 1965. Fundamentals of phytomorphology. New York: Ronald Press.
Meicenheimer, R. D. 1979. Relationships between shoot growth and changing phyllotaxy of *Ranunculus*. Amer. J. Bot. 66:557–569.
Meinhardt, H. 1982. Models of biological pattern formation. London: Academic Press.
Meinke, D. W. 1985. Embryo-lethal mutants of *Arabidopsis thaliana:* Analysis of mutants with a wide range of lethal phases. Theo. Appl. Genet. 69:543–552.
Meinke, D. W., and I. M. Sussex. 1979. Embryo-lethal mutants of *Arabidopsis thaliana:* A model system for genetic analysis of plant embryo development. Dev. Biol. 72:50–61.
Meins, F., and A. N. Binns. 1979. Cell determination and plant development. Bioscience 29:221–225.
Merxmuller, H., and P. Leins. 1967. Die verwandtschaftsbeziehungen der krenzblutter und Mohngewachse. Bot. Jahrb. Syst. 86:113–129.
Metzger, J. D. 1987. Hormones and reproductive development. In P. J. Davies [ed.], Plant hormones and their role in plant growth and development (Dordrecht: Martinus Nijhoff), pp. 431–462.
———. 1988. Localization of the site of perception of thermoinductive temperature in *Thlaspi arvense* L. Plant Physiol. 88:424–428.

Metzger, J. D., and J. A. D. Zeevaart. 1980. Comparison of the levels of six endogenous gibberellins in roots and shoots of spinach in relation to photoperiod. Plant Physiol. 66:679–683.

———. 1985. *Spinacia oleracea.* In A. H. Halevy [ed.], CRC handbook of flowering, vol 4 (Boca Raton, FL: CRC Press), pp. 384–392.

Meyer, V. G. 1966. Flower abnormalities. Bot. Rev. 32:165–218.

Meyer, V. G., and J. R. Meyer. 1964. Cytoplasmic effects on the differentiation of anthers and ovules of cotton. Amer. J. Bot. 51:693–696.

Meyerowitz, E. M. 1987. *Arabidopsis thaliana.* Annu. Rev. Gen. 21:93–111.

Meyerowitz, E. M.; J. L. Bowman; L. L. Brockman; G. N. Drews; T. Jack; L. E. Seiburth; and D. Weigel. 1991. A genetic and molecular model for flower development in *Arabidopsis thaliana.* Dev. Supp. 1:157–167.

Meyerowitz, E. M.; D. R. Smyth; and J. L. Bowman. 1989. Abnormal flowers and pattern formation in floral development. Development 106:209–217.

Miginiac, E. 1978. Some aspects of regulation of flowering: Role of correlative factors in photoperiodic plants. Bot. Mag. (Tokyo), Special Issue No. 1:159–173.

Mitchison, G. J. 1977. Phyllotaxis and Fibonacci series. Science 196:270–275.

Moeliono, B. M. 1970. Cauline or carpellary placentation among dicotyledons. 2 vols. Assen, Netherlands: Koninklijke Van Gorcum.

Molder, M., and J. N. Owens. 1973. Ontogeny and histochemistry of the intermediate and reproductive apices of *Cosmos bipinnatus* var. 'Sensation' in response to gibberellin A_3 and photoperiod. Can. J. Bot. 51:535–551.

———. 1985. *Cosmos.* In A. H. Halevy [ed.], CRC handbook of flowering, vol. 2 (Boca Raton, FL: CRC Press), pp. 341–349.

Moncur, M. W. 1981. Floral initiation in field crops: An atlas of scanning electron micrographs. Melbourne: Division of Land Crops, CSIRO.

Morel, G. 1963. Leaf regeneration in *Adiantum pedatum.* J. Linn. Soc. Lond. (Bot.) 58:381–383.

Morris, R. O. 1987. Genes specifying auxin and cytokinin biosynthesis in prokaryotes. In P. J. Davies [ed.], Plant hormones and their role in plant growth and development. (Dordrecht: Martinus Nijhoff), pp. 636–655.

Mosley, M. F., Jr. 1967. The value of the vascular system in the study of the flower. Phytomorphology 17:159–164.

Müller, H. 1883. The fertilization of flowers. London: Macmillan.

Murakami, Y. 1973. The role of gibberellins in the growth of floral organs of *Pharbitis nil.* Plant & Cell Physiol. 14:91–102.

———. 1975. The role of gibberellins in the growth of the floral organs of *Mirabilis jalapa.* Plant & Cell Physiol. 16:337–345.

Murfet, I. C. 1985. *Pisum sativum.* In A. H. Halevy [ed.], CRC handbook of flowering, vol. 4 (Boca Raton, FL: CRC Press), pp. 97–126.

———. 1989. Flowering genes in *Pisum.* In E. Lord and G. Bernier [eds.], Plant reproduction: From floral induction to pollination (Rockville, MD: American Society of Plant Physiologists), pp. 10–18.

Napier, R. M., and M. A. Venis. 1991. From auxin-binding protein to plant hormone receptor? TIBS 16:72–75.

Napp-Zinn, K. 1969. *Arabidopsis thaliana* (L.) Heynh. In L. T. Evans [ed.], The induction of flowering: Some case histories. (Ithaca, NY: Cornell University Press), pp. 291–304.

———. 1985. *Arabidopsis thaliana.* In A. H. Halevy [ed.], CRC handbook of flowering, vol. 1 (Boca Raton, FL: CRC Press), pp. 492–503.

Narayanaswamy, S., and L. George. 1982. Anther culture. In B. M. Johri [ed.], Experimental embryology of vascular plants. (Berlin: Springer-Verlag), pp. 79–103.
Nasrallah, J. B. and M. E. Nasrallah. 1984. Electrophoretic heterogeneity exhibited by the S-allele specific glycoproteins of *Brassica*. Experientia 40:279–281.
Natarella, N. J., and K. C. Sink. 1971. The morphogenesis of double flowering *Petunia hybrida* Hort. J. Amer. Soc. Hort. Sci. 96:600–602.
Naylor, F. L. 1941. Effect of length of inductive period on floral initiation and development in *Xanthium pennsylvanicum*. Bot. Gaz. 103:146–154.
Neuffer, M. C. 1982a. Growing maize for genetic purposes. In W. F. Sheridan [ed.], Maize for biological research (Charlottsville, VA: Plant Molecular Biology Association), pp.19–30.
———. 1982b. Mutant induction in maize. In W. F. Sheridan [ed.], Maize for biological research (Charlottsville, VA: Plant Molecular Biology Association), pp. 61–64.
Neuffer, M. G.; L. Jones; and M. S. Zuber. 1968. The mutants of maize. Madison, WI: Crop Science Society of America.
Nicholls, P. B. 1986. Induction of sensitivity to gibberellic acid in developing wheat caryopses: Effect of sugars in the culture medium. Aust. J. Plant Physiol. 13:795–801.
Nickerson, N. H. 1954. Morphological analysis of the maize ear. Amer. J. Bot. 41:87–92.
———. 1959. Sustained treatment with gibberellic acid of five different kinds of maize. Ann. Mo. Bot. Gard. 46:19–37.
———. 1960a. Studies involving sustained treatment of maize with gibberellic acid. II. Responses of plants carrying certain tassel modifying genes. Ann. Mo. Bot. Gard. 47:243–262.
———. 1960b. Sustained treatment with gibberellic acid of maize plants carrying one of the dominant genes Teopod and Corngrass. Amer. J. Bot. 47:809–815.
Nickerson, N. H., and E. E. Dale. 1955. Tassel modifications in *Zea mays*. Ann. Mo. Bot. Gard. 42:195–211.
Nickerson, N. H., and T. N. Embler. 1960. Studies involving sustained treatment of maize with gibberellic acid. I. Further notes on responses of races. Ann. Mo. Bot. Gard. 47:227–242.
Nijhout, H. F. 1990. Metoplors and the role of genes in development. Bioessays 12:441–446.
Nishino, E. 1978. Corolla tube formation in four species of Solonaceae. Bot. Mag. Tokyo 91:263–277.
———. 1982. Corolla tube formation in six species of Apocynaceae. Bot. Mag. Tokyo. 95:1–17.
Nogler, G. A. 1984. Gametophytic Apomixis. In B. M. Johri [ed.], Embryology of angiosperms (Berlin: Springer-Verlag), pp. 475–518.
Nojuku, E. 1956. The effect of light intensity on leaf shape in *Ipomoea caerulea*. New Phytol. 55:91–110.
———. 1957. The effect of mineral nutrition and temperature on leaf shape in *Ipomoea caerulea*. New Phytol. 56:154–171.
Nougarède, A. 1967. Experimental cytology of the shoot apical cells during vegetative growth and flowering. Int. Rev. Cyt. 21:203–351.
O'Brien, T. P.; M. E. Sammut; J. W. Lee; and M. G. Smart. 1985. The vascular system of the wheat spikelet. Aust. J. Plant Physiol. 12:1173–1187.
Ockendon, D. J. 1982. An S-allele survey of cabbage *(Brassica oleracea* var. *capitata)*. Euphytica 31:325–331.
Oka, H. I. 1991. Genetic diversity of wild and cultivated rice. In G. S. Khush and G. H.

Toenniessen [eds.], Rice biotechnology (Wallingford, UK: C·A·B International), pp. 55–81.
Opatrna, J.; F. Seidlova; and K. Benes. 1964. The anatomy of the shoot apex of wheat (*Triticum aestivum* L.) during transition from the vegetative to the reproductive state and the determination of the primordia. Biol. Plant. 6:219–225.
Ouyang, Q., and H. L. Swinney. 1991. Transition from a uniform state to hexagonal and striped Turing patterns. Nature 352:610–612.
Palmer, J. H., and J. Marc. 1982. Wound-induced initiation of involucral bracts and florets in the developing sunflower inflorescence. Plant & Cell Physiol. 23:1401–1409.
Palmer, J. H., and B. T. Steer. 1985. The generative area as the site of floret initiation in the sunflower capitulum and its integration to predict floret number. Field Crops Res. 11:1–12.
Pareddy, D. R. 1990. Studies on development and attempted chemical reversion of cultured tassels of two genic male steriles of maize (ms14 and ms24). Maydica 35:203–208.
Pareddy, D. R.; R. I. Greyson; and D. B. Walden. 1987. Fertilization and seed production with pollen from *in-vitro*-cultured maize tassels. Planta 170:141–143.
Paterson, K. E. 1984. Shoot tip culture of *Helianthus annuus*. Flowering and development of adventitious and multiple shoots. Amer. J. Bot. 71:925–931.
Patterson, E. B. 1975. An alternative approach for producing corn hybrids. Illinois Res. 17:16–18.
Payer, J. B. 1857. Traité d'organogénie comparée de la fleur. Paris: Masson.
Percival, J. 1921. The wheat plant: A monograph. London: Duckworth (reprint edition 1974).
Peterson, R. L.; M. G. Scott; and S. L. Miller. 1979. Some aspects of carpel structure in *Caltha palustris* L. (Ranunculaceae). Amer. J. Bot. 66:334–342.
Pharis, R. P., and R. W. King. 1985. Gibberellins and reproductive development in seed plants. Annu. Rev. Plant Physiol. 36:517–568.
Philipson, W. R. 1946. Studies in the development of the inflorescence. I. The capitulum of *Bellis perennis* L. Ann. Bot. 10:257–270.
———. 1947*a*. Studies in the development of the inflorescence. II. The capitula of *Succisa pratensis* Moench. and *Dipsacus fullonum* L. Ann. Bot. 11:285–297.
———. 1947*b*. Studies in the development of the inflorescence. III. The thyrse of *Valeriana officinalis* L. Ann. Bot. 11:409–416.
———. 1948*a*. Studies in the development of the inflorescence. IV. The capitula of *Hieracium boreale* Fries and *Dahlia gracilis* Ortg. Ann. Bot. 12:65–75.
———. 1948*b*. Studies in the development of the inflorescence. V. The raceme of *Lobelia Dortmanna* L. and other Campanulaceous inflorescences. Ann. Bot. 12:147–156.
Phillips, I. D. J. 1969. Apical dominance. In M. B. Wilkins [ed.], The physiology of plant growth and development (London: McGraw-Hill), pp.165–202.
———. 1975. Apical dominance. Annu. Rev. Plant Physiol. 26:341–367.
Phinney, B. O. 1984. Gibberellin A_1 dwarfism and the control of shoot elongation in higher plants. In A. Crozier and J. R. Hillman [eds], The biosynthesis and metabolism of plant hormones (Cambridge: Cambridge University Press). S.E.B. Seminar Series 23:17–45.
Phinney, B. O., and C. Spray. 1982. Chemical genetics and the gibberellin pathway in *Zea mays* L. In P. F. Wareing [ed.], Plant growth substances. (New York: Academic Press), pp. 101–110.
———. 1983. Gibberellin biosynthesis in *Zea mays:* the 3-hydroxylation step GA_{20} to GA_1. In J. Miyamoto et al. [eds.], IUPAC pesticide chemistry: Human welfare and the environment (Oxford: Pergamon Press), pp. 81–86.

———. 1985. Gibberellins (GAs), gibberellin mutants and their future in molecular biology. Curr. Topics Plant Biochem. Physiol. 4:67–74.
Phipps, I. F. 1928. Heritable characters in maize. XXX1. Tassel seed 4. J. Hered. 19:399–404.
Pinthus, M. J. 1985. *Triticum*. In A. H. Halevy [ed.], CRC handbook of flowering, vol. 4 (Boca Raton, FL: CRC Press), pp. 418–443.
Pizzolato, T. D. 1993. Vascular system of the fertile floret of *Panicum* (Gramineae: Panicoideae: Paniceae). Amer. J. Bot. 80:53–64.
Plack, A. 1957. Sexual dimorphism in Labiatae. Nature. 180:1218–1219.
Plantefol, L. 1948. La theorie des helices foliares multiples. Paris: Masson et Cie.
Plickert, K. 1936. Die Zuchtung der grossblutigen superbissima-Petunien. Zuechter 8:255–260.
Poethig, R. S. 1984. Cellular parameters of leaf morphogenesis in maize and tobacco. In R. A. White and W. C. Dickinson [eds.], Contemporary problems in plant anatomy (New York: Academic Press), pp. 235–259.
———. 1985. Homeotic mutations in maize. In M. Freeling [ed.], Plant genetics (New York: A. R. Liss), pp. 33–43.
———. 1987. Clonal analysis of cell lineage patterns in plant development. Amer. J. Bot. 74:581–594.
———. 1990. Phase changes and the regulation of shoot morphogenesis in plants. Science 250:923–930.
Polowick, P. L., and V. K. Sawhney. 1986. A scanning electron microscope study on the initiation and development of floral organs of *Brassica napus* (cv. Westar). Amer. J. Bot. 73:254–263.
Popham, R. A., and A. P. Chan. 1950. Zonation in the vegetative stem tip of *Chrysanthemum morifolium* Bailey. Amer. J. Bot. 37:476–483.
———. 1952. Origin and development of the receptacle of *Chrysanthemum morifolium* Bailey. Amer. J. Bot. 39:329–339.
Porter, E. K.; D. Parry; J. Bird; and H. G. Dickinson. 1984. Nucleic acid metabolism in the nucleus and cytoplasm of angiosperm meiocytes. In C. W. Evans and H. G. Dickinson [eds.], Controlling events in meiosis: Soc. Exp. Biology Symp. 38:363–379.
Posluszny, U.; M. G. Scott; and R. Sattler. 1980. Revisions in the technique of epi-illumination light microscopy for the study of floral and vegetative apices. Can. J. Bot. 58:2491–2495.
Postlethwait, S. N., and O. E. Nelson. 1964. Characterization of development in maize through the use of mutants. 1. The polytypic (Pt) and ramosa-1 (ra-1) mutants. Amer. J. Bot. 51:238–243.
Prusinkiewicz, P., and A. Lindenmayer. 1990. The algorithmic beauty of plants. New York: Springer-Verlag.
Puri, V. 1941. Studies in floral anatomy. I. Gynaecium constitution in the cruciferae. Proc. Ind. Acad. Sci. 14:166–187.
———. 1951. The role of floral anatomy in the solution of morphological problems. Bot. Rev. 17:471–553.
———. 1952. Floral anatomy and inferior ovary. Phytomorphology 2:122–129.
Purvis, O. N. 1961. The physiological analysis of vernalization. Encycl. Plant Physiol. 16:76–122.
Raab, M. M., and R. E. Koning. 1987. Interacting roles of gibberellin and ethylene in corolla expansion of *Ipomoea nil* (Convolvulaceae). Amer. J. Bot. 74:921–927.
———. 1988. How is floral expansion regulated? BioScience 38:670–674.

Raff, R. A. and T. C. Kaufman. 1983. Embryos, genes and evolution: The developmental-genetic basis of evolutionary change. New York: Macmillan.

Raghavan, V. 1976. Experimental embryogenesis in vascular plants. London: Academic Press.

———. 1986. Embryogenesis in angiosperms—A developmental and experimental study. Cambridge: Cambridge University Press.

———. 1988. Anther and pollen development in rice *(Oryza sativa)*. Amer. J. Bot. 75:183–196.

Rajeevan, M. S., and A. Lang. 1987. Comparison of de-novo flowerbud formation in a photoperiodic and day neutral tobacco. Planta 171:560–564.

Raman, K., and R. I. Greyson. 1978. Further observations on the differential sensitivities to plant growth regulators by cultured "single" and "double" flower buds of *Nigella damascena* (Ranunculaceae). Amer. J. Bot. 65:180–191.

Raman, K.; D. B. Walden; and R. I. Greyson. 1980. Propagation by shoot tip culture of *Zea mays* L.—A feasibility study. Ann. Bot. 45:183–189.

Rao, V. S. 1951. The vascular anatomy of flowers—A bibliography. J. Univ. Bombay 19:38–63.

———. 1961. Floral anatomy. New York: Scholar's Library.

Rastogi, R., and V. K. Sawhney. 1986. *In vitro* culture of young floral buds of tomato *(Lycopersicon esculentum* Mill.). Plant Sci. 47:221–227.

———. 1988. Flower culture of a male sterile stamenless-2 mutant of tomato *(Lycopersicon esculentum)*. Amer. J. Bot. 75:513–518.

———. 1989. *In vitro* development of angiosperm floral buds and organs. Plant Cell, Tissue and Organ Culture 16:145–174.

———. 1990a. Polyamines and flower development in the male sterile stamenless-2 mutant of tomato *(Lycopersicon esculentum* Mill.). I. Levels of polyamines and their biosynthesis in normal and mutant flowers. Plant Physiol. 93:439–445.

———. 1990b. Polyamines and flower development in the male sterile stamenless-2 mutant of tomato *(Lycopersicon esculentum* Mill.). II. Effects of polyamines and their biosynthetic inhibitors on the development of normal and mutant floral buds cultured *in vitro*. Plant Physiol. 93:446–452.

Raven, P. H.; R. F. Evert; and S. E. Eichorn. 1986. Biology of plants, 4th ed. New York: Worth Publishers.

Ray, P. M., and W. E. Alexander. 1966. Photoperiodic adaptation to latitude in *Xanthium strumarium*. Amer. J. Bot. 53:806–816.

Redei, G. P. 1975. Induction of auxotrophic mutations in plants. In L. Ledoux [ed.], Genetic manipulations with plant material (New York: Plenum Press), pp. 183–209.

Reid, J. B. 1980. Apical senescence in *Pisum:* A direct or indirect role for the flowering genes. Ann. Bot. 45:195–201.

Reid, J. B., and I. C. Murfet. 1984. Flowering in *Pisum:* A fifth locus, *Veg.* Ann. Bot. 53:369–382.

Reiner, J. M. 1968. The organism as an adaptive control system. Englewood Cliffs, NJ: Prentice-Hall.

Reynolds, J., and J. Tampion. 1983. Double flowers: A scientific study. New York: Van Nostrand Reinhold.

Reznickova, S. A., and M. T. M. Willemse. 1980. Formation of pollen in the anther of *Lilium*. II. The function of the surrounding tissues in the formation of pollen and pollen wall. Acta Bot. Neerl. 29:141–156.

Rick, C. M. 1977. Linkage summary. Report of Tomato Gen. Coop. 27:3–7.

Rickett, H. W. 1944. The classification of inflorescences. Bot. Rev. 10:187–231.
———. 1955. Materials for a dictionary of botanical terms. III. Inflorescences. Bull. Torrey Bot. Club 82:419–445.
Rijven, A. H. G. C., and L. T. Evans. 1967a. Inflorescence initiation in *Lolium temulentum* L. IX. Some chemical changes in the shoot apex at induction. Aust. J. Biol. Sci. 20:1–12.
———. 1967b. Inflorescence initiation in *Lolium temulentum* L. X. Changes in ^{32}P incorporation into nucleic acids of the shoot apex at induction. Aust. J. Biol. Sci. 20:13–24.
Riley, R., and R. B. Flavell. 1977. A first view of the meiotic process. Phil. Trans. Roy. Soc. Lond. Ser. B. 277:191–199.
Ritterbusch, A. 1976. Die Organopoiese der Blute von *Calceolaria tripartita* R. et P. (Scrophulariaceae). Bot. Jahrb. Syst. 95:267–320.
———. 1977. A new technique for the study of floral histology and morphogenesis: Tissue strips made from oriented and cleared thick microtome sections. Can. J. Bot. 55:1373–1382.
———. 1980a. Size or age related transformations and sliding scales for the representation of spatial and temporal ordered developmental sequences. Flora 169:299–308.
———. 1980b. The normalization of developmental sequences for the representation of relative growth. Flora 169:309–315.
———. 1980c. The spatio-temporal patterns of growth and development in floral ontogenesis as visualized by Bildscharen and trajectories. Flora 169:405–423.
Ritterbusch, A., and U. Wunderlin. 1989. On growth and development—A spatio-temporal analysis of flower ontogenesis. Env. and Exp. Bot. 29:111–124.
Rogers, M. N. 1980. Snapdragons. In R. A. Larson [ed.], Introduction to floriculture (San Diego: Academic Press), pp. 107–131.
Roman, H. 1947. Mitotic non-disjunction in the case of interchanges involving the B-type chromosome in maize. Genetics 32:391–409.
———. 1948. Directed fertilization in maize. Proc. Nat. Acad. Sci. USA 34:36–42.
Rood, S. B.; R. P. Pharis; and D. J. Major. 1980. Changes of endogenous gibberellin-like substances with sex reversal of the apical inflorescence of corn. Plant Physiol. 66:793–796.
Rosenberg, S. M., and H. T. Bonnett. 1983. Floral organogenesis in *Nicotiana tabacum:* A comparison of two cytoplasmic male sterile cultivars with a male-fertile cultivar. Amer. J. Bot. 70:266–275.
Roth, V. L. 1988. The biological basis of homology. In C. J. Humphries [ed.], Ontogeny and systematics. (London: British Museum of Natural History), pp. 1–26.
Robinson-Beers, K.; R. E. Pruitt; and C.S. Gasser. 1992. Ovule development in wild-type *Arabidopsis* and two female sterile mutants. Plant cell 4:1237–1249.
Rudich, J. 1985. *Cucumis sativus*. In A. H. Halevy [ed.], CRC handbook of flowering, vol. 2 (Boca Raton, FL: CRC Press), pp. 365–374.
Rudich, J.; A. H. Halevy; and L. R. Kedar. 1972. The level of phytohormones in monoecious and gynoecious cucumbers as affected by photoperiod and ethephon. Plant Physiol. 50:585–590.
Russell, S. D. 1991. Isolation and characterization of sperm cells in flowering plants. Annu. Rev. Plant Physiol. Plant Mol. Biol. 42:189–204.
Sachs, R. M. 1965. Stem elongation. Annu. Rev. Plant Physiol. 16:73–96.
———. 1968. Control of intercalary growth in the scape of *Gerbera* by auxin and gibberellic acid. Amer. J. Bot. 55:62–68.

———. 1977. Nutrient diversion: An hypothesis to explain the chemical control of flowering. Hort. Science 12:220–222.
Sachs, R. M., and W. P. Hackett. 1977. Chemical control of flowering. Acta Horticultura 68:29–49.
Sachs, T. 1969. Regeneration experiments on the determination of the form of leaves. Israel J. Bot. 18:21–30.
———. 1981. The control of the patterned differentiation of vascular tissues. In H. W. Woolhouse [ed.], Advances in botanical research, vol. 9 (London: Academic Press), pp. 151–262.
———. 1986. Cellular interactions in tissue and organ development. In D. H. Jennings and A. J. Trewavas [eds.], Plasticity in plants: Soc. Exp. Biology. Symp. 40:181–210.
Sakate, T. 1976. Determination of the most sensitive stage to sterile-type cool injury in rice plants. Res. Bull. Hokkaido Nat. Agric. Exp. Stn. 113:1–35.
Salisbury, F. B. 1963. The flowering process. Oxford: Pergamon Press.
———. 1969. *Xanthium strumarium* L. In L.T. Evans [ed.], The induction of flowering: Some case studies (Ithaca, NY: Cornell University Press), pp. 14–61.
———. 1985. *Xanthium strumarium*. In A. H. Halevy [ed.], CRC handbook of flowering, vol. 4 (Boca Raton, FL: CRC Press), pp. 473–522.
Sanchez-Martinez, D.; P. Puigdomench; and M. Pages. 1986. Regulation of gene expression in developing *Zea mays* embryos. Plant Physiol. 82:543–549.
Sangwan, R. S., and B. S. Sangwan-Norreel. 1987. Biochemical cytology of pollen embryogenesis. Int. Rev. Cyt. 107:221–272.
Sass, J. E. 1977. Morphology. In G. F. Sprague [ed.], Corn and corn improvement, 2d ed. (Madison, WI: American Society of Agronomy), pp. 89–110. Agronomy Monograph 18.
Satina, S. 1945. Periclinal chimeras in *Datura* in relation to the development of the ovule. Amer. J. Bot. 32:72–81.
Satina, S., and A. F. Blakeslee. 1941. Periclinal chimeras in *Datura stramonium* in relation to development of leaf and flower. Amer. J. Bot. 28:862–871.
———. 1943. Periclinal chimeras in *Datura* in relation to the development of the carpel. Amer. J. Bot. 30:453–462.
Sattler, R. 1966. Towards a more adequate approach to comparative morphology. Phytomorphology 16:417–429.
———. 1967. Petal inception and the problem of pattern detection. J. Theor. Biol. 17:31–39.
———. 1968. A technique for the study of floral development. Can. J. Bot. 46:720–722.
———. 1973a. Organogenesis of flowers: A photographic text-atlas. Toronto: University of Toronto Press.
——— 1973b. Centrifugal primordial inception in floral development. In Y. S. Murty, B. M. Johri, H. Y. Mohan Ram and T. M. Varghese [eds.], Advances in plant morphology, Professor V. Puri commemoration volume (Meerut, India: Sarita Prakashan), pp. 170–178.
———. 1974. A new approach to gynoecial morphology. Phytomorphology 24:22–34.
———. 1978. Fusion and continuity in floral morphology. Notes Roy. Bot. Gard. Edinburgh 36:397–405.
———. 1988*a*. A dynamic multidimensional approach to floral morphology. In P. Leins, S. C. Tucker and P. K. Endress [eds.], Aspects of floral development (Berlin: Lubrecht & Cramer), pp.1–6.

———. 1988b. Homeosis in plants. Amer. J. Bot. 75:1606–1617.
Sattler, R., and C. Lacroix. 1988. Development and evolution of basal cauline placentation: *Basella rubra*. Amer. J. Bot. 75:918–927.
Sattler, R., and R. Rutishauser. 1990. Structural and dynamic descriptions of the development of *Utricularia folisa* and *U. australis*. Can. J. Bot. 68:1989–2003.
Saunders, E. R. 1913. Double flowers. J. Roy. Hort. Soc. 38:469–482.
———. 1926. A reply to comments on the theory of the solid carpel and carpel polymorphism. New Phytol. 25:294–306.
———. 1928. Further studies on inheritance in *Matthiola incana*. II. Plastid colour and doubling. J. Genet. 20:53–76.
Sawhney, V. K. 1974. Morphogenesis of the stamenless-2 mutant in tomato. III. Relative levels of gibberellins in the normal and mutant plants. J. Exp. Bot. 25:1004–1009.
———. 1983. Temperature control of male sterility in a tomato mutant. J. Hered. 74:51–54.
Sawhney, V. K.; S. K. Bhadula; P. L. Polowick; and R. Rastogi. 1989. Regulation and development of male sterility in tomato and rapeseed. In E. Lord and G. Bernier [eds.], Plant reproduction: From floral induction to pollination. (Rockville, MD: American Society of Plant Physiologists), pp. 114–120.
Sawhney, V. K., and D. H. Dabbs. 1978. Gibberellic acid induced multilocular fruits in tomato and the role of locule number and seed number in fruit size. Can. J. Bot. 56:2831–2835.
Sawhney, V. K., and R. I. Greyson. 1973a. Morphogenesis of the stamenless-2 mutant in tomato. I. Comparative description of the flowers and ontogeny of stamens in the normal and mutant plants. Amer. J. Bot. 60:514–523.
———. 1973b. Morphogenesis of the stamenless-2 mutant in tomato. II. Modifications of sex organs in the mutant and normal flowers by plant hormones. Can. J. Bot. 51:2473–2479.
———. 1979. Interpretations of determination and canalization of stamen development in a tomato mutant. Can. J. Bot. 57:2471–2477.
Sawhney, V. K., and E. B. Nave. 1986. Enzymatic changes in post-meiotic anther development in *Petunia hybrida*. II. Histochemical localization of esterase, peroxidase, malate- and alcohol dehydrogenase. J. Plant Physiol. 125:467–473.
Sawhney, V. K., and P. L. Polowick. 1985. Fruit development in tomato: The role of temperature. Can. J. Bot. 63:1031–1034.
———. 1986. Temperature-induced modifications in the surface features of stamens of a tomato mutant: An SEM study. Protoplasma 131:75–81.
Schaeffer, G. W.; M. D. Lazar; and P. S. Baenziger. 1984. Wheat. In W. R. Sharp, D. A. Evans, P. V. Ammirato and Y. Yamada [eds.], Handbook of plant cell culture, vol. 2, Crop species (New York: Macmillan), pp. 108–136.
Schaeverbeke, J. 1965. Actions compareés de l'acide gibbérellique et de l'acide indolyl-B-acétique sur l'allongement des filets staminaux isolés du *Zea mays* L. C. R. Acad. Sci. Ser. D 260:4580–4582.
Schmid, R. 1976. Filament histology and anther dehiscence. Bot. J. Linn. Soc. 73:303–315.
———. 1988. Reproductive versus extra-reproductive nectaries—Historical perspective and terminological recommendations. Bot. Rev. 54:179–232.
Schock-Bodmer, H. 1939. Beitrage zur Kenntnis des Streckungswachstums der Gramineen-Filamente. Planta 30:168–204.
Schultz, E. A., and G. W. Haughn. 1991. LEAFY, a homeotic gene that regulates inflorescence development in *Arabidopsis*. Plant Cell 3:771–781.

Schuster, W. H. 1985. *Helianthus annuus*. In A. H. Halevy [ed.], CRC handbook of flowering, vol. 3 (Boca Raton, FL: CRC Press), pp. 98–121.

Schwabe, W. W., and A. H. Al-Doori. 1973. Analysis of juvenile-like condition affecting flowering in the black currant *(Ribes nigrum)*. J. Exp. Bot. 24:969–981.

Schwarz-Sommer, Z.; P. Huijser; W. Nacken; H. Saedler; and H. Sommer. 1990. Genetic control of flower development by homeotic genes in *Antirrhinum majus*. Science 250:931–936.

Scott, J. W., and L. R. Baker. 1975. Inheritance of sex expression from crosses of dioecious cucumber *(Cucumis sativus* L.). J. Am. Soc. Hortic. Sci. 100:457–461.

Seigel, B. A., and J. A. Verbeke. 1989. Diffusible factors essential for epidermal cell redifferentiation in *Catharanthus roseus*. Science 244:580–582.

Selker, J. M., G. L. Steucek; and P. B. Green. 1992. Biophysical mechanisms for morphogenetic progressions at the shoot apex. Dev. Biol. 153:29–43.

Shannon, S., and D. R. Meeks-Wagner. 1991. A mutation in the *Arabidopsis* TFL1 gene affects inflorescence meristem development. Plant Cell 3:877–892.

Sharma, H. C.; B. S. Gill; and R. G. Sears. 1984. Inflorescence culture of wheat-*Agropyron* hybrids: Callus induction, plant regeneration, and potential in overcoming sterility barriers. Plant Cell, Tissue and Organ Culture 3:247–255.

Sharman, B. C. 1967. Interpretation of the morphology of various naturally occurring abnormalities of the inflorescence of wheat *(Triticum)*. Can. J. Bot. 45:2073–2080.

———. 1978. Morphogenesis of 2,4-D induced abnormalities of the inflorescence of bread wheat *(Triticum aestivum* L.). Ann. Bot. 42:145–53.

———. 1983. Developmental anatomy of the inflorescence of bread wheat *(Triticum aestivum* L.) during normal initiation and when affected by 2,4-D. Ann. Bot. 52:621–639.

Sheridan, W. F. [ed.], 1982. Maize for biological research. Charlottesville, VA: Plant Molecular Biology Association.

———. 1988. Maize developmental genetics: Genes of morphogenesis. Annu. Rev. Genet. 22:353–385.

Sheridan, W. F., and M. G. Neuffer. 1982. Maize developmental mutants. J. Hered. 73:318–329.

Sherry, R. A.; K. J. Eckard; and E. M. Lord. 1993. Flower development in dioecious *Spinacia oleracea* (Chenopodiaceae). Amer. J. Bot. 80:283–291.

Shewry, P. R. [ed.]. 1991. Barley: Genetics, biochemistry, molecular biology and biotechnology. Wallingford, UK: C·A·B International.

Shifriss, O. 1961. Sex control in cucumbers. J. Hered. 52:5–12.

Shininger, T. L. 1979. The control of vascular development. Annu. Rev. Plant Physiol. 30:313–337.

Shivanna, K. R., and B. M. Johri. 1985. The angiosperm pollen: Structure and function. New Delhi: Wiley Eastern Ltd.

Silk, W. 1984. Quantitative descriptions of development. Annu. Rev. Plant Physiol. 35:479–518.

Singer, S. R.; C. H. Hannon; and S. C. Huber. 1992. Acquisition of competence for floral development in *Nicotiana* buds. Planta 188:546–550.

Singer, S. R., and C. N. McDaniel. 1986. Floral determination in the terminal and axillary buds of *Nicotiana tabacum* L. Dev. Biol. 118:587–592.

———. 1987. Floral determination in internode tissues of day-neutral tobacco first occurs many nodes below the apex. Proc. Nat. Acad. Sci. 84:2790–2792.

Singh, B. K., and C. F. Jenner. 1983. Culture of detached ears of wheat in liquid culture: Modification and extension of the method. Aust. J. Plant Physiol. 10:227–36.

Singh, V., and D. K. Jain. 1979. Floral organogenesis in *Antirrhinum majus* (Scrophulariaceae). Proc. Indian Acad. Sci. B. 88:183–188.

Singh, V., and R. Sattler. 1974. Floral development of *Butomus umbellatus*. Can. J. Bot. 52:223–230.

Sink, K. C., Jr. 1973. The inheritance of apetalous flower type in *Petunia hybrida* Vilm. and linkage tests with the genes for flower doubleness and grandiflora characters and its use in hybrid seed production. Euphytica 22:520–526.

———. [ed.]. 1984. *Petunia*. Berlin: Springer-Verlag.

Sinnott, E. W. 1958. The genetic basis of organic form. Annu. New York Acad. Sci. 71:1223–1233.

———. 1960. Plant morphogenesis. New York: McGraw-Hill.

Skoog, F. 1955. Growth factors, polarity and morphogenesis. Annee. Biol. 31:1–11.

Slack, F. M. V. 1983. From egg to embryo: Determinative events in early development. Cambridge: Cambridge University Press.

Sladky, Z. 1986. The role of growth regulators in the differentiation of flowers and inflorescences. Biol. Plant. (Praha) 28:31–37.

Smith, A. G.; C. S. Gasser; K. A. Budelier; and R. T. Fraley. 1990. Identification and characterization of stamen- and tapetum-specific genes from tomato. Mol. Gen. Genet. 222:9–16.

Smith, B. G., and P. H. Rubery. 1981. The effects of infection by *Phytophthera infestans* on the control of phenylpropanoid metabolism in wounded potato tissue. Planta 151:665–668.

Smith, F. J.; J. H. de Jong; and J. L. Oud. 1975. The use of primary trisomics for the localization of genes on the seven different chromosomes of *Petunia hybrida*. Triplo V. Genetica 45:361–370.

Smith, L. G., and S. Hake. 1992. The initiation and determination of leaves. Plant Cell 4:1017–1027.

Smith, S. E. 1990. Plant biology and social responsibility. Plant Cell 2:367–368.

Smith, S. E. M., and C. N. McDaniel. 1992. The *Maryland Mammoth* allele and rooting both perturb the fate of florally determined apices in *Nicotiana tabacum*. Dev. Biol. 153:176–184.

Smyth, D. R.; J. L. Bowman; and E. M. Meyerowitz. 1990. Early development of *Arabidopsis*. Plant Cell 2:755–767.

Smyth, D. R., and E. M. Meyerowitz. 1988. Research on the molecular genetics of flower development at the California Institute of Technology. Flowering Newslett. 6:3–8.

Snow, M., and R. Snow. 1931. Experiments on phyllotaxis. 1. The effect of isolating a primordium. Phil. Trans. Roy. Soc. Lond. Ser. B 221:1–43.

Soetiarto, S. R., and E. Ball. 1969. Ontogenetical and experimental studies of the floral apex of *Portulaca grandiflora*. 2. Bisection of the meristem in successive stages. Can. J. Bot. 47:1067–1076.

Sommer, H.; J.–P. Beltran; P. Huijser; H. Pape; W.–E. Lonnig; H. Saedler; and Z. Schwarz-Sommer. 1990. Dieficiens, a homeotic gene involved in the control of flower morphogenesis in *Antirrhinum majus:* The protein shows homology to transcription factors. EMBO 9:605–613.

Sommer, H.; R. Carpenter; B. J. Harrison; and H. Saedler. 1985. The transposable element Tam3 of *Antirrhinum majus* generates a novel type of sequence alteration upon excision. Mol. Gen. Genet. 199:225–231.

Sommer, H., and H. Saedler. 1986. Structure of the chalcone synthase gene in *Antirrhinum majus*. Mol. Gen. Genet. 202:429–434.

Sporne, K. R. 1958. Some aspects of floral vascular systems. Proc. Linn. Soc. London. 169:75–84.
Spribille, R., and G. Forkmann. 1982. Genetic control of chalcone synthase activity in flowers of *Antirrhinum majus*. Phytochemistry 21:2231–2234.
Stanton, M. L. 1987a. The reproductive biology of petal color variants in wild populations of *Raphanus sativus* L. I. Pollinator response to color morphs. Amer. J. Bot. 74:178–187.
———. 1987b. The reproductive biology of petal color variants in wild populations of *Raphanus sativus* L. II. Factors limiting seed production. Amer. J. Bot. 74:188–196.
Stanton, M. L., and R. E. Preston. 1988. Ecological consequences and phenotypic correlates of petal size variation in wild radish, *Raphanus sativus* (Brassicaceae). Amer. J. Bot. 75:528–539.
Stebbins, G. L. 1964. Four basic questions of plant biology. Amer. J. Bot. 51:220–230.
———. 1974. Flowering plants—Evolution above the species level. Cambridge, MA: Belknap Press of Harvard University Press.
———. 1992. Comparative aspects of plant morphogenesis: A cellular, molecular and evolutionary approach. Amer. J. Bot. 79:589–598.
Stebbins, G. L., and E. Yagill. 1966. The morphogenetic effects of the hooded gene in barley. I. The course of development in hooded and awned genotypes. Genetics 54:727–741.
Steeves, T. A. 1966. On the determination of leaf primordia in ferns. In E. G. Cutter [ed.], Trends in plant morphogenesis (London: Longmans), pp. 200–219.
Steeves, T. A.; M. A. Hicks; J. M. Naylor; and P. Rennie. 1969. Analytical studies on the shoot apex of *Helianthus annuus*. Can. J. Bot. 47:1367–1375.
Steeves, T. A.; M. W. Steeves; and R. Olson. 1991. Flower development in *Amelanchier alnifolia* (Maloideae). Can. J. Bot. 69:844–857.
Steeves, T. A., and I. M. Sussex. 1989. Patterns in plant development, 2d ed. Cambridge: Cambridge University Press.
Steffen, J. D.; R. M. Sachs; and W. P. Hackett. 1988a. Growth and development of reproductive and vegetative tissues of *Bougainvillea* cultured *in vitro* as a function of carbohydrate. Amer. J. Bot. 75:1219–1224.
———. 1988b. *Bougainvillea* inflorescence meristem development: Comparative action of GA_3 *in vivo* and *in vitro*. Amer. J. Bot. 75:1225–1227.
Stern, H., and Y. Hotta. 1973. Biochemical controls of meiosis. Annu. Rev. Gen. 7:37–66.
Stern, K. R. 1991. Introductory plant biology, 5th ed. Dubuque, IA: W. C. Brown.
Stork, H. E. 1956. Epiphyllous flowers. Bull. Torrey Bot. Club 85:338–341.
Stubbe, H. 1966. Genetik und zytolgie von *Antirrhinum* L. sect. *Antirrhinum*. Jena: G. Fischer.
Suge, H. 1984. Hormonal control of growth and development. In S. Tsunoda and N. Takahashi [eds.], Biology of rice (Tokyo: Japanese Scientific Societies Press, and Amsterdam: Elsevier), pp.133–151.
Sumner, M. J., and L. van Caeseele. 1990. The development of the central cell of *Brassica campestris* prior to fertilization. Can. J. Bot. 68:2553–2563.
Sundberg, M. D. 1982a. Floral ontogeny in *Cyclamen persicum* "F-1 Rosemunde Rose" (Primulaceae). Amer. J. Bot. 69:380–388.
———. 1982b. Petal-stamen initiation in the genus *Cyclamen* (Primulaceae). Amer. J. Bot. 69:1701–1709.

Sundberg, M. D., and A. R. Orr. 1986. Early inflorescence and floral development in *Zea diploperennis*, diploperennial teosinte. Amer. J. Bot. 73:1699–1712.

Sunderland, N. 1974. Anther culture as a means of haploid production. In K. J. Kasha [ed.], Haploids in higher plants: Advances and potential (Guelph: University of Guelph Press), pp. 91–222.

Sunderland, N., and B. Huang. 1987. Ultrastructural aspects of pollen dimorphism. Int. Rev. Cyt. 107:175–220.

Sussex, I. M. 1955. Morphogenesis in *Solanum tuberosum* L.: Experimental investigation of leaf dorsiventrality and orientation in the juvenile shoot. Phytomorphology 5:286–300.

Svensson, L. 1988. Inbreeding, crossing and variation in stamen number in *Scleranthus annuus* (Caryophyllaceae), a selfing annual. Evol. Trends in Plants 2:31–37.

———. 1990. Distance-dependent regulation of stamen number in crosses of *Scleranthus annuus* (Caryophyllaceae) from a discontinuous population. Amer. J. Bot. 77:889–896.

Takeba, G., and A. Takimoto. 1966. Translocation of the floral stimulus in *Pharbitis nil*. Bot. Mag. (Tokyo) 70:811–814.

Takehashi, N. 1984. Differentiation of ecotypes in *Oryza sativa* L. In S. Tsunoda and N. Takahashi [eds.], Biology of rice (Tokyo: Japan Scientific Societies Press, and Amsterdam: Elsevier), pp. 71–88.

Takeoka, Y.; A. A. Mamun; T. Wada; and P.B. Kaufman. 1992. Reproductive Adaptation of rice to enviromental stress. Tokyo: Japan Scientific Societies Press, and Amsterdam: Elsevier.

Tamas, I. A. 1987. Hormonal regulation of apical dominance. In P. J. Davies [ed.]; Plant hormones and their role in plant growth and development (Dordrecht: Martinus Nijhoff), pp. 393–410.

Tanimoto, S., and H. Harada. 1985. Hormonal regulation of flowering. In S. S. Purohit [ed.], Hormonal regulation of plant growth and development (Advances in agricultural biotechnology) (Dordrecht: Martinus Nijhoff/Dr. W. Junk Publishers), pp. 41–93.

Tanksley, S. D., and M. A. Mutschler. 1990. Linkage map of tomato *(Lycopersicon esculentum)* (2N = 24). In S. J. O'Brien [ed.], Genetic maps, 5th ed. (New York: Cold Spring Harbor Press), pp. 6.3–6.15.

Taylor, T. N. 1988. The origin and evolution of angiosperms. ISI Atlas of Science: Animal and Plant Sciences 1:47–50.

Tennant, N. W. 1986. Reductionism and holism in biology. In T. J. Horder, J. A. Witkowski and C. C. Wylie [eds.], A history of embryology (Cambridge: Cambridge University Press), pp. 407–433.

Tepfer, S. S. 1953. Floral anatomy and ontogeny in *Aquilegia formosa* var. *truncata* and *Ranunculus repens*. Univ. Calif. Publ. Bot. 25(7):513–648.

Tepfer, S. S.; R. I. Greyson; W. R. Craig; and J. L. Hindman. 1963. *In vitro* culture of floral buds of *Aquilegia*. Amer. J. Bot. 50:1035–1045.

Thomas, J. F.; C. E. Anderson; C. D. Raper; and R. J. Downs, Jr. 1975. Time of floral intitiation in tobacco as a function of temperature and photoperiod. Can. J. Bot. 53:1400–1410.

Tilton, V. R., and H. T. Horner, Jr. 1980. Stigma, style and obdurator of *Ornithogalum caudatum* (Liliaceae) and their function in the reproductive process. Amer. J. Bot. 67:1113–1131.

Tomato Genetics Cooperative Reports 37. 1987. Information available from TGC Reports, Cornell University Departments of Plant Breeding and Biometry and Agronomy and the USDA-ARS, 1017 Bradfield Hall, Cornell University, Ithaca, NY 14853–1901.

Toxopeus, H. J. 1927. Erblechkeitsuntersuchungen an *Nigella damascena*. Genetica 9:341–440.

Tran Thanh Van, K. M. 1981. Control of morphogenesis in *in vitro* cultures. Annu. Rev. Plant Physiol. 32:291–311.

Tran Thanh Van, M.; N. Thi Dien; and A. Chlyah. 1974. Regulation of organogenesis in small explants of superficial tissue of *Nicotiana tabacum* L. Planta (Berlin) 115:149–159.

Trees, S. C., and J. N. Rutger. 1978. Inheritance of four genetic male steriles in rice. J. Hered. 69:270–272.

Trewavas, A. 1982. Possible control points in plant development. In H. Smith and E. D. Grierson [eds.], The molecular biology of plant development (Berkeley: University of California Press), pp. 1–27. Botanical Monographs. Vol. 18.

———. 1987. Sensitivity for hormone involvement in growth substance responses. In G. V. Hoad, J. R. Lenton, M. B. Jackson and R. K. Aitkin [eds.], Hormone action in plant development—A critical appraisal (London: Butterworths), pp. 19–38.

Trewavas, A., and R. E. Cleland. 1983. Is plant development regulated by changes in the concentration of growth substances or by changes in the sensitivity to growth substances? Trends in Biol. Sci. 8:354–357.

Trione, E. J., and V. O. Stockwell. 1989. Development of detached wheat spikelets in culture. Plant Cell, Tissue and Organ Culture 17:161–170.

Troll, W. 1964. Die Infloreszenzen, vol. I. Stuttgart: Gustav Fischer.

———. 1969. Die Infloreszenzen, vol. II. Stuttgart: Gustav Fischer.

Tucker, S. C. 1984*a*. Origin of symmetry in flowers. In R. A. White and W. C. Dickison [eds.], Contemporary problems in plant anatomy (New York: Academic Press), pp. 351–395.

———. 1984*b*. Unidirectional organ initiation in legumious flowers. Amer. J. Bot. 71:1139–1148.

———. 1987. Floral initiation and development in legumes. In C. H. Stirton [ed.], Advances in legume systematics, vol. III (Kew: Royal Botanic Garden), pp. 83–239.

———. 1988*a*. Dioecy in *Bauhinia* resulting from organ suppression. Amer. J. Bot. 75:1584–1597.

———. 1988*b*. Loss versus suppression of floral organs. In P. Leins, S. C. Tucker and P. K. Endress [eds.], Aspects of flower development (Berlin: Cramer), pp. 69–82.

———. 1989. Overlapping organ formation and common primordia in flowers of *Pisum sativum* (Leguminosae: Papilionoideae). Amer. J. Bot. 76:714–729.

———. 1991. Helical floral organogenesis in *Gleditsia,* a primitive caesalpinoid legume. Amer. J. Bot. 78:1130–1149.

Turing, A. M. 1952. The chemical basis of morphogenesis. Phil. Trans. Roy. Soc. Ser. B 237:37–72.

Uyldert, I. E. 1928. The influence of growth promoting substances on decapitated stalks of *Bellis perennis*. Proc. Kon. Ned. Acad. Wetensch. 31:59–61.

van Heel, W. A. 1966. Morphology of the androecium in Malvales. Blumea 13:177–394.

———. 1978. Morphology of the pistil in Malvaceae-Ureneae. Blumea 24:123–127.

———. 1981. A SEM investigation on the development of free carpels. Blumea 27:499–522.

———. 1983. The ascidiform early development of free carpels: A SEM-investigation. Blumea 28:231–270.

van Holst, G.-J., and A. E. Clarke. 1986. Organ-specific arabinogalactan-proteins of *Ly-*

copersicon peruvianum (Mill) demonstrated by crossed electrophoresis. Plant Physiol. 80:786–789.
van Nigtevecht, G. 1966a. Genetic studies in dioecious *Melandrium*. I. Sex-linked and sex-influenced inheritance in *Melandrium album* and *Melandrium dioicum*. Genetica 37:281–306.
———. 1966b. Genetic studies in dioecious *Melandrium*. II. Sex determination in *Melandrium album* and *Melandrium dioicum*. Genetica 37:307–344.
Van't Hof, J. 1985. Control points within the cell cycle. In J. A. Bryant and D. Francis [eds.], The cell division cycle in plants (Cambridge: Cambridge University Press), pp. 1–13.
van Went, J. L., and M. T. M. Willemse. 1984. Fertilization. In B. M. Johri [ed.], Embryology of angiosperms (Berlin: Springer-Verlag), pp. 273–317.
Vasil, I. K. 1967. Physiology and cytology of anther development. Biol. Rev. 42:327–373.
———. 1987. The physiology and culture of pollen. Int. Rev. Cytol. 107:127–174.
Veen, A. H., and A. Lindenmayer. 1977. Diffusion mechanism for phyllotaxis: Theoretical physico-chemical and computer study. Plant Physiol. 60:27–139.
Venis, M. A. 1985. Hormone binding in plants. New York: Longmans.
Vergara, B. S. 1985. *Oryza sativa*. In A. H. Halevy [ed.], CRC handbook of flowering, vol. 3 (Boca Raton, FL: CRC Press), pp. 435–441.
Vergara, B. S., and T. T. Chang. 1985. The flowering response of the rice plant to photoperiod: A review of literature, fourth ed. International Rice Research Institute. P.O. Box 933. 1099 Manila, Philippines.
Veuskens, J.; D. Ye; M. Oliviera; D. D. Ciupercescu; P. Installe; H. A. Verhoeven; and I. Negrutiu. 1991. Sex determination in the dioecious *Melandrium alba:* Androgenic embryogenesis requires the presence of the X chromosome. Genome 35:8–16.
Vince-Prue, D.; B. Thomas; and K. E. C. Cockshull [eds.]. 1984. Light and the flowering process. London: Academic Press.
Vithanage, H. I. M. V., and R. B. Knox. 1980. Periodicity of pollen development and quantitative cytochemistry of exine and intine enzymes in the grasses *Lolium perenne* and *Phalaris tuberosa*. Ann. Bot. 45:131–142.
Vollbrecht, E.; B. Veit; N. Sinha; and S. Hake. 1991. The developmental gene Knotted-1 is a member of a maize homeobox gene family. Nature 350:241–243.
Waddington, C. H. 1956. Principles of embryology. New York: Macmillan.
———. 1966a. Fields and gradients. In M. Locke [ed.], Symposium of the Society for Developmental Biology 25:105–124.
———. 1966b. Principles of development and differentiation. New York: Macmillan.
Walbot, V. 1989. Is anatomy destiny in plants? In R. Goldberg [ed.], The molecular basis of plant development (New York: Alan R. Liss), pp. 37–47.
Walden, D. B.; R. I. Greyson; V. R. Bommineni; D. R. Pareddy; J.-P. Sanchez; E. Banasikowska; and D. Kudirka. 1989. Maize meristem culture and recovery of mature plants. Maydica 34:263–275.
Walker, D. B. 1975. Postgenital carpel fusion in *Catharanthus roseus*. III. Fine structure of the epidermis during and after fusion. Protoplasma 86:43–63.
———. 1978. Morphogenetic factors controlling differentiation and dedifferentiation of epidermal cells in the gynoecium of *Catharanthus roseus*. 1. The role of pressure and cell confinement. Planta 142:181–186.
Walters, M. S. 1976. Variation in preleptotene chromosome contraction among three cultivars of *Lilium longiflorum*. Chromosoma 57:51–80.

———. 1980. Premeiosis and meiosis in *Lilium* "Enchantment." Chromosoma 80:119–146.
Ward, M. 1963. Developmental patterns of adventitious sporophytes in *Phlebodium aureum* J. Sm. J. Linn. Soc. Lond. (Bot.) 58:377–380.
Wardell, W. L., and F. Skoog. 1969. Flower formation in excised tobacco segments. I. Methodology and effects of plant growth hormones. Plant Physiol. 44:1402–1406.
Wardlaw, C. W. 1956–57. The floral meristem as a reaction system. Proc. Roy. Soc. Edinb. B. 66:394–408. Reprinted in C. W. Wardlaw, Essays on form in plants (Manchester, Manchester University Press, 1968), pp. 224–238.
———. 1961. Growth and development of the inflorescence and flower. In M. X. Zarrow [ed.], Growth in living systems. (New York: Basic Books), pp. 491–523.
———. 1965a. Organization and evolution in plants. London: Longmans.
———. 1965b. The organization of the shoot apex. In W. Ruhland [ed.], Encyclopedia of plant physiology, vol. XV/1 (Berlin: Springer/Verlag), pp. 967–1076.
———. 1968. Problems of organization in plants: A practical approach. In C. W. Wardlaw [ed.], Essays on form in plants (Manchester: University of Manchester Press), pp. 348–355.
Wareing, P. F. 1978. Determination in plant development. Bot. Mag. (Tokyo) Special Issue No. 1:3–17.
Warm, E. 1984. Changes in the composition of *in vitro* translated leaf mRNA caused by photoperiodic flower induction in *Hyoscyanus niger*. Physiol. Plant. 61:344–350.
Warmke, H. E. and S.-L. J. Lee. 1977. Mitochondrial degeneration in Texas cytoplasmic male-sterile corn anthers. J. Hered. 68:213–222.
Watson, L.; H. T. Clifford; and M. J. Dallwitz. 1985. The classification of the Poaceae: Subfamilies and supertribes. Aust. J. Bot. 33:433–484.
Weatherwax, P. 1916. Morphology of the flowers on *Zea mays*. Bull. Torrey Bot. Club 43:127–144.
———. 1955. Structure and development of reproductive organs. In G. F. Sprague [ed.], Corn and corn improvement (New York: Academic Press), pp. 89–121. Agronomy Monograph 5. American Society of Agronomy. Madison, WI.
Weaver, R. F., and P. W. Hedrick. 1989. Genetics. Dubuque, IA: Wm. C. Brown.
Weaver, S. E., and M. J. Lechowicz. 1983. The biology of Canadian weeds. 56. *Xanthium strumarium* L. Can. J. Plant Sci. 63:221–225.
Weber, H. 1938. Gramineen-studien 1. ber das Verhalten des Gramineen-Vegetationskegels beim bergang zur Infloreszenzbildung. Planta 28:275–289.
Weberling, F. 1965. Typology of inflorescences. J. Linn. Soc. (Bot.) 59:215–221.
———. 1983. Fundamental features of modern inflorescence morphology. Bothalia 14:917–922.
———. 1989. Morphology of flowers and inflorescences. Cambridge: Cambridge University Press.
Weibel, E. R. 1973. Stereological techniques for electron microscopy morphometry. In M. A. Hayat [ed.], Principles and techniques of electron microscopy: biological applications. vol. 3 (New York: Van Nostrand Rheinhold), pp. 237–296.
Weigel, D.; J. Alvarez; D. R. Smyth; M. F. Yanofsky; and E. M. Meyerowitz. 1992. LEAFY controls floral meristem identity in *Arabidopsis*. Cell 69:843–859.
Weiss, P. 1963. Cell interactions. Can. Cancer Conf. 5:241–276.
Wellensiek, S. T. 1961. Leaf vernalization. Nature 192:1097–1098.
Wenzel, G., and E. Thomas. 1974. Observations on growth in culture of anthers of *Secale cereale*. Zeit. fur Pflanzenzucht. 72:89–94.

Westergaard, M. 1985. The mechanism of sex determination in dioecious flowering plants. Adv. Genet. 9:217–281.

Wetmore, R. H.; E. M. Gifford, Jr.; and M. C. Green. 1959. Development of vegetative floral buds. In R. B. Withrow [ed.], Photoperiodism and related phenomena in plants and animals (Washington: American Association for the Advancement of Science), pp. 255–273.

Whelan, E. D. P. 1978. Cytology and interspecific hybridization. In J. F. Carter [ed.], Sunflower science and technology. No. 19 in series Agronomy (Madison, WI: American Society of Agronomy), pp. 339–369.

Whitaker, T. W., and G. N. Davis. 1962. Cucurbits. New York: Interscience Publishers.

White, P. R. 1933. Plant tissue cultures: Results of preliminary experiments on the culturing of isolated stem-tips of *Stellaria media*. Protoplasma 19:97–116.

Wiering, H.; P. de Vlaming; P. Cornu; and A. Maizonnier. 1979. *Petunia* genetics. I. List of genes. Ann. Amerior. Plant. (Paris) 29:611–622.

Wilkins, A. S. 1986. Genetic analysis of animal development. New York: John Wiley & Sons.

Willemse, M. T. M., and S. A. Reznickova. 1980. Formation of pollen in the anther of *Lilium*. 1. Development of the pollen wall. Acta. Bot. Neerl. 29:127–140.

Willemse, M. T. M., and J. L. van Went. 1984. The female gametophyte. In B. M. Johri [ed.], Embryology of angiosperms (Berlin: Springer-Verlag), pp. 159–196.

Williams, P. H., and C. B. Hill. 1986. Rapid cycling populations of *Brassica*. Science 232:1385–1389.

Williams, R. F. 1960. The physiology of growth of the wheat plant. 1. Seedling growth and the pattern of growth at the shoot apex. Aust. Jour. Biol. Sci. 13:401–428.

———. 1966a. Development of the inflorescence in Gramineae. In F. L. Milthorpe and J. D. Ivins [eds.], The growth of cereals and grasses (London: Butterworths), pp. 74–87.

———. 1966b. The physiology of growth in the wheat plant. III. Growth of the primary shoot and inflorescence. Aust. J. Biol. Sci. 19:949–966.

———. 1975. The shoot apex and leaf growth. Cambridge: Cambridge University Press.

Wilms, H. J. 1980a. Development and composition of the spinach ovule. Acta Bot. Neerl. 29:243–260.

———. 1980b. Ultrastructure of the stigma and style of spinach in relation to pollen germination and pollen tube growth. Acta Bot. Neerl. 29:33–47.

———. 1981a. Ultrastructure of the developing embryo sac of spinach. Acta Bot. Neerl. 30:75–99.

———. 1981b. Pollen tube penetration and fertilization in spinach. Acta Bot. Neerl. 30:101–122.

Wolpert, L. 1971. Positional information and pattern formation. Curr. Top. Dev. Biol. 6:183–224.

———. 1989. Introduction. In Cellular basis of morphogenesis. Ciba Foundation Symposium 144 (Chichester: John Wiley & Sons), pp. 1–4.

Wolpert, L., and W. D. Stein. 1984. Positional information and pattern formation. In G. M. Malacinski and S. V. Bryant [eds.], Pattern formation: A primer in developmental biology (New York: Macmillan), pp. 3–21.

Woodger, J. H. 1967. Biological principles: A critical study. London: Routledge & Kegan.

Worsdell, W. C. 1916. The principles of plant-teratology, vol. 2. London: Ray Society.

Wyatt, R. 1982. Inflorescence architecture: How flower number, arrangement and phenology affect pollination and fruit-set. Amer. J. Bot. 69:585–594.

Xinggui, L., and W. Jilin. 1988. Fertility transformation and genetic behaviour of Hubei photoperiod-sensitive genic male sterile rice. In Hybrid rice: Proceedings of the International Symposium on Hybrid Rice. (Manila, Philippines: International Rice Research Institute), pp. 129–138.

Xu, X.-B, and B. S. Vergara. 1986. Morphological changes in rice panicle development: A review of literature. IRRI Research Paper Series No. 117. (Manila, Philippines: International Rice Research Institute), pp. 1–13.

Yampolsky, C., and H. Yampolsky. 1922. Distribution of sex forms in the phanerogamic flora. Bibl. Genet. Leipzig 3:1–62.

Yanofsky, M. F.; H. Ma; J. L. Bowman; G. N. Drews; K. A. Feldmann; and E. M. Meyerowitz. 1990. The protein encoded in the *Arabidopsis* homeotic gene agamous resembles transcription factors. Nature 346:35–39.

Yoshida, S. 1981. Fundamentals of rice crop science. Manila, Philippines: International Rice Research Institute.

You, R., and W. A. Jensen. 1984. Ultrastructural observations of the mature megagametophyte and the fertilization in wheat *(Triticum aestivum)*. Can. J. Bot. 63:163–178.

Young, P. A., and J. W. MacArthur. 1947. Horticultural characters in tomatoes. Tex. Agric. Exp. Stn. Bull. No. 698, pp. 1–61.

Zeevaart, J. A. D. 1958. Flower formation as studied by grafting. Meded. Landbourwhogesch. (Wageningen) 58:1–88.

Index

Abnormal flowers, 77, 115
Abortion of meristems, 181
Actinomorphic, 69
Adnation, 16, 67
Aestivation, 7
Amygdalus, 137
Anagalis arvensis, flower reversion, 138
Androecium, 15, 87
Andromonoecious dwarfs of maize, 208
Angiosperms
   life cycle, 19, 20
   number of species, genera and families, 13
Anther, 15
   embryos from anther culture, 126
Anthocyanins, 84
Antipetalous, 46
Antirrhinum majus
   culture protocol, 261
   Cycloidia, 71
   deficiens, 71, 251
   nivea, 84
   source of genetic stocks, 264
   symmetry, 71
Antisepalous, 46
Apical dominance, 181
Apomixis, 169
Aquilegia
   double flowers, 76
   gynoecium initiation, 131
   sepal growth in vitro, 65, 67
   stamen initiation, 88
   vasculature, 82
Arabidopsis thaliana, 78
   agamous and apetala mutants, 78, 251
   apetala-2 and apetala-3, 117
   carpellody, 166
   culture protocol, 261
   homeotic genes, 251
   inflorescence genes, 171, 182
   ovule mutants, 151
   pistillata, 78, 91
   source of genetic stocks, 264
   stamen initiation, 89
Ascidiate carpel, 131
Autoradiography, induced and uninduced meristems of *Sinapis,* 31
Axillary buds, 46

Begonia
   in vitro culture of flower buds, 163
   monoecy, 162
   role of PGS and flower form, 163
Bellis, inflorescence stalk, 179
Betalains, 84
Bilateral symmetry, 69
Biophysical components of development, 243
Bisymmetry, 69
Bougainvillea, control of flowering, 42
Bract, 15
Bracteole, 15
Brassica campestris
   carpellody, 166
   flower induction, 25
   initiation of flower primordia, 48, 50
   self-incompatability, 113
   source of genetic stocks, 265
   stamen initiation, 89
Bryophyllum, control of staminate flowers, 164
Butomus, carpel initiation, 131

Calceolaria, corolla growth, 71
Calyx, 15
   growth in *Fittonia,* 61
Canabis sativa
   flower formation, 159
   staminate flower promotion, 164
   temperature and perianth, 74
Carotenoids, 84
Carpel, involute and conduplicate types, 146
Carpellody, 166
Catharanthus, fusion of organs, 69, 142
Cell differentiation, 6
   in anthers, 100
   in gynoecia, 145
   in sepals and petals, 80
   in stamen filaments, 96
Centrifugal initiation, 51, 90
Centripetal initiation, 91
Chalcone synthase, 84
Chasmogamy, 60, 63
Cheiranthus, organ initiation, 48
Choripetalous, 16
Cleistogamy, 60, 63
Cleome, filament growth, 94

Clonal analysis, 30
　in sunflower, 33
　in *Zea*, 30, 32
Coalescence of stamens, 17
Cohesion, 16, 67
Common primordium, in *Cyclamen*, 51
　in *Pisum*, 51
Comparative morphology, 8
Competence, 27, 237
Compound gynoecia, 18
Computer simulation of inflorescence structure, 176
Confocal microscopy, 54
Congenital fusion, 142
Corolla, 15
　cell differentiation, 80
　growth and role of PGS, 62
　meristic variation, 74
Cosmos, flower meristem development, 178
Crossing over, 22
Cruciferae, gynoecial vasculature, 146
Cucumis sativus
　genetic basis of flower form, 160
　role of PGS in flower form, 161
Cyclamen orbiculatum, sepal and petal initiation, 51
Cytoplasmic male sterility, 124

Darlingtonia, ovule development, 153
Darwin, Charles, 8, 67, 71
Datura
　embryogenesis in cultured anthers, 126
　gynoecium initiation, 136
　periclinal chimeras, 54
　stamen initiation, 91
Day-neutral plants, 25
Delphinium elatum, gynoecium initiation, 131
Determination, 6, 237
　assessment, 34
　Waddington's interpretation, 238, 240
Development, iii
　definition, 4
　factors involved, 44
　variation, 115
Developmental sequence, 4
Diadelphous, 17
Differentiation. *See* cell differentiation
Diffusion-reaction hypothesis, 232
Dioecy, in *Silene* and *Spinacia*, 159
Distylly, 168
Double fertilization, 22, 157
Double flowers, 75
　in *Aquilegia* 75
　in *Begonia*, 162
　in *Petunia*, 79

Egg, 20, 156
　egg apparatus, 155
Embryo, 20
Embryo sac, 20, 154, 155
Endosperm, 20, 23
Endothecium, 101
Epigenetic, 9, 242
Epigynous flower, 19
Epistase, 153
Eucnide, flower development, 141. *See also* Inferior ovary
Evocation, 29, 30
Evocation period, 36
Expressivity, 249

Fate map
　in sunflower, 33
　in *Zea*, 32
Female gametophyte, 155
Female germ unit, 156
Fertilization. *See* double fertilization
Field interpretation of meristem, 45
Filament, 17
　anatomy, 96
Filiform apparatus, 156
Fittonia, calyx growth, 61
Flavonoids, 84
Floral determination, 30
　in *Lolium temulentum*, 32
　in *Nicotiana tabacum*, 32
　in *Silene armeria*, 32
　in *Xanthium strumarium*, 32
Floral formula and diagram, 15
Floral stimulus, 28
Flower
　carpellate, gynoecial, 16
　complete, 16
　definition, 13, 14
　gynoecious, 158
　imperfect, unisexual, diclinous, 16, 158
　incomplete, 16
　in vitro culture, 127
　perfect, bisexual, monoclinous, 16
　staminate, 16, 158
Fritillaria, 22, 179
Funiculus, 152
Fusion of organs, 67, 142

Gaillardia grandiflora
　corolla growth, 65
　filament elongation, 94, 95
　PGS and corolla growth, 66
Gametophyte, gametophytic phase, 22
Genic male sterility, 124
　in maize, 125

Gerbera, inflorescence growth, 179
Gibberellins. *See also* plant growth substances
  and corolla growth in *Gaillardia*, 67
  endogenous content in maize inflorescences, 210
  and filament growth in *Ipomoea*, 97
  and floral organ development in *Zea*, 203
Goebel, 9
Goethe, 8
Grasses (Gramineae, Poaceae), 195
Growth of organs, 6
Gynoecium, 16, 130
  growth, 139
  initiation, 130, 134
  syncarpous, 18
  vasculature, 145

Helianthus annuus,
  flower, 191
  morphology of inflorescence, 188
Hermaphroditic flower, 157
Heteroblasty, 115, 181
Heteromorphic flowers, 113
Heterospermy, 157
Heterospory, 22
Heterostyly, 167
  pin and thrum forms, 168
Hoffmeister, 9
Holistic, 10
Homeobox in angiosperms, 251
Homeotic mutants, homeosis, 43, 251
  in *Antirrhinum*, 251
  in *Arabidopsis*, 251
  in *Zea*, 208
Homology, 8
Homology, developmental, 14
Homomorphic flowers, 113
Humulus, staminate flower promotion, 164
Hyoscyamus, gynoecial flower promotion, 164
Hypogynous flower, 19
Hypostase, 153

Impatiens balsamina, flower reversion, 138
Imperfect flower
  in *Cucumis*, 160
  environmental factors, 166
  genetic basis, 159
  in *Melandrium*, 159
  PGS involvement, 164
  in *Spinacia*, 159
In vitro culture, flower meristems and buds, 65, 127
  petals, 65
Independent assortment, 22
Induction, photoperiodic, 25

Inferior ovary, 19, 139
Inflorescence, 13, 171
  classification, 174
  determinate vs indeterminate, 172
  polytelic and monotelic, 172
Initiation, organ, 6. *See also* Gynoecium, initiation; Leaf initiation and growth; Perianth, initiation; Stamen, initiation
Integuments, 152
Ipomoea nil (*Pharbitis*)
  GAs and filament growth, 96
  PGS and organ growth, 96
Ipomopsis aggregata, petal number, 74

Lamium, 65
  GA and petal growth, 65
Leaf initiation and growth, 44
  factors involved, 45
Leaves, arrangement, 45
Levels of biological organization, 5
Levisticum officinale, ovary formation, 143. *See also* Inferior ovary
Lilium, filament growth, 94
Lilium longiflorum
  anther growth, 99
  gynoecium growth, 139
  microsporogenesis, 107
Linanthus, corolla lobe number, 116
Linum, leaf initiation, 45
Lolium temulentum
  culture protocol, 261
  evocation period, 36
  floral stimulus, 32, 38
  inflorescence morphology, 227
  microsporogenesis, 106
  shoot apex culture, 34
  spikelet initiation, 228
  stamen initiation, 92
Long-day plants, 25
Lunaria annua, 36
Lycopersicon esculentum
  carpellody, 166
  culture protocol, 262
  enzyme differentiation, 254
  floral genes, 121
  meristic variation, 166
  source of genetic stocks, 265
  stamen initiation, 89
  stamenless-2, 120
  staminate flower promotion, 164

Magnoliophyta, 13
Male gametophyte, 111
Marginal meristems, 58
Matthiola incana, doubleness, 77

Megaspore, 22, 23, 154
Megasporocyte, 154
Megasporogenesis, 23, 154
Meiosis, 22
Melandrium alba, dioecy, 159
Mercurialis annua
   cytokinin involvement, 164
   enzyme differentiation, 254
Meristem d'attente, 188
Meristic variation, 74
   in *Gleditsia,* 74
   in gynoecia, 166
   in *Microseris* inflorescence, 193
   in stamens, 116
Merosity, 74
Microseris, meristic variation, 193
Microsporangium, differentiation, 99
   in *Oryza,* 100
   in *Zea,* 105
Microspore, 22, 23, 108
Microsporogenesis, 23, 106
Mirabilis jalapa, petal growth and GA content, 64
Mitosis
   in *Helianthus* meristems, 192
   in leaves, 58
   in meristems, 30, 31
   in petals, 59, 60
Monadelphous, 17
Monoecy, 158
Monophylesis, 13
Mosaic development, 85, 239

Nectaries, floral and extra-floral, 148
Nepeta, flower symmetry, 73
Nicotiana alata, self-incompatability, 113
Nicotiana sylvestris, 41
Nicotiana tabacum
   corolla growth, 59
   enzyme differentiation, 254
   flower induction, 27
   fusion of organs, 68
   gynoecium initiation, 136
   in vitro culture of corolla, 66
   meristem proliferation, 34
   ovule initiation, 151
   sepal and petal initiation, 54
   source of genetic stocks, 265
   stamen initiation, 91
Nigella damascena
   doubleness, 77
   in vitro flower bud culture, 128
Nigella hispanica, filament growth, 94
Nuphar, carpel number, 167
Nymphaea, 46

Obdurator, 148
Ontogeny, 4, 9, 10, 11
Oryza sativa
   anther growth, 99
   culture protocol, 262
   flower morphology, 222
   genetic variability, 225
   inflorescence initiation, 220
   male sterility, 226
   microsporangia, 100
Ovary, inferior and superior, 19
Overlapping initiation, 52
Ovules, 18
   endothelium, 153
   initiation, 149
   morphology, 152
   in *Nicotiana,* 151
   in *Papaver rhoeas,* 149

Panicle, 173
   of rice, 223
Papaver, ovule initiation, 149
Parastichies, 48, 189
Pedicel, 15, 179
Peduncle, 179
Peloric corolla, 71
Peltate carpel, 131
Penetrance, 249
Perfect flower, 157. *See also* Imperfect flower
Perianth, definition 15
   growth, 57
   initiation, 54
   vasculature, 80
Periclinal chimeras, 54
Periclinal divisions, 44
Perigynous flower, 19
Perilla crispa, induction, 27
Petal, 15
   growth, 59
   initiation, 48, 54
Petalody, 76
Petunia
   chalcone synthase, 84
   doubleness, 79
   floral genes, 80
Phalaris, microsporogenesis, 106
Pharbitis nil (Ipomoea)
   evocation period, 36
   floral stimulus, 28
   induction to flower, 27
   petal growth and GA content, 63
   shoot apex culture, 34;
Photoperiodic induction, 25
   developmental patterns, 26
Phylogeny, 8, 9, 11

Phytochrome, 27
Pigmentation, 83
Pistil, 18
Pisum sativum
  carpel initiation, 130
  common primordia, 51
  flower competence, 41
  inflorescence genes, 183
  leaf initiation, 44
  source of genetic stocks, 266
  unidirectional initiation, 52
Placenta, 18, 149
Placentation, 18, 149
Plant growth substances, 9
  influence of development, 244
Plasticity, 239
Poaceae (Gramineae), 195
Pollen
  germination, 112
  in vitro culture, 114
  wall structure, 110
Polyadelphous, 18
Polygonum-type megasporogenesis, 154
Polyphylesis, 13
Positional information, 241
Primula vulgaris, heterostyly, 168
Proliferation, 137
Proliferation of flowers *in vitro*, 128
Prophyll, 200

Ranunculus repens
  organ initiation, 46
  stamen initiation, 88
  vasculature, 81
Receptacle, 15
Reductionism, 10
Regular symmetry, 71
Regulative development, 85, 239
Reversion, 137
Ricinus, gynoecial flower stimulation, 164

Schleiden, 9
Scleranthus annuus, stamen variation, 116
Seed, 18
Self-incompatability, 113
Senescence, 5, 7
Sepal, 15
  growth, 57
  growth in *Aquilegia*, 67
  initiation, 48
Shoot apex culture, 33
Short-day plants, 25
Silene armeria, induction, 27
Silene coeli-rosa, sepal initiation, 48
Silene dioica, dioecy, 159

Sinapis alba, autoradiography of induced and uninduced meristems, 31
Sinnott, E. W., 9
Solanum dulcamara, growth of perianth, 58
Spikelet branch in *Zea*, 199
Spikelet in grasses, 196
Spinacia oleracea, flower form, 159
Sporogenesis, 21. See also Megasporogenesis; Microsporogenesis
Sporophyte, sporophytic phase, 21
Stamen, 17
  genetically based variability, 116
  growth, 93
  initiation, 87
Staminodia, 18
  in *Aquilegia*, 88
Stellaria media, flower symmetry, 73
Stigma, 18
Style, 18
Sunflower, fate map, 33. See also *Helianthus annuus*
Superior ovary, 19
Surgical bisection, 238
Symmetry, perianth, 69
Sympetalous, 16

Tapetum, 104
Taraxicum, inflorescence stalk, 179
Tassle-seed mutants in maize, 206
Temperature and flowering, 26
Tepal, 15
Teratomas, 77
Thin-tissue explants, 176
Thlaspi arvense, flowering stimulus, 36
Tradescantia, petal growth, 60
Transfer cell wall projections, 156
Transmitting tissue, 147
Tristylly, 168
Triticum
  culture protocol, 262
  growth of lemma and palea, 62
  gynoecium initiation, 134
  in vitro inflorescence culture, 219
  morphology of spikelet and flower, 217
  origin of gynoecium, 218
  studies of speltoid genes, 219
  transition to flowering, 211
Tunica layers, 44

Unidirectional initiation, 52
Utricularia, axillary buds, 46

Vernalization, 36
Vinca, stamen growth, 94

Wardlaw, 9

Xanthium strumarium
  induction, 27, 184
  monoecy, 185
  morphology of inflorescences, 184
Zea mays
  andromonoecious dwarfs, 208
  culture protocol, 263
  cytoplasmic male sterility, 203
  developmental basis of monoecy, 164, 201
  fate map from clonal analysis, 32
  genetic basis of flower form, 203
  genic male sterility, 205
  microsporogenesis, 106
  PGS and flower form, 209
  source of genetic stocks, 266
  tassel and ear morphology, 197
  transition to flowering, 197
Zygomorphy, 69
Zygote, 23

WITHDRAWN

WAKE TECHNICAL COMMUNITY COLLEGE LIBRARY
9101 FAYETTEVILLE ROAD
RALEIGH, NORTH CAROLINA 27603

WITHDRAWN

DATE DUE

JUN 2 4 '02			
GAYLORD			PRINTED IN U.S.A.

WITHDRAWN